APPLIED SYSTEMS ENGINEERING

APPLIED SYSTEMS ENGINEERING

ADRIAN GHEORGHE
Department of Management Science
Bucharest Polytechnic Institute,
ROMANIA

A Wiley-Interscience Publication

EDITURA ACADEMIEI
București

JOHN WILEY & SONS
Chichester • New York • Brisbane
Toronto • Singapore

Copyright © 1982 by John Wiley & Sons, Ltd.

All rights reserved.

No part of this book may be reproduced by any means, nor transmitted, nor translated into a machine language without the written permission of the publisher.

British Library Cataloguing in Publication Data

Gheorghe, Adrian
  Applied systems engineering.
  1. Systems engineering
  I. Title
  620.7   TA168

ISBN 0 471 09997 X

This edition is the revised English version of the Romanian book
"INGINERIA SISTEMELOR. METODE ŞI TEHNICI DE CALCUL"
published in 1979 by EDITURA ACADEMIEI
Calea Victoriei 125, București, 79717

All rights reserved.

PRINTED IN ROMANIA

Table of contents

	Introduction	9
PART I.	BASIC PROBLEMS IN SYSTEMS ENGINEERING	11

Chapter 1. **Fundamentals and concepts in systems engineering** 13

    1.1. Field and boundaries of systems engineering 13
    1.2. Systems definitions, decision processes, information processes and hierarchies   19
    1.3. Cost-effectiveness models in systems engineering. Fundamentals and algorithms 27

Chapter 2. **The practice of systems engineering** 37

    2.1. State of the art 40

Chapter 3. **Multifunctional systems and extensions** 42

    3.1. Concepts of polyfunctional structures and systems 42
    3.2. Polyfunctional structures and systems: basic definitions 43
    3.3. Coherent structures and extensions 46
    3.4. Modular decomposition of polyfunctional structures 48
    3.5. Boolean coherent structures 49
    3.6. Structures of fixed time. A general comparison 58
    3.7. Economic aspects concerning multifunctional systems design . . . 65
    3.8. General optimization of boolean hierarchical dependent structures . . 65
    3.9. Conception and measurement of system complexity 71
    3.10. Diagnostic questionnaires for complex systems 79
    3.11. Resilience of complex systems 83
    3.12. Systems engineering and systems resilience 86
        3.12.1. What is a resilient design? 87
        3.12.2. Safety engineering — risk embedding — resilience 87

Chapter 4. **A probabilistic safety modelling methodology for large-scale technical systems** . 89

    4.1. Basic definitions 89
    4.2. A mathematical approach to "How safe is safe enough?" 94
    4.3. How safe is "too" safe? 96
    4.4. The model of consequences and utilities 102

4.5. Basic structure of a pilot model for systems safety evaluation 103
4.6. A quantitative model for safety analysis 106

PART II. MODELS, ALGORITHMS AND OPTIMIZATION TECHNIQUES IN SYSTEMS ENGINEERING 113

Chapter 5. **Modelling aspects of reliability, maintenance and safety strategies using probabilistic systems analysis** 115

5.1. Markovian decision processes. Basic definitions 117
5.2. Basic definitions in operational engineering technical systems 120
5.3. Hybrid OR methods for maintenance and safety modelling 122
5.4. Reliability prediction of complex systems with Markov behaviour having many failed states 124
5.5. Reliability and availability prediction of polyfunctional technical systems with semi-Markov structure 130
    5.5.1. Completely observable complex systems 131
    5.5.2. The exit probability matrix 132
    5.5.3. Partially observable systems and availability prediction functions . . 136
5.6. Short-term reliability of a complex operating system 142
5.7. Mathematical foundations for safety analysis 148
    5.7.1. Definition of a fault tree 148
    5.7.2. Algorithms for evaluating min cut sets in a fault tree analysis . . . 151
    5.7.3. Modular representation of fault trees 153
    5.7.4. Fault tree analysis using techniques of boolean difference 154
    5.7.5. Fault tree analysis and measures of event importance 156
    5.7.6. A linguistic representation of a fault tree 159
    5.7.7. Automated construction of fault trees 166
    5.7.8. Vesely model for fault tree evaluation 168
    5.7.9. Error analysis and fault trees 175

Chapter 6. **On risk-sensitive Markovian decision models for complex systems maintenance** . . 177

6.1. Preliminaries on the risk-preference phenomenon 177
6.2. A risk-sensitive Markov decision model for maintenance strategies in technical systems 179
6.3. A risk-sensitive semi-Markov decision model for maintenance strategies in technical systems. Finite-horizon decision-making 189
6.4. A special case: partially observable risk-sensitive Markov decision models for maintenance strategies 194
6.5. Risk-sensitive MDP with probabilistic observation of states; an infinite-horizon case 200
    6.5.1. A Branch and Bound algorithm 201
    6.5.2. A Fibonnacci search method for Branch and Bound algorithm of a Markovian decision process with logical conditions 202
    6.5.3. Algorithm description 203
6.6. Inspection-maintenance-replacement model for technical systems 216
6.7. Maintenance processes where "cannibalization" is the only repair activity . 220

6.8. Generalized cannibalization policies in multicomponent systems	225
6.9. A Markov decision model for multicomponent system maintenance	231

Chapter 7. Inspection, diagnosis, reliability and maintenance policies for coherent structures — 234

7.1. Cause-effect models and Markovian decision processes 234
7.2. Model formulation 237
    7.2.1. The core process 238
    7.2.2. The observation space 239
7.3. Computing probabilities in cause-effect models and overall system dynamics 239
    7.3.1. State-space dynamics for Markov processes with logical conditions . 241
    7.3.2. The value of a complete analysis for maintenance policies 242
7.4. Inspection, diagnosis and maintenance policies 244
    7.4.1. A finite-horizon maintenance planning model 244
    7.4.2. An infinite-horizon maintenance planning model 250

Chapter 8. Semi-Markov maintenance and safety models for polyfunctional systems — 263

8.1. A semi-Markov population model for maintenance planning 263
8.2. Semi-Markov process statistics for maintenance modelling 265
8.3. An invariant index for polyfunctional systems 272
8.4. The vector semi-Markov process 273
8.5. Maintenance policies in polyfunctional technical systems experiencing semi-Markov deterioration. General formulation and "policy iteration" solution 275
8.6. Linear programming formulation for replacement strategies 279
8.7. Systems with a complex maintenance: an LP formulation 280
8.8. Degree of decomposition for maintenance strategies in large-scale technical systems 282
8.9. An implicit enumeration program for optimization and management of maintenance strategies in polyfunctional systems 286
8.10. The implicit enumeration (0—1) program model for maintenance management strategies 288

Part III. APPLIED STUDIES — 291

Chapter 9. Applications of systems engineering in maintenance-safety models and medical diagnosis — 293

9.1. Description of a BWR 293
9.2. Risk-sensitive Markovian policies for detection and maintenance of fatigue cracks 294
9.3. A probabilistic safety analysis for a power station equipped with a BWR . 299
9.4. Computer systems safety 310
9.5. Dynamic decision models for clinical diagnosis 317

List of notation 327

References 331

Introduction

Applied Systems Engineering deals with systems reliability, maintenance and safety analysis of complex systems. It was considered in our investigation that the behaviour of systems is described by a Markov or semi-Markov process, but whenever systems of practical interest (e.g. nuclear reactors — BWR) are considered, the concept of (complete or partial) observability of systems was incorporated into maintenance decision models.

For the case when the decision-maker is a risk-sensitive person and repair replacement maintenance actions must be taken, an appropriate risk-preference index is used in RSMDP. For the case when each individual function of the polyfunctional technical system has a multilevel performance index, a semi-Markov population model was developed in order to produce safety figures under the assumption that the system is coherent.

The main developments to be followed by the reader are:
— the class of MDP with logical conditions whose applications range from systems engineering to medical diagnosis (one pass algorithm, and Branch and Bound); a semi-Markov population model for maintenance planning and safety strategy of polyfunctional coherent systems; the operation of risk-sensitive Markovian models for modelling maintenance strategies in large technical systems; combined OR techniques for maintenance management planning in polyfunctional technical systems; a probabilistic methodology for safety analysis.

The identification of distinct problems in systems engineering concerns reliability and safety of complex systems evolving in a dynamic environment (see Chapter 1).

Chapter 2 discusses polyfunctional coherent structures and gives a formal mathematical representation for such structures. In the general case presented, a finite number of functions are assumed to operate at different levels of performance.

Chapter 3 is a study of MDP. As technical systems are allowed different degrees of observability, the theoretical models are discussed in consideration of the concept of partial and general observability. A full account of the well-known problem of real *versus* model world is given for the maintenance modelling of large and complex technical systems.

Chapter 4 introduces maintenance models for the case when the decision-maker is a risk-sensitive person and the deterioration process of the equipment is governed by a Markov or semi-Markov chain.

Chapter 5 is a study of a class of Markovian models for partially observable systems, whose underlying behaviour could be implicitly represented by the use of cause-effect models. The theoretical and practical value of such models is duly emphasized. Maintenance models using this class of MDP are studied for the case of finite and infinite horizon and appropriate solution algorithms are given (i.e. the "one pass algorithm" for the finite horizon case and the Branch and Bound solution for the infinite horizon process).

Chapter 6 studies the maintenance policies and safety analysis for polyfunctional technical systems when each function is allowed a multilevel behaviour described by a semi-Markov Decision Process (SMDP) and by hybrid OR techniques. One also has to include a set of constraints for the system operation. A semi-Markov population model is introduced for safety analysis in polyfunctional coherent structures, using results from Chapter 2.

Chapter 7 gives a methodology for probabilistic safety analysis on large-scale engineering systems such as nuclear power stations.

The third section, Chapter 8 in particular, deals with two applications of the theoretical models given before. Thus, a pressure vessel maintenance policy is given for a BWR system, under the assumption that the decision-maker is a risk-sensitive person. The safety methodology presented in Chapter 7 was applied to a nuclear power station, and complete practical results and conclusions are given.

The application of the class of models (MDP with l.c.) introduced in Chapter 5 extends beyond systems engineering and, among these converging areas, the problem of medical diagnosis applied to the respiratory system is considered (Chapter 9).

Part I

Basic problems in systems engineering

Chapter 1

Fundamentals and concepts in systems engineering

General systems theory conjoins many distinct areas of knowledge — from physical, biological, social, behavioural sciences to engineering management, mathematics and computer sciences. All these areas provide foundations and extensions for new interdisciplinary efforts in the field of systems science and cybernetics, bionics, human engineering, systems engineering, operations research.

As was emphasized by M'Pherson [119], "The breadth of vision and the unification of concepts afforded by systems science is timely because of the interdisciplinary framework that it provides both for comprehending the increasing complexity of today's social-ecological problems and for unravelling such problems by the better design of man-organized and man-made systems. It is here that systems engineering should come into its own for (with a wealth of engineering design experience behind it) it has been expanding its compass rapidly of late to become a (the?) methodology for the macro-design of man-made systems".

In what follows, we shall explore the fundamentals and concepts used in systems engineering. The field of systems engineering is viewed as "the creative realisation and maintenance of effective structures for operating-man-machine systems. It contributes to the realization of systems in the real world and provides its own brand of unification between disciplines. It needs assistance from many of its sister disciplines, perhaps most importantly from biocybernetics where the study of the control and organization of biological systems is beginning to provide concepts for the better design of complex man-machine systems" [119].

1.1. Field and boundaries of systems engineering

Systems engineering has been developed over the last 25 years with the view of elaborating and implementing appropriate methodologies for the study of complex man-machine systems. Such systems required a special treatment; they raised multiple problems with respect to their design, and special questions pertaining to economics-engineering, social-ecological constraints and environmental protection.

In the literature, the concept of systems engineering was given many concise definitions. But though numerous, these definitions are not disentangled in content. Warfield and Hall [120] consider that "the term 'systems engineering' has been chosen as the basis for structuring the approach and incorporating the methodology. The inclusion of the word 'engineering' has both benefits and disadvantages. The principal benefit is that the word carries with it a connotation of more than simply the study of a problem — engineering has always been oriented toward application and implementation. Analysis, an approach, the use of a particular methodology — none of these carries the powerful implication of *implementation* for the benefit of mankind. The principal disadvantage is that some will probably believe that the choice of the word implies that only people who are educated in and engaged in one of the recognized branches of engineering engage in implementation. This disadvantage, hopefully, can be dispelled by stating here that such is not the intent". The same authors, in an effort for a unified systems engineering concept, define the field of systems engineering as the engineering management, direction, control, and technical effort applied to a total system to ascertain and maintain overall technical integrity and integration of that specific system as related to design configuration, reliability and performance.

Another author (Rau [11]) considers that systems engineering "consists of the application of scientific methods in integrating the definition, design, planning, development, manufacture, and evaluation of systems. It encompasses such terms as systems approach, system functional analysis, system reliability analysis, task analysis, maintenance analysis and operation analysis. It is fundamentally concerned with deriving a coherent total system to achieve a stated set of objectives subject to physical, environmental, state-of-the art, and economical constraints".

The complex process of system engineering is schematically plotted in fig. 1.1 and represents a sequential process of system definition, system development and system evaluation.

(1) *System definition* specifies what a specific system is required to do, the attributes and functional relationships between its components and its basic design.

(2) *System development* (according to Rau [11]) "consists of finalizing the system design, planning and implementing its development process, and actual system production".

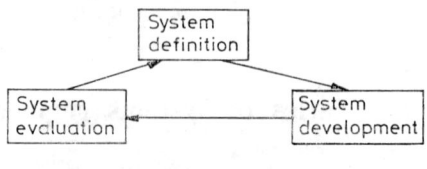

Fig. 1.1

(3) *System evaluation* places the system in operation, it observes its behaviour, compares its observed behaviour with that which is required, recommends means for extending both the life-cycle and usefulness of the system and evaluates the overall system performance.

In what follows, we shall adopt the definition given by M'Pherson [119], which presents the field of systems engineering as "the over-all realization of man-machine systems such that the resources allocated to the system satisfy stated objectives during the whole of the system life-cycle".

The key words (e.g. realization, man-machine system) in the above definition require some explanation. Throughout the book, realization is understood to denote the design planning, implementation, evaluation and management of the complex system engineering process. According to M'Pherson, a man-machine system is a system in which (1) human and technological-machine functions are either coordinated or integrated to a greater or lesser extent in the achievement of system goals and constraints and where (2) human and technological systems interact. The concept of system is considered to be the complex of man-machine objects interrelated in a purposeful and creative structure open to learning, capable of structural stability, development and adaptability. In fact, such a system is finally considered to be resilient, robust and flexible either on a short or a long time horizon.

The birth-to-scrap time sequence through which a specific (e.g., engineering or biomedical) system evolves and operates is known as *system life-cycle*. It can be partitioned, for specific reasons concerning systems engineering, into three or four main phases, each of them with several stages. A diagrammatical representation of the concept of systems engineering is given in fig. 1.2.

It is clear from the above that the process of systems engineering refers mainly to the optimal design and appropriate management of a large variety of man-machine systems. In this chapter, machine is understood to denote "any technological complex ranging from machines *qua* machines", through industrial plant

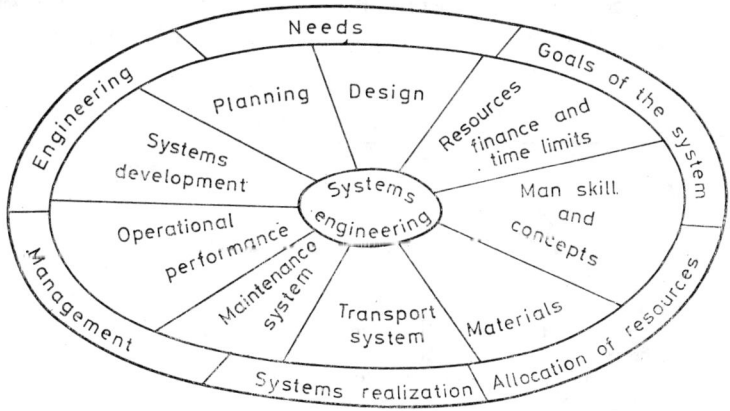

Fig. 1.2

to world communication systems, "as well as any man-made environment, from buildings through urban environments to ecological systems in which man's social needs and his technological desires interact either directly or through the natural environment" [119]. Concerning the concept of system life-cycle (see fig. 1.3), one can mention that the design phase is characterized by an iterative-cycle process and the operational phase is itself considered to be a cyclic process. The feedback system described in the figure below indicates either design modification, operational system modification in the light of experience and learning or design updating and the penetration of innovation in any of the next generation of the system.

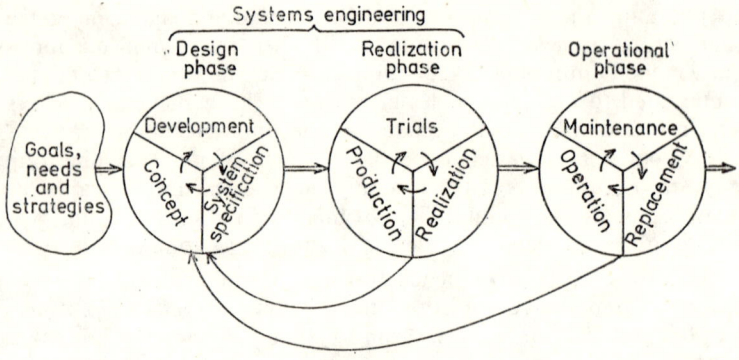

Fig. 1.3

The whole process is very time-consuming and "typically for a large technological system, the design and acquisition phases may take 10 years while the operational phase may last for 20 years" [119]. The substitution of different technologies is a time-consuming (see table 1.1) operation and the systems engineer must be very careful in doing this throughout the phases of existence of technological systems.

Table 1.1

Substitution for different technologies	The takeover time of the substitution (years)
1. Synthetic/Natural Rubber	58
2. Plastic/Natural Leather	57
3. Electric Arc/Open Hearth Specialty Steels	47
4. Open Hearth/Bessemer Steel	42
5. Plastic/Hardwood Residence Floors	25
6. Synthetic/Natural Tyre Fibres	17.5
7. Plastic/Metal Cars	16
8. Detergent/Natural Soap (US)	8.75

In systems engineering, man-machine systems and the concept of total systems design must be considered explicitly in the operation of machines and equipment, this, of course, in the presence of the operating personnel. However, the latter's professional knowledge and skill, capacity for adaptability and learning concerning the operation and behaviour of complex machines and systems are implicit in the overall design process. This class of systems could be represented using the hierarchy of man-machine communication systems (see fig. 1.4).

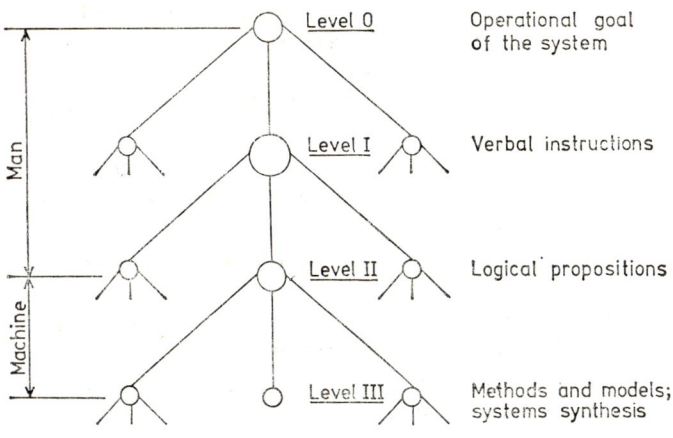

Fig. 1.4

The problems of systems engineering must be integrated in the interplaying man-machine systems (see fig. 1.5) [119]: "Overall systems design requires the proper assignment of functions to men and machines with human knowledge

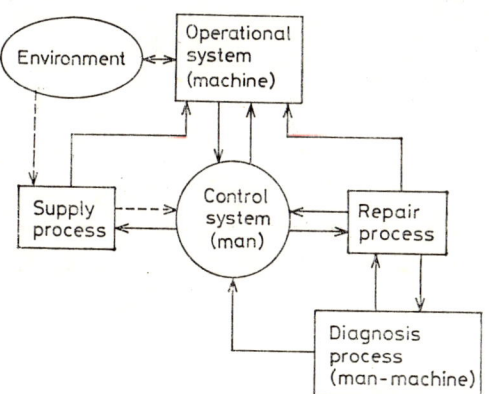

Fig. 1.5

and skills as a proper resource component for system design. The allocation takes into account the factors as inherent machine-human capabilities, environment, job satisfaction, efficiency, interfacing, costs".

Systems engineering deals with the design and acquisition phases.

Design means the design development and specification of the machine system (e.g. the equipment and systems of equipment for the operational system, the technological support maintenance and logistic systems to maintain the operational system in an acceptable state of efficiency and performance) and human system (e.g. the tasks for the human personnel by skill, trade and responsibility, training selection, promotion procedures for system personnel).

The *acquisition* phase deals with the production testing and trials of the equipment plus the selection and training of system personnel.

The complex management process of the system engineering and operational phase is shown in fig. 1.6 [119]. On examining this figure, several pieces of practical advice for a systems engineer may be derived:

1) design for the operational phase must include the design or operational control and information systems;

2) systems engineering phases are concerned with the design of their own processes as well as with their own management;

3) the optimal time-cost schedule must be designed to achieve a delivered system on time and within the stated budget;

4) research and development processes in the technological and human engineering field will forecast substitution, penetration and diffusion of technologies and economic environments in which the system may operate;

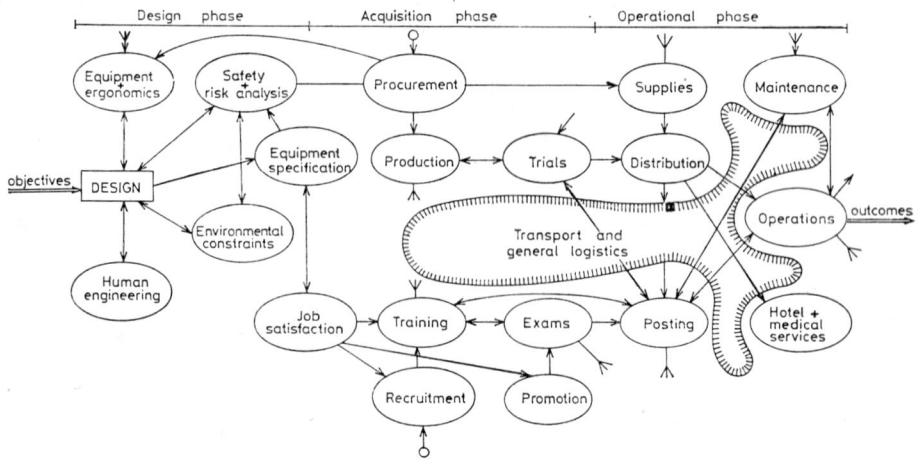

Fig. 1.6

5) many operational organizations involved in the systems engineering process are embedded as subsystems of the larger organization or institution which "sets the design and operational strategies and provides the resources to create and operate the system" [119].

In ref. [119], it is considered that the field of systems engineering has the following three major dimensions:

a) *the knowledge dimension* — encompassing specialized knowledge from the disciplines and profession (e.g. engineering, medicine, business, management, mathematics, social sciences, etc.);

b) *the time dimension*, extending from initial conception through retirement or major modification of the system;

c) *the logic dimension* which deals with the steps that are carried out in the field of systems engineering.

Figure 1.7 gives a portrayal of the three categories as elements of the "tool space" of systems engineering.

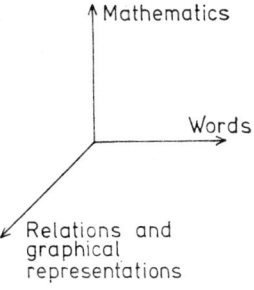

Fig. 1.7

1.2. Systems definitions, decision processes, information processes and hierarchies

In this paragraph, we shall present set-theoretic definitions of systems and systems elements in order to conjugate the general descriptive approach to the concept of systems engineering given above and more precise mathematical models used in the practice of management science and systems engineering. These definitions and formalized concepts will be used in the next chapters in view of elaborating models specific to the field of systems engineering with special application to reliability, safety and maintenance.

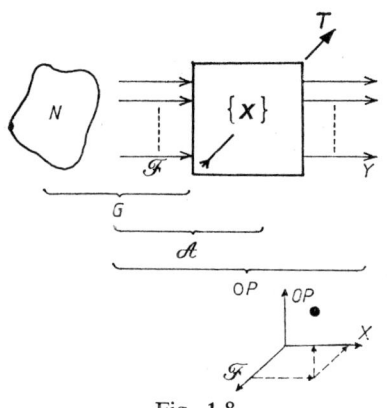

Fig. 1.8

Throughout this book, a system is considered to be open; the system is realized in the view of having a given set of functions operate in time in the presence of the human operator. The basic definition of systems in the context of systems engineering is taken from M'Pherson [119]. It provides a comprehensive framework from which other subsequent system models used will be developed. This definition also incorporates notions from system definitions given elsewhere by Klir [121], Wymore [124], Lee [122] and Mesarovic [123].

DEFINITION 1.1. A system $\mathscr{S}$ is represented by a set (see also fig. 1.8)

$$\mathscr{S} = \{N, G, \mathscr{F}, \mathscr{A}, X, PO, Y, T\}, \qquad (1.1)$$

where N represents the set of environment objects, $G: N \to \mathscr{F}$ is the mapping of environment states into system inputs, $\mathscr{F}$ represents the set of admissible system input functions, X represents the allowable state space for the system, $\mathscr{A}: \mathscr{F} \to X$

is the state transition function for the system, $OP: \mathscr{F} \otimes X \to Y$ is the operational process of the system, Y is the set of possible output functions from the system and T represents the time scale.

DEFINITION 1.2. The OP as the system process structure is given by relation

$$OP: F \otimes Q \to X, \qquad (1.2)$$

where Q is the set of crosscouplings between the processes and results of the decomposition

$$\mathscr{B}: F \otimes X \to Q \qquad (1.3)$$

and $\mathscr{B}$ is the crosscoupling function.

The *decision process* in systems engineering is illustrated in fig. 1.9 for the case when the decision-making process is considered under uncertainty. It is also assumed that the decision-maker could be a risk-sensitive person and an appropriate measure of his sensitiveness is given by a utility function.

We choose an alternative $d_i^* \in \mathscr{D}$, $\forall i$ from a set of possible decision alternatives $\mathscr{D}$ such that the utility measure $u_{ij} \in \mathscr{U}$ from the possible outcome $m_i \in M$ be maximal in the context of the stated set of utilities $\mathscr{U}$ and the expected future environment $p[n_j]$. The decision taken is that which maximizes the expected utility such that

$$\langle \mathscr{U} | d_i, m_i, p[n_j] \rangle = \max_{i} \sum_{j} u_{ij} p[n_j], \qquad (1.4)$$

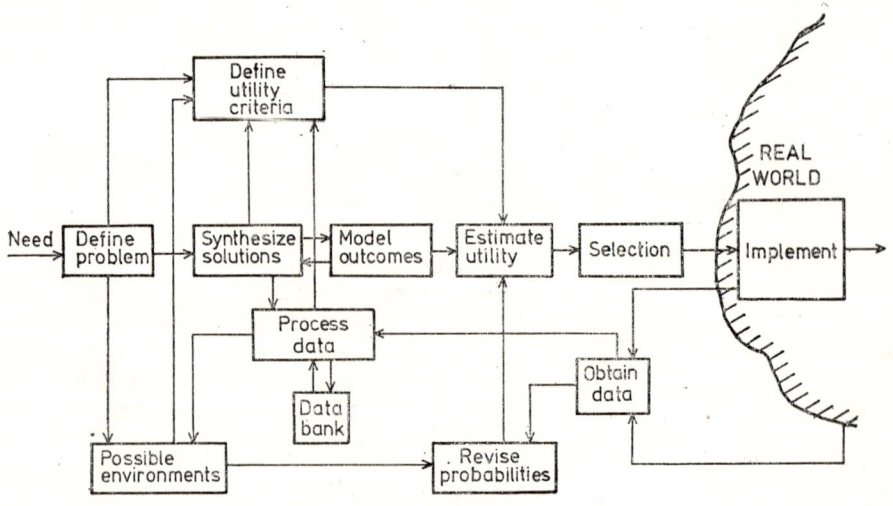

Fig. 1.9

where $\langle . | . \rangle$ indicates the expected utility for the decision process. It is clear that the decision process is subjectively biased as the human operator is in an uncertain and fuzzy decision environment: "decision-making is subjective, based on estimates of reality and coloured by the decision-maker's attitudes" [119].

DEFINITION 1.3. A *decision unit* for a decision process is represented by the set

$$DU = \{\mathscr{D}, \mathscr{I}, M, \mathscr{U}, G\}, \quad (1.5)$$

where $\mathscr{D}$ is the set of admissible decisions, $\mathscr{I}$ represents the information input set, M is the set of possible outcomes, $\mathscr{U}$ represents the utility criteria set, and G is the performance function

COROLLARY 1.1. $DU: \mathscr{I} \otimes \mathscr{D} \to M.$ (1.6)

COROLLARY 1.2. $G: \mathscr{I} \otimes M \to \mathscr{U}.$ (1.7)

COROLLARY 1.3. *Selection of an action/decision d_i is governed by the relation*

$$g(d_i, M_i, \mathscr{I}) \geqslant g(d_j, m_j, \mathscr{I}); \quad i \neq j. \quad (1.8)$$

COROLLARY 1.4. *A policy D is the set of individual decisions selected for each possible information input such that $D = \{(d_{ij}, i_j)\}$, where $d_{ij} \in \mathscr{D}$ and $i_j \in \mathscr{I}$.*

The management of systems engineering phases and operations implies an *information process (IP)* which essentially consists of a data processor (DP) and an information distributor (ID). Raw data H(N) from the system and environment is processed either directly or *via* a model for a future quantitative or a qualitative approach. For large quantities of data or for efficient management results using OR or other mathematical models, we need computers and appropriate system and human support.

COROLLARY 1.5. $IP = H(N) \otimes H(X, Y) \to \mathscr{I}.$ (1.9)

COROLLARY 1.6. $ID = \mathscr{I} \to \{\mathscr{I}_i\}.$ (1.10)

A *control system (CS)* includes a decision unit (DU), an appropriate information process/system and an implementation process (PM). The task of the control system is to determine and evaluate an optimal policy. Using the above comments and notation, the control system in the framework of systems engineering is represented as in [119].

COROLLARY 1.7.

$$CS: \{\underbrace{(N \otimes X \to \mathscr{I})}_{IP} \to D\} \xrightarrow{\max G} M \quad (1.11)$$

$$CS: N \otimes X \underset{G}{\to} \widetilde{M}, \quad (1.12)$$

where $\widetilde{M}$ is the optimal control set.

DEFINITION 1.4. The performance function G of a controlled system is expressed for the following three cases of interest.

a) Continuous control

$$G(t) = \max_{m(t)} \int_{t_1}^{t_f} g(x, n, y, m, t) \, dt; \quad (1.13)$$

b) Discrete control (e.g. policy iteration, dynamic programming, game theory);

$$G(r) = \max_{m(r)} \{G_{\max}(r-1) + g(r)(x, n, u, m, r)\}; \quad (1.14)$$

c) Optimal design (e.g. multivariable search in automation and control engineering):

$$G = \max_{m} \{E(x, n, u, m, t)| L(x, m, t)\}, \quad (1.15)$$

where generally $E(.)$ is for system effectiveness and $L(.)$ for the system cost.

In the literature [119], a total (whole) system is characterized by four types of functional system: operational system, maintenance system, transport system and procurement system. According to M'Pherson, (1) an operational system will be specific in its nature and structure to its specific class of operation (e.g. power generation, production), (2) the maintenance system is designed to keep the previous system at an acceptable level of operational performance, (3) the transport system conveys resources within the whole system and (4) the procurement system is responsible for storing the resources of the total system.

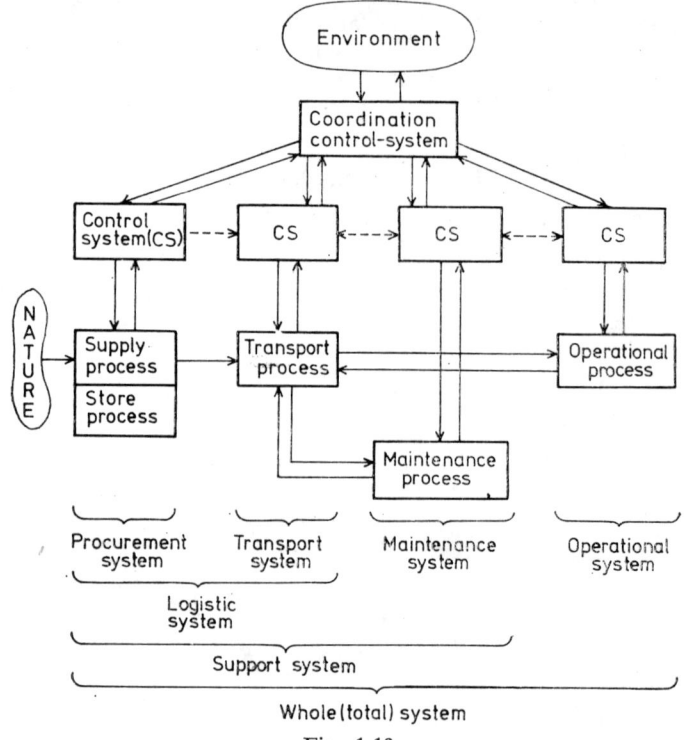

Fig. 1.10

A diagrammatical representation for a total system is given in fig. 1.10. At the higher level, the coordination control system (CCS) acts on an effective coordination between the different systems of the total system. The control systems, on the other hand, are hierarchically structured at the functional and level stages. The functional hierarchy depends on self-organization, learning, adaptiveness, and feedback of the system. The level in the hierarchy is described by the strategic, tactical, coordination, operational and process control.

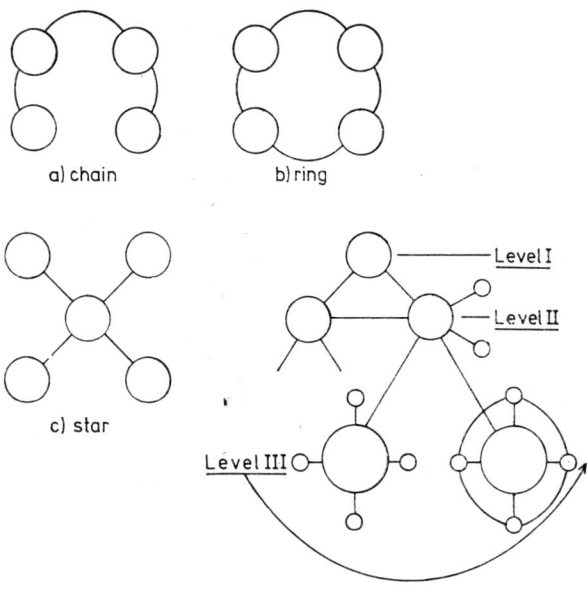

Fig. 1.11

An important role in system design is played by the degree of organization of such entities. The recognition and evaluation of simple structures play an important role in practical applications of systems engineering. In fig. 1.11 a few basic structures in the analysis and design of complex systems are given. These have multiple applications, ranging from the design of technological production lines to the management of industrial processes, computer networks and the evaluation of international cooperation strategies.

The chain structure could describe a sequence of operations, a production line or a chain of commands and instructions; the ring structure represents cyclic operations, feedback systems. Complex structures could be viewed as basic structures for many physical and artificial (man-made) processes up to political systems with cross-checks and balances. Conventional management trees and control operating systems are well represented by a star structure. As an example for a simple coalition, we may take the maintenance system or coordination control of a transport system, for a complex coalition, which is illustrated in [119], the central control imposed on a large and complex technological and/or social system. All these structures can be integrated in a hierarchical organization.

Systems design takes into consideration the multiple aspects emphasized so far in the framework of systems engineering. It is not infrequent that such aspects may interact and appear conflictual in nature; systems analysts and designers have to transform them into equipments and operating systems able to achieve the design objectives as well as other sets of constraints (e.g. cost investment and operational costs, reliability, etc.).

DEFINITION 1.5. The *systems engineering process* is given by the mapping

$$SE: N \otimes \mathscr{F} \xrightarrow{G^*_{WS}} WS, \qquad (1.16)$$

where N and $\mathscr{F}$ are environmental and resource inputs and WS is the whole process.

The *objective function* for the whole system G_{WS} is expressed by $G_{WS} = g$ [G (Schedule Control); G (Operational System); G (Support System)].

Optimal value for an SE problem is given by the functional

$$G^*_{WS} = \min_{M_{SE}} \left\{ \int_{t_0}^{t_1} g_{SC}(X_{SC}, M_{SE}, N) \, dt \right\} +$$
$$+ \max_{M_{OS}, M_{SS}} \{E | L \ (X_{OS}, X_{SS}, M_{OS}, M_{SS}, N, U)(t_f - t_c)\}, \qquad (1.17)$$

where g_{SC} is the performance function for the schedule control (SC), X_{SC} represents the trajectory of SC, M_{SE} are the model outcomes from SE. The second term in the above functional relation represents the measure of the system cost-effectiveness.

In the above adapted model we have considered that real systems operate in environments with natural or man-made perturbations (e.g. human error), uncertainty in the evolution of the system, multiple interactions with the environment, penalties for an improper operation and low quality performances. In the realization of the total system, the planning process is dependent on additional cost and time constraints such that

$$\sum_i r_i \leqslant t_c; \quad \sum_j L_{ij}(r_j) \leqslant L_{\max}, \quad (i = 1, 2, \ldots). \qquad (1.18)$$

In the above relation r_j represents the time for which a system can be in state j and $L_{ij}(.)$ represents the cost of a transition between two states of the system. Parameters t_c and $L_{\max}$ represent the start time for the system and the upper limit of available resources, respectively.

The following steps are involved in the design of an operational total system. a) Make a separate design of the operational and support systems; b) balance these so that the total system is optimal with respect to one or more decision criteria.

For practical reasons "whole-system design should achieve an overall optimum with respect to the two criteria that matter most: performance and cost" [119]. In fig. 1.12 we give a flow diagram for the cost-effectiveness design process. We can identify that each system design intensively interacts with the other and the common denominator which has to be taken into consideration is that of effectiveness. The system design process is considered an art and a science. "... all good system design calls for artistry — artistry in the imaginative and insightful selection of entities and structuring of their attributes".

By coupling his artistic skill and the operational tools of system design available to him, the designer develops a "feeling" for the "key" attribute relationships among the "key" entities in a system. Thus, the system design process is a search to find the "key" entities and establish the "key" relationships which will produce the objectives sought [119].

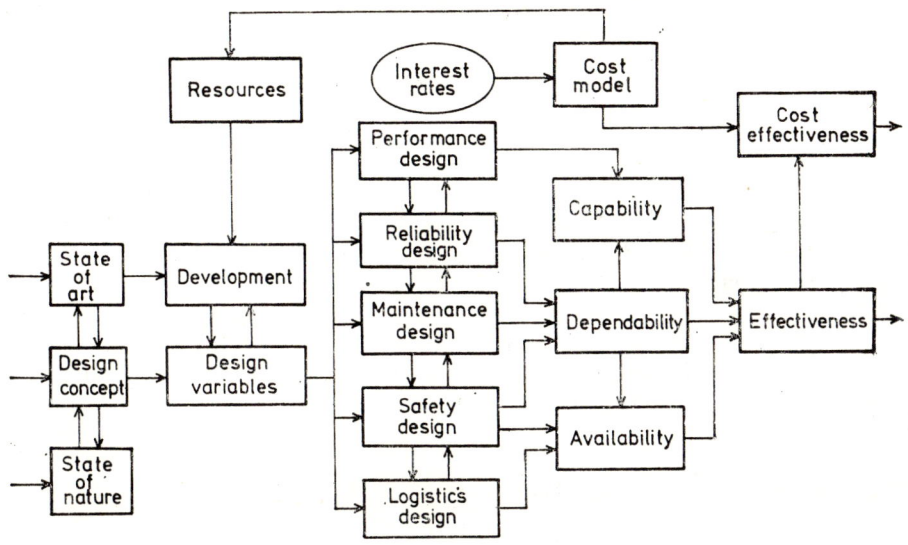

Fig. 1.12

With regard to system design process, Nugent and Vollmann consider the following steps:

a) identifying goals;

b) imagining attributes that would help to achieve these goals;

c) imagining entities that have the attributes which are sought;

d) imagining a structure among those atributes that will help to achieve the goals.

As far as the system designer is concerned, he must be able to identify appropriate criteria and thereafter distinguish "valuable" attributes and entities which possess those attributes. Also, he must have the ability to include and structure, as measured by effects on the criteria set and the skill in knowing when to structure or to make a synthesis.

The systems engineer must be creative and innovative in his effort to define goals and criteria for the system. According to Hall, "perhaps the most creative act of all is the initial one, the ability to ask a new question...", and "as an extension of the same idea, the hardest job is not finding answers, it is asking pertinent (or even impertinent) questions. The creative process of inclusion and structuring follows from the asking of pertinent questions".

For technological and engineering systems of a series-parallel type (see fig. 1.13) with k independent and not necessarily identical components, where p_i and c_i are the reliability and cost, respectively, of the i-th part ($i = 1, 2, \ldots, k$), the system reliability is of the form

$$e(r_1, r_2, \ldots, r_k) = \prod_{i=1}^{k} [1 - (1 - p_i)^{r_i}]. \tag{1.19}$$

The appropriate value for the total cost of such a system is given by

$$c(r_1, r_2, \ldots, r_k) = \sum_{i=1}^{k} c_i r_i. \tag{1.20}$$

For non-series-parallel systems, the problem of computing system reliability is more difficult to handle and few algorithms are given in the literature (see Kim [196], Cârlan [260]).

Now, a simple computational procedure will be presented for allocating redundancy among subsystems. The multistage system has to achieve maximum reliability subject to multiple constraints, either linear or nonlinear. Examples are shown. The algorithm given below is due to Sharma and Ventkateswaran [116].

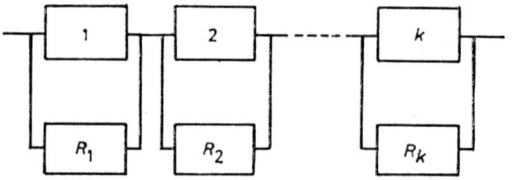

Fig. 1.13

Consider an engineering system with k stages (subsystems) connected in series. At stage i, r_i, ($i = 1, 2, \ldots, k$), represents the number of redundant (parallel) elements, each having probability of failure $q_i = 1 - p_i$. It is assumed that failure of each individual element occurs independently and the probability of failure is $q_i \neq q_j$ for $i \neq j$. The system unreliability is calculated by the relation

$$Q(r) = e(r_1, r_2, \ldots, r_k) = 1 - \prod_{i=1}^{k} (1 - q_i^{r_i}), \tag{1.21}$$

where r is a vector of non-negative integers such that $r = (r_1, r_2, \ldots, r_k)$. The allocation of redundancies obeys the following constraints which are not necessarily linear

$$\sum_{i=1}^{k} G_{ij}(r) \leq b_j, \quad (j = 1, 2, \ldots, n), \tag{1.22}$$

where n represents the total number of constraints associated with the system. For $q_i \to 0$, we notice that

$$Q(r) \cong q(r) = \sum_{i=1}^{k} q_i^{r_i}. \tag{1.23}$$

1.3. Cost-effectiveness models in systems engineering. Fundamentals and algorithms

As was already mentioned, the selection of an optimal design solution for complex technical systems implies consideration of a complete set of parameters, such as reliability, value, penalties, weight, volume, system's efficiency, utility, etc. In a simplified model for analysis, we can use such parameters as cost and effectiveness. However, "effectiveness by itself is just a number; all one may infer is that a larger number indicates increased effectiveness" [119]. It has become customary in the field of systems engineering to develop a cost model simultaneously with the effectiveness model. In fig. 1.14 a few ways of dealing with cost-effectiveness comparisons are indicated. In all these models the systems engineer has to take due account of the available resources and how effectiveness and cost depend upon resources [11].

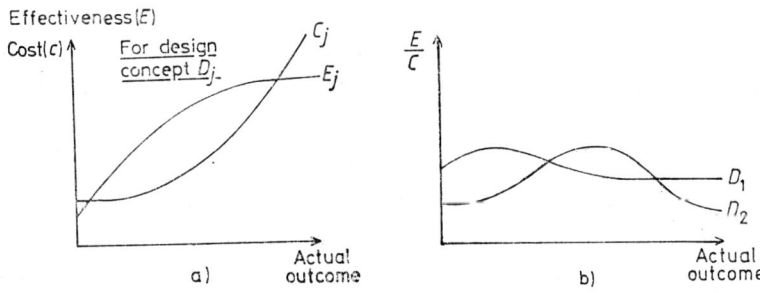

Fig. 1.14

Consider k types of resource; in this case, system effectiveness is of the form $e(r) = e(r_1, r_2, \ldots, r_k)$ and the system cost is $c(r) = c(r_1, r_2, \ldots, r_k)$, where r_i denotes the amount of the i-th resource used in the realization of the system. We can formulate the following three problems:

PROBLEM I: max $e(r_1, r_2, \ldots, r_k)$ subject to $c(r_1, r_2, \ldots r_k) \leqslant C$; $0 \leqslant r_i \leqslant R_i$ for $1 \leqslant i \leqslant k$, where R_i is a constraint on each resource of type i and C is an upper cost allowed for the system.

PROBLEM II: min $c(r_1, r_2, \ldots, r_k)$ subject to $e(r_1, r_2, \ldots, r_k) \geqslant E, 0 \leqslant r_i \leqslant R_i$ for $1 \leqslant i \leqslant k$, where E is a minimum performance level for the system.

PROBLEM III: max $\dfrac{e(r_1, r_2, \ldots, r_k)}{c(r_1, r_2, \ldots, r_k)}$, subject to $0 \leqslant r_i \leqslant R_i$ for $1 \leqslant i \leqslant k$.

As a result of applying one of the optimization processes given above, an alternative corresponds to an expenditure of resources and, thus, the optimal alternative is the k-tuple $(r_1^*, r_2^*, \ldots, r_k^*)$ which satisfies the desired criterion formulation.

The problem consists of maximizing the system reliability subject to a set of pertinent constraints (e.g. cost, weight, volume) and then we can write

$$\min q(r) = \sum_{i=1}^{k} q_i^{r_i}, \tag{1.24}$$

subject to

$$\sum_{i=1}^{k} G_{ji}(r) \leqslant b_j, \quad (j = 1, 2, \ldots, n). \tag{1.25}$$

As Sharma et al. [116] appreciate, "a good way to reduce $q(r)$ is to pick out the maximum term on the right-hand side and reduce it by adding a redundant unit to the corresponding stage". In successive iterations, the systems analyst can reduce $q(r)$. In doing this, we have to add one redundant element to the stage with the highest q^{r_i}, assuming that the system constraints are not violated. Although the nature of this algorithm is extremely intuitive, we may have good practical results without using special mathematical techniques which are sometimes difficult to apply. The algorithm [116] is carried out in the following steps.

Step 1: Assign $r_i = 1$ for all i, $(i = 1, 2, \ldots, k)$. Because we deal with a series system, there must be at least one element in each stage. Check to see that no constraints are violated.

Step 2: Find the stage that has the highest unreliability and add a redundant component to that stage (subsystem).

Step 3: a) If any constraint is violated, go to Step 4.

b) If no constraint has been reached, go to Step 2.

c) If any constraint is exactly satisfied, STOP. The current r is the optimum number of redundant components.

Step 4: Remove the redundant component added in Step 2. The resulting number is optimum for the stage. Remove this stage from further consideration.

Step 5: If all stages have been removed from consideration, STOP. The current r is optimum. Otherwise, go to Step 2.

Example 1 [116]. Consider a system with five subsystems connected in series. The operation probabilities p_i are given in table 1.1. The constraint set is of a linear form. Find the allocation vector $r = (r_1, r_2, r_3, r_4, r_5)$ which will maximize system reliability.

We shall apply the above given algorithm starting with $r = (1, 1, 1, 1, 1)$ and add one element at a time, as is shown in table 1.2. The optimum number of redundant components is given by $r_{opt} = (3, 4, 5, 4, 3)$. The system reliability for r_{opt} is 0.985.

Table 1.2

Number of events					System non-reliability					Total weight	Total cost
r_1	r_2	r_3	r_4	r_5	1	2	3	4	5		
1	1	1	1	1	0.1	0.25	0.35[a]	0.20	0.15	38	32
1	1	2	1	1	0.1	0.25[a]	0.1225	0.20	0.15	44	41
1	2	2	1	1	0.1	0.0625	0.1225	0.20[a]	0.15	53	45
1	2	2	2	1	0.1	0.0625	0.1225	0.04	0.15[a]	60	54
1	2	2	2	2	0.1	0.0625	0.1225[a]	0.04	0.0225	68	59
1	2	3	2	2	0.1[a]	0.0625	0.042875	0.04	0.0225	74	68
2	2	3	2	2	0.01	0.0625[a]	0.042875	0.04	0.0225	82	73
2	3	3	2	2	0.01	0.015625	0.042875[a]	0.04	0.0225	91	77
2	3	4	2	2	0.01	0.015625	0.015006	0.04[a]	0.0225	97	86
2	3	4	3	2	0.01	0.015625	0.015006	0.008	0.0225[a]	104	93
2	3	4	3	3	0.01	0.015625[a]	0.015006	0.008	0.003375	112	100
2	4	4	3	3	0.01	0.003906	0.015006[a]	0.008	0.003375	121	104
2	4	5	3	3	0.01[a]	0.003906	0.005252	0.008	0.003375	127	113
2	4	5	3	3	0.001	0.003906	0.005252	0.008[a]	0.003375	135	118
3	4	5	4	3	0.01	0.003906	0.005252	0.0016	0.003375	142	125

[a] This is the stage to which a redundant component is to be added.

Example 2 [*116*]. Consider a system with five subsystems having the reliability characteristics given below.

Subsystem i	1	2	3	4	5
p_i	0.80	0.85	0.90	0.65	0.75

The set of restrictions is nonlinear and has the form

$g_1(r) = r_1^2 + 2r_2^2 + 3r_3^2 + 4r_4^2 + 2r_5^2 \leq 110$

$g_2(r) = 7(r_1 + e^{r_1/4}) + 7(r_2 + e^{r_2/4}) + 5(r_3 + e^{r_3/4}) + 9(r_4 + e^{r_4/4}) + 4(r_5 + e^{r_5/4}) \leq 175$ (1.26)

$g_3(r) = 7r_1 e^{r_1/4} + 8r_2 e^{r_2/4} + 8r_3 e^{r_3/4} + 6r_4 e^{r_4/4} + 9r_5 e^{r_5/4} \leq 200$.

Using the procedure given by Sharma and Venkateswaran, the optimal solution is $r_{opt} = (3, 2, 2, 3, 3)$ with reliability 0.9045 (see also table 1.3).

The second algorithm uses a computational method and techniques of variational calculus. The purpose of the model is to obtain an optimal redundancy allocation of series-parallel systems and the objective is to maximize the system profit. The procedure is developed for k-stage parallel systems. The problem was intensively investigated by Fan, Wang, Tillman and Hwang [125]. The total cost of a system with parallel redundancy is of the form $\sum_{i=1}^{k} C^i r^i$ and the net profit NP for the entire system is

$$NP = PR_s - \sum_{i=1}^{k} C^i r^i, \quad (1.27)$$

where P represents the profit provided that the system operates successfully, R_s is the reliability of the series-parallel system, C^i is the component's cost from stage i and r^i represents the redundancy level at stage i, $(i = 1, 2, \ldots, k)$.

We shall present now the computational algorithm, as well as other details about the model. General computational relations are given by the following equations:

a) $U^i = 1 - R^i$ and represents the operational unreliability of the i-th component in the series-parallel system;

b) $y_i^i = (U^i)^{r^i}, \quad (i = 1, 2, \ldots, k)$ (1.28)

c) $a^i = \dfrac{-C^i}{\ln U^i}, \quad (i = 1, 2, \ldots, k)$ (1.29)

d) $NP = PR_s - \sum_{i=1}^{k} C^i r^i$ (1.30)

Fan et al. [125] gave the following computational algorithm.

Table 1.3

Number of events					System non-reliability					Constraints		
r_1	r_2	r_3	r_4	r_5	1	2	3	4	5	$g_1(r)$	$g_2(r)$	$g_3(r)$
1	1	1	1	1	0.2	0.15	0.1	0.35a)	0.25	12	73.1	48.8
1	1	1	2	1	0.2	0.15	0.1	0.1225	0.25a)	24	85.4	60.8
1	1	1	2	2	0.2a)	0.15	0.1	0.1225	0.0625	30	90.8	79.0
2	1	1	2	2	0.04	0.15a)	0.1	0.1225	0.0625	33	100.4	93.1
2	2	1	2	2	0.04	0.0225	0.1	0.1225a)	0.0625	39	109.9	109.2
2	2	1	3	2	0.04	0.0225	0.1a)	0.04875	0.0625	59	123.1	127.5
2	2	2	3	2	0.04	0.0225	0.01	0.04875	0.0625a)	68	130.0	143.6
2	2	2	3	3	0.04a)	0.0225	0.01	0.04875b)	0.015625	78	136.0	171.1
3	2	2	3	3	0.008	0.0225	0.01	0.04875b)	0.015625	83	146.1	192.5

a) This is the stage to which a redundant component is to be added.
b) This indicates that the stage has been removed from further consideration.

Step 1: Let $y^1(1)$ and $y^1(2)$ be assumed with the condition that $1 > y^1 > 0$ because $y^1 = (U^1)^{r^1}$, where $1 > U^1 = (1 - R^1) > 0$ and r^1 is a positive unknown integer.

Step 2: The value of E is computed such that

$$E = \frac{a^1(1 - y^1)}{y^1}. \qquad (1.31)$$

Step 3: The values of y^i, $(i = 2, 3, \ldots, k)$, are computed from the following equations

$$y^i = \frac{a^i}{E + a^i}, \quad (i = 2, 3, \ldots, k), \text{ with the specification that}$$

$$E = \frac{a^1(1 - y^1)}{y^1} = \frac{a^i(1 - y^i)}{y^i}. \qquad (1.32)$$

Step 4: A computation procedure is made for determining $S = P \prod_{i=1}^{k} (1 - y^i)$.

Step 5: The error ER is computed such that $ER = S - E$.

Step 6: The value of $y^1(3)$ is computed from the equation

$$y^1(3) = \frac{y^1(2) - ER(2) \, y^1(1)/ER(1)}{1.0 - ER(2)/ER(1)}. \qquad (1.33)$$

Step 7: Steps 2 ÷ 5 are iterated to obtain $ER(3)$.

Step 8: A check is made to verify if

$$\{|ER(3)| - (ER)_{\max}\} \leq 0. \qquad (1.34)$$

Step 9: If inequality (1.34) from step 8 is satisfied, then $y^1(3)$ is the required y^1 and so are $y^i(3)$ for $i = 2, 3, \ldots, k$. r^i, $(i = 1, 2, \ldots)$, are computed by the equation

$$r^i = \ln y^i / \ln U^i, \quad (i = 1, 2, \ldots, k) \qquad (1.35)$$

Stop.

Step 10: If inequality (1.34) from step 8 is not satisfied, then $y^1(1)$ and $y^1(2)$ are replaced by $y^1(2)$ and $y^1(3)$, respectively, and go to Step 6.

Example 3 [125]. Consider two technical systems of a series type, the first with $k = 3$ and the second with $k = 8$ subsystems. In order to improve the reliability performance of these systems, parallel redundancy may be used.

Computational data for the system $k = 3$ are presented in table 1.4. The net profit for the system with zero redundancies is $NP = P \prod_{i=1}^{3} R^i - \sum_{i=1}^{3} C^i = -0.95$.

The optimum redundancy obtained by following the above optimization algorithm is $r^1 = 6.7$, $r^2 = 2.6$, $r^3 = 1.7$ (see also table 1.5). Since all r^i, $i = 1, 2, 3$, should be positive integers, then

$r^1 = 7$, $r^2 = 3$, $r^3 = 2$ and the optimum net profit is $NP = 1.323$ monetary units. Using a similar methodology for the complex system $k = 8$ with data from table 1.6, the net system profit $NP = P \prod_{i=1}^{8} R^i - \sum_{i=1}^{8} C^i = 8.74$ when no redundancies are allowed.

Table 1.4

Subsystem	R	C
1	0.333	0.200
2	0.500	1.000
3	0.750	1.000

$P = 10.00$;   3 subsystems

Table 1.5

Subsystem	Optimal value	Number of parallel components
1	6.720816	7
2	2.553489	3
3	1.713077	2

Maximum profit = 1.4208.
Real profit = 1.3230.

Table 1.6

Subsystem	R	C
1	0.900	0.500
2	0.750	0.400
3	0.650	0.900
4	0.800	0.700
5	0.850	0.400
6	0.950	1.000
7	0.750	0.800
8	0.600	

$P = 100.00$;   8 subsystems

Table 1.7

Subsystem	Optimal value	Number of parallel components
1	2.646672	3
2	4.191516	4
3	4.503256	5
4	3.356344	3
5	2.933708	3
6	2.196330	2
7	3.533785	4
8	5.139768	5

Maximum profit = 86.0585.
Real profit = 75.6421.

Optimal redundancy for $k = 8$ is presented in table 1.7. Since all r^i, $1 \leqslant i \leqslant 9$, should be positive integers, then $r^1 = 3$, $r^2 = 4$, $r^3 = 5$, $r^4 = 3$, $r^5 = 3$, $r^6 = 2$, $r^7 = 4$, $r^8 = 5$ and the optimum net profit $NP = 75.64$.

Let us present now a different computational procedure for maximizing reliability of multistage parallel systems subject to multiple nonlinear constraints (see Tillman et al. [168]).

The problem is stated as follows

$$\text{Max } R_s = \prod_{i=1}^{k} (1 - (1 - R^i)^{r^i}) \tag{1.36}$$

subject to

$$\sum_{i=1}^{k} g_l^i(r^i) \leq b_l, \quad (l = 1, 2, \ldots, s), \tag{1.37}$$

where R_s is the system reliability, k is the total number of stages in a series-parallel system, $g_l^i(r^i)$ is a function representing the amount of l-th resource consumed at stage i as a function of r^i, s is the number of constraints and b_l is the total amount of the l-th resource available. If x_l^i is the l-th resource corresponding to the l-th constraint which is consumed in the first i stages, the performance equations for the complex system is written as $x_l^i = x_l^{i-1} + g_l^i(r^i)$ for $i = 1, 2, \ldots, k$ and $l = 1, 2, \ldots, s$, and $x_l^0 = 0$, $x_l^k \leq b_l$.

The objective function which has to be optimized is written as

$$S = \ln R_s = x_{s+1}^k - \sum_{l=1}^{s+1} c_l x_l^k, \tag{1.38}$$

where

$$c_l = 0, \quad (l = 1, 2, \ldots, s) \tag{1.39}$$

$$c_{s+1} = 1$$

$$x_{s+1}^i = x_{s+1}^{i-1} + \ln[1 - (1 - R^i)^{r^i}], \quad (i = 1, 2, \ldots, k)$$

$$x_{s+1}^0 = 0.$$

For optimization purposes, a Hamiltonian will be introduced such that

$$H^i = \sum_{l=1}^{s+1} z_l^i x_l^i = \sum_{l=1}^{s} z_l^i \{x_l^{i-1} + g_l^i(r^i)\} + z_{s+1}^i \{x_{s+1}^{i-1} + \ln[1 - (1-R^i)^{r^i}]\}, \quad (i=1,\ldots,k) \tag{1.40}$$

$$z_l^{i-1} = \frac{\partial H^i}{\partial x_l^{i-1}} = z_l^i, \quad (i = 1, 2, \ldots, k; l = 1, 2, \ldots, s, s+1) \tag{1.41}$$

$$z_{s+1}^i = c_{s+1} = 1. \tag{1.42}$$

The necessary condition for the local optimality of resource allocation in series-parallel systems can be obtained such that

$$\frac{\partial H^i}{\partial r^i} = 0 = \sum_{l=1}^{s} z_l^i \frac{\partial g_l^i(r^i)}{\partial r^i} + \frac{-(1 - R^i)^{r^i} \ln(1 - R^i)}{1 - (1 - R^i)^{r^i}}. \tag{1.43}$$

For the evident assumptions on the nature of r^i, $\forall_i$, we have $z_l^k = c_l = 0$, $(l = 1, 2, \ldots, s, l \neq j)$ and $z_l^i = 0$, $(l = 1, 2, \ldots, s; l \neq j; i = 1, 2, \ldots, N)$, and, finally, the above relation becomes

$$z_m^i \frac{\partial g_m^i(r^i)}{\partial r^i} - \frac{(1 - R^i)^{r^i} \ln(1 - R^i)}{1 - (1 - R^i)^{r^i}} = 0. \tag{1.44}$$

We give below the algorithm to solve the problem of reliability allocation subject to nonlinear constraints.

Step 1: Assuming a value for r^1 and using relation (1.44) we obtain z_m^1 and (see relation (1.41)), $z_m^1 = z_m^i$, $(i = 2, 3, \ldots, k)$.

Step 2: Find r^i, $(i = 2, 3, \ldots, k)$, from (1.44) by using the values of z_m^i obtained at Step 1.

Step 3: Compute x_l^k, $(l = 1, 2, \ldots, s)$.

Step 4: One of the following conditions occurs:

4.1. If $x_l^k < b_l$ for all $l = 1, 2, \ldots, s$, then we assume a higher value for r^1 and go to Step 1.

4.2. If $x_m^k > b_m$ and $x_l^k < b_l$ for $l \neq m$, $(l = 1, 2, \ldots, s)$, then we assume a smaller value for r^1 and return to Step 1.

4.3. If $x_p^k > b_p$, $p \neq m$ and $x_l^k < b_l$ for $l = 1, 2, \ldots, s$, $l \neq m$, where m is the active constraint, then go to Step 5.

4.4. If $x_m^k = b_m$ and $x_l^k < b_l$, for $l = 1, 2, \ldots, s$, $l \neq p$, then we have a candidate for the optimal solution.

Step 5: Replace constraint m by constraint p. Similarly replace m by p in equation (1.44) and Steps 1 and 2, and repeat the procedure given in Steps 1–4.

It is not infrequent in many applications of systems engineering that constraints of a technical system could be the total weight, total volume or total cost, and all these appear in a nonlinear form. If c^i is the cost per element at stage i, w^i represents the weight per element at the i-th stage, v^i is the volume per element at stage i and r^i denotes the number of elements in parallel at stage i, then the system constraints could be imagined under the following form

(a) $\quad \sum_{i=1}^{k} g_1^i(r^i) = \sum_{i=1}^{k} t^i(r^i)^2 \leq P$, where $t^i = w^i r^i$, $\qquad (1.45)$

(b) $\quad \sum_{i=1}^{k} g_2^i(r^i) = \sum_{i=1}^{k} c^i(r^i + \exp(r^i/4)) \leq C$; $\qquad (1.46)$

(c) $\quad \sum_{i=1}^{k} g_3^l(r^l) = \sum_{i=1}^{k} w^i r^i \exp(r^i/4) \leq W$. $\qquad (1.47)$

The objective function which has to be maximized is given by the functional

$$S = \sum_{i=1}^{k} \ln[1 - (1 - R^i)^{r^i}] = \sum_{l=1}^{4} c_l x_l^4 = x_4^k, \qquad (1.48)$$

where $c_l = 0$, $(l = 1, 2, 3)$, $c_4 = 1$. We solve the problem using the general framework above and obtain

$$r^i = (U^i)^{r^i}\left(r^i + \frac{\ln U^i}{2 z_1^i t^i}\right) \qquad (1.49)$$

$$z_1^i = \frac{1}{2 t^i r^i} \frac{(U^i)^{r^i} \ln U^i}{1 - (U^i)^{r^i}}, \qquad (1.50)$$

if only the first constraint is active. The values of r^i, $(i = 1, 2, \ldots, k)$, are obtained by solving the equation $f(r^i) = r^i - (U^i)^{r^i}(r^i + A^i) = 0$, $A^i = \ln U^i/2z_1^i t^i$, using Newton's method. In a similar way, we can calculate all r^i if the second constraint is active by solving the equation

$$f(r^i) = \left[1 + \frac{1}{4}(e)^{r^i/4}\right] - (U^i)^{r^i} \cdot \left[\left(1 + \frac{1}{4}(e)^{r^i/4}\right) + \ln U^i/z_2^i c^i\right] = 0, \quad (1.51)$$

where

$$z_2^i = \frac{1}{c^i\left[1 + \frac{1}{4}(e)^{r^i/4}\right]} \left[\frac{(U^i)^{r^i} \ln U^i}{1 - (U^i)^{r^i}}\right]. \quad (1.52)$$

If the third constraint is active, then

$$f(r^i) = (e)^{r^i/4}\left(1 + \frac{1}{4}r^i\right) - (U^i)^{r^i}\left((e)^{r^i/4}\left(1 + \frac{1}{4}r^i\right) + \frac{\ln U^i}{z_3^i w^i}\right) = 0,$$

where

$$z_3^i = \frac{1}{w^i\left[(e)^{r^i/4} + \frac{1}{4}r^i(e)^{r^i/4}\right]} \left[\frac{(U^i)^{r^i} \ln U^i}{1 - (U^i)^{r^i}}\right].$$

Example 4 [*168*]. Consider a system with $k = 5$ and the constants presented in table 1.8. By using the above model, we obtain $r^1 = 3$, $r^2 = 2$, $r^3 = 2$, $r^4 = 3$, $r^5 = 3$. The system reliability is 0.9045.

Table 1.8

i	R^i	t^i	P	c^i	C	w^i	W
1	0.80	1		7		7	
2	0.85	2		7		8	
3	0.90	3	100	5	175	8	200
4	0.65	4		9		6	
5	0.75	2		4		9	

Chapter 2

The practice of systems engineering

The last decade witnessed a considerable increase of interest in mathematical modelling for designing and implementing large and complex technical systems. As a result, a new field of management science — systems engineering — has developed. Its aim was to structure, synthetise and implement the design and operation of technical systems. The development of large industrial processes, offshore engineering equipment, new alternatives for producing power (i.e. nuclear power stations) and advanced satellite systems showed that adequate maintenance strategies must be followed in order to reduce failure rates and maximize the system profit or benefit. It is difficult to obtain data for analysing maintenance costs; the results of a 10-years review of cost ratios covering 63 of the largest U.S. companies (prepared by Albert Raymond and Associates, Chicago) are presented in ref. [70]. It is shown that in the period 1960—1969, there has been a real growth of maintenance expenditures even though net sales and investment in plants and equipment have risen at an even greater rate. Among the factors which determined the structure

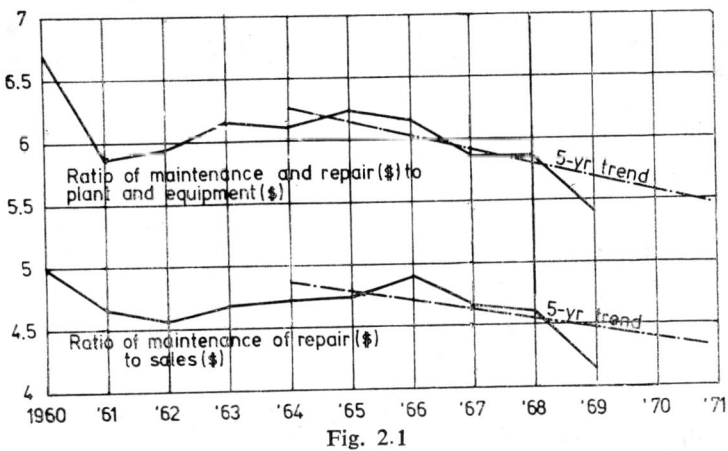

Fig. 2.1

of maintenance and repair costs, mention should be made of increased mechanization, increased complexity of equipment, tighter control over production, increased quality requirements, rising cost for maintenance labour, supplies and material costs. Figure 2.1 gives composite data for maintenance cost trends (see ref. [70]).

Maintenance and repair expenditures (percentage per year) have increased in the period 1965—1969 vs. 1960—1964 as stated in ref. [70]: in radio-T.V. companies (4) (the bracketed figure represents the number of companies participating in the above survey), the figure was 45.2%, for aircraft industry (5) — 16%, for electrical companies (4) — 12.5%. The composite figure for 63 companies participating in the survey over the above period was 8%. Among the conclusions reached in ref. [70], the following deserve special mention.

"Many factors have worked to increase maintenance costs, some important factors have been at work to reduce the cost of maintaining buildings and machinery...

... There is more interest in maintenance at management levels today...

... There is much more invested in new plants and expansions. The study showed that the value of plant and equipment increased by 40% in the second five-year period (1965—1969). Once they have been debugged, new plants and equipment require less maintenance".

On the other hand, new techniques have been brought in to assist in the structuring of the maintenance management process, such as computer-aided maintenance or mathematical modelling of complicated maintenance systems. Advanced operations research (OR) techniques such as mathematical programming, stochastic processes have been used for solutions to maintenance problems of single and multicomponent systems. Turban [17] describes and discusses the results of a U.S. survey on the use of mathematical models in plant maintenance decision-making. Among the conclusions of his study, he pointed out that "The use of Maintenance Management Models (MMM) may be seen in terms similar to those involved in the use of any another commodity or service: the model builder produces a commodity (or service) for the use of the management (as buyer or client)...

... The properties of most existing models need improvement. It is no secret that industrial users invest very little in the construction of models, and that most support is given by the military, university, research foundations, or the Government. However industry can, and some large corporations do, promote the construction of models...

... MMM must face the competition of substitute commodities. Comparing MMM with alternative methods (such as "rules of thumb") involves evaluating utilities (gains) and costs, but the relationship between use of MMM (or any other method) and their utility to the user is quite complex. We can ask: utility to whom? To the organisation as a whole? To the maintenance unit? Or to the maintenance manager?...

... Development of data processing will probably have the greatest effect on the evaluation of alternatives in the future. All indications suggest a radical change in this area of maintenance".

One may wonder "Why is the gap between theory and practice becoming wider in maintenance analysis?" Part of the answer is that "the vast majority of maintenance management personnel have no education in OR and MMM" [17]. However, "Theoretical knowledge and understanding of OR and MMM, although essential to successful implementation of the models, is not sufficient. Even the manager who is aware of the theory and is willing to try, will usually be very cautious. An excellent way to convince a manager that MMM are potentially successful is to let him talk, in his own language, to his counterpart in a plant where MMM are in use".

The reader could easily ask himself: "If so, then do we need to build up more complicated MMM and why?" The answer is positive. The decision-maker and the systems analyst can help by working together in improving the operational status of the equipment, its availability and finally maximize the system's output (see fig. 2.2).

Turban [17] noted that "Large maintenance units will continue to be a principal target for model implementation because of high absolute potential saving the availability of professional and clerical help required for the implementation, and because of the unit's high status in the organization."

As a conclusion of a comprehensive literature survey concerning the maintenance problem for complex systems and appropriate mathematical modelling, the following comments may be derived.

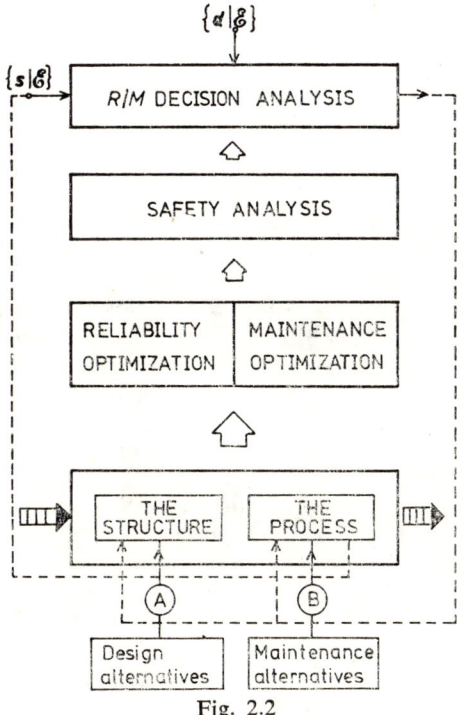

Fig. 2.2

1. Plants have become more sophisticated, units are larger, and cost systems have developed to the point where they will reveal the appalling cost of stoppages.

2. There are therefore two aims, cheaper maintenance and better maintenance, and they are not necessarily mutually exclusive, provided that the maintenance systems can be improved.

3. The systems analyst is then faced with the following questions regarding the maintenance modelling:

"What kind of philosophy has resulted from the experience?" and,

"To what kind of modelling approach does this philosophy lead?" (see also [69]).

Starting from the above statements, research developed on the analytical studies and practical applications of reliability, maintenance and safety of large technical systems.

One of the objectives of this book was to use the study of optimal maintenance strategies for technical systems when their behaviour is described by a Markovian Decision Process (MDP) to the special cases when the system is polyfunctional and when it can be completely or partially observable. Appropriate decomposition methods for the state space representation of the deterioration-maintenance (repair or replacement) process for the case of large and complex systems (coherent structures) have also to be investigated.

Another objective was to investigate maintenance strategies when the decision-maker is a risk-sensitive person, and to see how the structure of optimal repair/replacement strategies changes when the maintenance engineer is either risk-averse or risk-preference. There are technical systems which are not completely observable; in this case, we must build adequate mathematical models to describe the behaviour of partially observable systems.

2.1. State of the art

A large body of literature is concerned with reliability, maintenance and safety analysis of technical systems. Analysis of systems with a Markovian behaviour has also been widely investigated. However, on closer inspection, it appears that the technical literature on reliability and maintenance of technical systems with Markovian changes of states is extremely redundant. Different authors present highly sophisticated models within the same philosophy of system description.

However, two major trends exist at present in this respect: the Anglo-Saxons, with major contributions in the American literature, and the Soviet. A comprehensive survey on the Soviet literature is given in ref. [92]; the authors investigated more than 700 papers. The following conclusions, derived after investigating such a large amount of information, are worth mentioning.

— Reliability and maintenance engineering has been recognized as a distinct discipline in the U.S.S.R. for more than fifteen years;

— Purely theoretical and statistical approaches to reliability, maintenance and replacement policies and confidence limits are due to the Soviets;

— Mathematical modelling for preventive maintenance and graceful degradation were investigated in the Soviet literature.

On the other hand, the Anglo-Saxon literature on reliability/maintenance/safety was surveyed by Barlow and Proschan [1], Jorgenson et al. [12], Rau [11], Gheorghe [55], Worrell et al. [243], Fussell et al. [262].

With respect to MDP applied to the reliability/maintenance problem, we may notice that completely observable Markov processes (Markov and semi-Markov) have been used in the analysis of optimal repair/replacement strategies for such simple cases as "machine replacement" (see Barlow and Proschan [1], Kolesar [18, 19], Klein [23], Howard [7, 8], Kao [27], Hobbs [37], Tahara et al. [29], Derman [59]), or for multicomponent systems (see Bonhomme [30], Luss and Kander [82], Rolfe [24], O'Neil [83]). Several authors ([18, 59]) considered randomized maintenance policies as a joint probability that the system is in a deterioration state and a maintenance action could be enforced for the system. When the system was not allowed a complete observation, partially observable Markovian processes (Markov and semi-Markov) have been used. A stochastic inspection process was explicitly introduced. Relevant work in this field was reported by Eckles [71], Satia [34], Sondik [13, 31] and Hobbs [37].

However, alongside sophistication in modelling optimal maintenance strategies using Markovian classes of models, adequate optimization techniques have been developed. Much interest was raised by the dynamic programming techniques given by Bellman [73] and Howard [7, 8], and linear programming [10], especially for completely observable systems. Several interesting approximation techniques were used for partially observable systems with Markov behaviour (e.g. the approximation technique given by Smallwood and Sondik [31]). Complicated systems maintenance techniques use hybrid OR models (more details can be found in ref. [55]).

Other stochastic models have been used for inspection and maintenance models applied to systems engineering (see Jardine [9], Munford [20], Stapleton [21], Quayle [22] and Simon [25]). Further literature comments on maintenance and safety modelling by means of MDP are given in this book where they are necessary.

Chapter 3

Multifunctional systems and extensions

3.1. Concepts of polyfunctional structures and systems

Broadly speaking, real systems (engineering, biological, etc.) have a complex topology. In order to decompose them for analysis and computation, appropriate rules must be found for design and maintenance decision making.

Throughout this chapter a distinction will be made between the concept of structure and that of system. We proceed from a formal mathematical representation of polyfunctional structures. The basic results presented in this chapter will be used throughout the remaining sections.

A formal mathematical representation of structures appeared in systems engineering in connection with aeroplane reliability during mission time. Birnbaum et al. [41] assumed that a structure can either perform or fail and they defined this as a *dichotomic reliability*, i.e. the probability that a structure will perform its task at any point of a given time interval. Later on, they [42] defined coherent structures and gave some of their properties for the case when the structure function can take only boolean values. However, real complex structures can perform their design objectives with partial or full satisfaction. The probability that the structure executes its task at different performance levels will be viewed as *multichotomic reliability*. For instance, let us consider a steam turbine which can meet its design objective (to produce electric power) at different performance levels, from zero up to the full power. If one removes some blades from the turbine, it will still work. However, the output will decrease in compliance with the new structure topology. Such examples could be found in many other systems engineering problems and related fields.

O'Brien [26] introduced *multifunctional structures* where each function is allowed to have dichotomic behaviour. He analysed different classes (hierarchical, symmetric and independent) of these structures. His prime interest was to calculate the reliability of these structures, i.e. the probability that a certain number of structure functions operate. O'Brien also developed a computing method for designing multifunctional hierarchical structures.

On closer analysis, real operating structures appear to be multicomponent, multichotomic and/or polyfunctional structures in general. By particularizing a multichotomic behaviour, we can find a dichotomic behaviour. A wide class of practical examples could be given in this respect, from engineering systems to marketing advertising for multiple products, management of organizational structures.

In our study, the formal representation of complex structures systems will rely on the methodology of systems engineering, with application to reliability and maintenance policies of polyfunctional coherent systems of Markovian behaviour (Howard [7, 8]).

3.2. Polyfunctional structures and systems: basic definitions

A *structure* represents a collection of components interconnected by a set of rules, which is assumed to be capable of performing one or more functions from a finite set Y. An oscilloscope, a hydraulic pump, a computer, a chemical plant or a nuclear power station may be considered to be structures.

For the sake of clarity, we shall introduce the following notation: ω represents the total number of functions for which the system has been designed; each function must perform at maximum operational level so as to ensure the perfect operation of the system; φ represents the number of functions which perform either at maximum or some intermediate levels of operation ($\varphi \leqslant \omega$); r represents the number of functions which perform at maximum level of operation ($r \leqslant \varphi \leqslant \omega$). Throughout this chapter, we shall consider that minimum r functions have to perform at maximum operational level for the system to perform its design task.

DEFINITION 3.1. *A polyfunctional structure* is a structure which can perform maximum ω functions; it is a structure with ω inputs and minimum r outputs ($\omega \times r$), where $r = \inf_{\omega} \{\omega; \omega \in Y\}$.

The above description is plotted in fig. 3.1. Practical examples of such structures refer to the safety system of a nuclear reactor (Emergency Core Cooling System for LWR) or a time-sharing computer network.

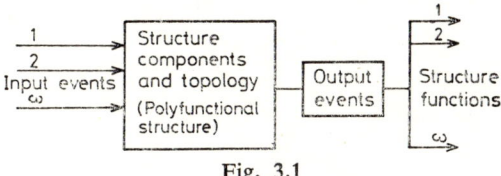

Fig. 3.1

Let Y be a finite set of functions in an E^n-Euclidian space for a given structure (a well-defined structure is a *non-null* subset of Y). From the above definition, a polyfunctional structure is a fixed subset Y_ω of Y such that

$$Y_\omega \subset Y \neq \emptyset; \quad \text{card } Y_\omega \geqslant 1; \quad \text{card } Y_\omega = \omega, \tag{3.1}$$

where $\emptyset$ is the empty set and ω is the cardinality of the fixed subset Y_ω.

Under normal operation, a structure can perform a number of φ functions (i.e. the cardinality of Y_φ is φ). Let r functions out of φ perform at maximum level. Following the above definitions, $Y_r = Y_\omega \cap Y_\varphi = \{r: r \in Y_\omega \text{ or } Y_\varphi \subseteq Y\}$; card $Y_\omega \geq$ card $Y_r \to \omega \geq r$, and consequently $r = \inf_\omega \{\omega: Y_\omega \supseteq Y_r\}$.

Let us consider a structure which consists of n components and the states of component i, $(i = 1, 2, \ldots, n)$, can be given by

$$x_i = \begin{cases} 0 - \text{ the component } i \text{ does not operate (a failed state)} \\ 1 - \text{ the component } i \text{ is in an operational state.} \end{cases}$$

The vector of the state components is

$$X = (X_1, X_2, \ldots, X_n). \tag{3.2}$$

In this chapter we shall follow a concept similar to that given by Hirsch et al. [43], and for the most general case we make the following assumptions.
— each component of a structure has a location in the structure; only one component can be installed in a given location (well defined topology); only components of the same type can be interchanged (parallel redundancy); in a structure all the locations must be occupied by components. A particular structure function $\Phi_\lambda(X)$, $\lambda = 1, 2, \ldots \omega$, can be in any of the finite number of states

$$\Phi_\lambda(X) = 0, 1, 2, \ldots, M, \quad (\lambda = 1, 2, \ldots, \omega), \tag{3.3}$$

and, consequently, one can have

$$\Phi_\lambda(X) = \begin{cases} 0 - \text{ the function } \lambda \text{ does not operate,} \\ 1 - \text{ the function } \lambda \text{ is just operating,} \\ M - \text{ the function } \lambda \text{ is fully operating.} \end{cases}$$

The set of the state variables of the components should be considered as an n-dimensional vector (relation (3.2)), which ranges over the set $\mathscr{V}^n$ consisting of the 2^n vertices of the unit n-cubs. The set $\mathscr{V}^n$ of all vertices describes all the possible states of locations in the structure. For a particular function λ, $(\lambda = 1, 2, \ldots, \omega)$, we can have the mapping relation

$$\Phi_\lambda(X): \mathscr{V}^n \to \mathscr{A}, \tag{3.4}$$

where $\mathscr{A}$ is the set of all possible states of each function

$$\mathscr{A} = \{0, 1, 2, \ldots, M\}. \tag{3.5}$$

A *total structure function* for the polyfunctional case is defined by

$$\Phi_\omega^*(X) = [r^0(X), r^1(X), \ldots, r^M(X)], \tag{3.6}$$

which represents the degree to which the ω-structure is capable of operating. The notation $r^j(X)$, $(j = 0, 1, \ldots, M)$, indicates how many functions are operating at the performance level j, and $r^j(X) = 0, 1, \ldots, \omega$ such that

$$\sum_{j=0}^{M} r^j(X) = \omega. \tag{3.7}$$

For the case of a polyfunctional structure with boolean values for each function λ, the total structure function is given by the summation of the performance value of the individual structure functions (O'Brien [26])

$$\Phi_\omega^*(X) = \sum_{\lambda=1}^{\omega} \Phi_\lambda(X). \tag{3.8}$$

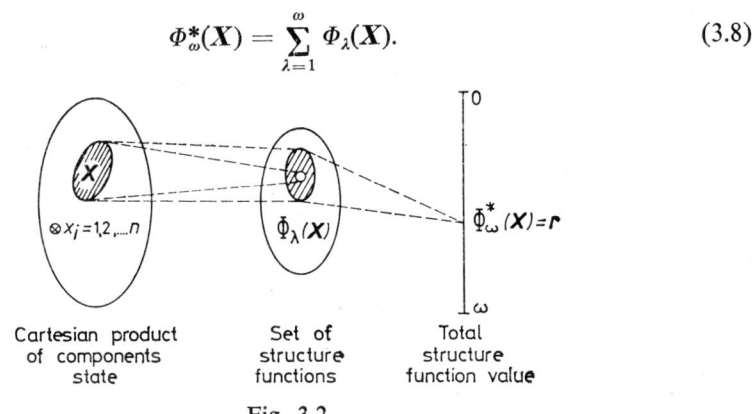

Cartesian product of components state | Set of structure functions | Total structure function value

Fig. 3.2

An alternative representation for the total structure function is $\Phi_\omega^*(X) = 0, 1, 2, \ldots, N$, which represents the overall index to which the ω-structure performs its design requirements (see fig. 3.2). Some properties of the ω-structure will be derived and used later. The following definitions are needed.

DEFINITION 3.2. A component i is an *inessential component* to a polyfunctional structure if

$$\Phi_\lambda(x_1, \ldots, x_{i-1}, 0, x_{i+1}, \ldots, x_n) = \Phi_\lambda(x_1, \ldots, x_{i-1}, 1, x_{i+1}, \ldots, x_n) \tag{3.9}$$

for all $\lambda = 1, 2, \ldots, \omega$.

PROPOSITION 3.1. *In a polyfunctional structure, if an independent component i is inessential to all $\lambda = 1, 2, \ldots, \omega$ functions, then it can be neglected in the structure.*

Proof. We can prove this proposition by induction. Let us suppose that the component i is present in the structure. But according to relation (3.9), we can see that the impact of component i over all the performances of the structure functions is null. Therefore component i can be neglected in the structure.

DEFINITION 3.3. A component i is *essential* to a polyfunctional structure if

$$\Phi_\lambda(x_1, \ldots, x_{i-1}, 0, x_{i+1}, \ldots, x_n) \neq \Phi_\lambda(x_1, \ldots, x_{i-1}, 1, x_{i+1}, \ldots, x_n), \tag{3.10}$$

for any $\lambda = 1, 2, \ldots, \omega$.

In the present work we consider only structures which have essential components.

For any real state vector $X = (x_1, x_2, \ldots, x_n)$ and $Y = (y_1, y_2, \ldots, y_n)$ a lexicographical property of the vectors of the state components is formally defined as $X \prec Y \Leftrightarrow X \neq Y$, and $y_k < x_k \Rightarrow x_j < y_j$ for some $j < k$, $(k = 2, 3, \ldots, n)$; in more general terms $X \preceq Y \Leftrightarrow x_1 < y_1$, or $[x_1 = y_1, x_2 < y_2]$, or, $\ldots$, or

$$[x_1 = y_1, \ldots, x_{n-1} = y_{n-1}, x_n \leq y_n].$$

In the above notation "$\preceq$" denotes a partial or a lexicographic ordering among the component state vectors. For a dichotomic component state vector X, a numerical (lexicographic) ordering is given by the integer value $2(X) = x_1 2^{n-1} + x_2 2^{n-2} + \ldots + x_n 2^0$. This is a refinement of the partial ordering of vectors for storage in the computer memory; if $X \preceq Y$, then $2(X) \leqslant 2(Y)$. However, $2(X) \leqslant 2(Y)$ does not imply that vectors X and Y are lexicographically ordered.

In the remainder of the book, we deal with polyfunctional systems which are structures; a time parameter is appropriate to describe their behaviour. We shall follow the definition of a system given by M'Pherson [45] (see Chapter 1).

3.3. Coherent structures and extensions

Birnbaum et al. [41, 42], Barlow and Proschan [1], O'Brien [24] and Vesely [14] studied this class of structures under the name of coherent structures. However, their approach was toward a boolean representation of X and $\Phi_\lambda(X)$, $(\lambda = 1, 2, \ldots, \omega)$. Hirsch et al. [43] introduced the notion of monotonic structures, when the structure has to perform one function which can take a finite number of performance levels.

DEFINITION 3.4. A polyfunctional structure is lexicographically semi-coherent for each function λ, $(\lambda = 1, 2, \ldots, \omega)$, if

$$\Phi_\lambda(X) - \Phi_\lambda(Y) \leqslant 0 \tag{3.11}$$

and if $X \prec Y$.

DEFINITION 3.5. A polyfunctional structure is lexicographically *coherent* for all functions λ, $(\lambda = 1, 2, \ldots, \omega)$, if it satisfies relation 3.11 and if

$$\Phi_\lambda(\boldsymbol{0}) = 0; \quad \Phi_\lambda(\boldsymbol{1}) = M. \tag{3.12}$$

In an alternative formulation of Definition 3.5., "A polyfunctional structure is coherent iff card $\Phi_\omega^*(X) \leqslant$ card $\Phi_\omega^*(Y)$, and card $\Phi_\omega^*(\boldsymbol{0}) = 0$; card $\Phi_\omega^*(\boldsymbol{1}) = M$." A pictorial representation is given in fig. 3.3. It is clear from the above definitions that by maintenance (repair, replacement) actions over a set of components in the structure, the structure function will eventually operate at some upper performance level of reliability and safety. The concept of lexicographical coherent structures is used in systems engineering for evaluating maintenance policies or performing safety analysis in complex technical systems. For systems endowed with a complicated topology, the bounds for the performance level of $\Phi_\lambda(.)$, $(\lambda = 1, 2, \ldots, \omega)$, could be given by introducing "path" and "cut" sets.

DEFINITION 3.6. A *cut set* for a function λ is a vector $\boldsymbol{v}_\lambda$ for which $\Phi_\lambda(\boldsymbol{v}_\lambda) = 0$; the vector $\boldsymbol{v}_\lambda$ is a cut set for the ω-structure if $\Phi_\lambda(\boldsymbol{v}_\lambda) = 0$ for all $\lambda = 1, 2, \ldots, \omega$.

A useful concept is that of a *minimal cut set* for a function λ. For any vector $\boldsymbol{v}_\lambda$ for which $\Phi_\lambda(\boldsymbol{v}_\lambda) = 0$, and for all $X_\lambda \succ \boldsymbol{v}_\lambda$, we can write the equation $\Phi_\lambda(X_\lambda) = \min \{1\}$.

DEFINITION 3.7. A *path set* for a function λ is a vector $\mathbf{Z}_\lambda$ for which $\Phi_\lambda(\mathbf{Z}_\lambda) = k$, $k \in [1, 2, \ldots, M]$.

A *minimal path set* for a function λ is a vector $\mathbf{S}_\lambda$ for which $\Phi_\lambda(\mathbf{S}_\lambda) = \min\{1\}$. For any vector $\mathbf{Z}_\lambda$, $\Phi_\lambda(\mathbf{Z}_\lambda) = 0$ such that $\mathbf{Z}_\lambda \prec \mathbf{S}_\lambda$. In a complex structure, however, there exist more than one minimal path. A minimal path $\mathbf{S}_\lambda j$

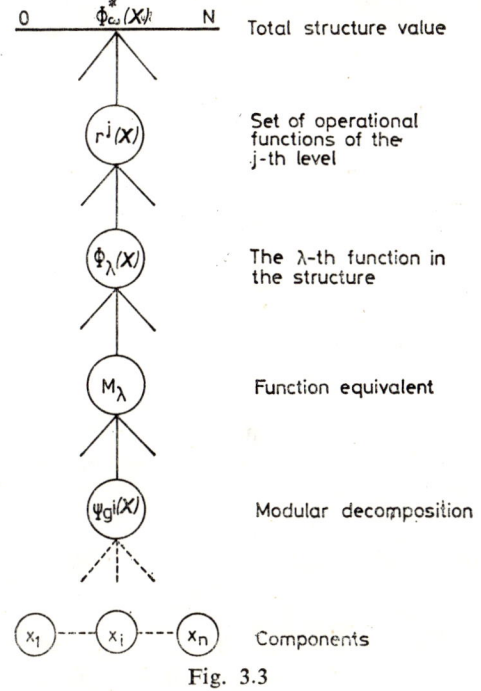

Fig. 3.3

for a function λ is

$$\mathbf{S}_\lambda j = [(\mathbf{S}_1)_\lambda j, (\mathbf{S}_2)_\lambda j, \ldots, (\mathbf{S}_n)_\lambda j], (j = 1, 2, \ldots, M_\lambda), \quad (3.13)$$

where M_λ indicates the number of minimal paths that exist in the structure. The concepts of minimal cut and path sets have been used by Vesely [16] and Fussell [46] in reliability analysis of complex systems by fault trees.

In order to identify the functions for which a component i is a number of a minimal path, define a ω-function system

$$\omega' = \sum_{\lambda=1}^{\omega} M_\lambda, \quad (3.14)$$

where each minimal path, for each function λ, $(\lambda = 1, 2, \ldots, \omega)$, will be considered to be an equivalent function (see O'Brien [26]). From the above comments, it is obvious that a component can be a member of more than one minimal path for a particular function λ, and it can also be a member of minimal paths for more than one function.

DEFINITION 3.8. If x_i, $i = 1, 2, \ldots, \hat{N}$, are multilevel variables, then $y = \int_{i=1}^{\hat{N}} x_i$, where y is also a multilevel variable such that

$$x_1 = x_2 = \ldots = x_{\hat{N}} = 0 \qquad y = 0$$

$$x_i = 1 \text{ for } i \in \{1, 2, \ldots, \hat{N}\} \qquad y = 1$$

$$x_i = \hat{N} \text{ for } i \in \{1, 2, \ldots, \hat{N}\} \qquad y = Z.$$

The next lemma shows that function λ operates at different performance levels if one or more of the pseudofunctions λ^j operate.

LEMMA 3.1. *The function $\Phi_\lambda(X)$ can be represented by relation*

$$\Phi_\lambda(X) = \int_{j=1}^{M_\lambda} \Phi_{\lambda^j}(X). \tag{3.15}$$

Proof. The above equation derives from Definition 3.8 for the case when the left-hand side vector is multilevel.

We can see the analogy between the results given here using the lexicographical property of coherent structures and the property of coherent structures and the "Max-flow, Min-cut"-Fulkerson theorem from the network theory.

3.4. Modular decomposition of polyfunctional structures

A modular decomposition of the ω-function structure is a partition in a finite set of modules; each module is identified by its own structure function. However, instead of looking at the system as a whole, by a modular decomposition we can analyse the system through the behaviour of its modules. In the structure, a module can be imagined as a subset of the basic components of the system, which are organized into some substructures of their own and which affect the system only through the performance of their substructure (Bodin [44]). A module is an assembly of components which can itself be treated as a component of the system.

In structures with complex topologies, functions depend upon each component; in case that one component fails, a group of minimal paths could also fail. Let us define the set $G: G = $ set $[g^i, i = 1, 2, \ldots, \omega']$, where g^i correspond to all combinations of the minimal paths for all functions and $Y^g = [g_1^{\gamma i}, g_2^{\gamma i}, \ldots, g_n^{\gamma i}]$ such that

$$g_k^{\gamma i} = \begin{cases} 1 - \text{if the } k\text{-th component is a member of exactly these minimal paths corresponding to } g^i, \\ 0 - \text{otherwise.} \end{cases}$$

Then we can see that

$$g_k^{\gamma^i} = \prod_{\lambda^j \in g^i}(Z_k)_{\lambda^j} \prod_{\lambda^j \notin g^i}(1-(Z_k)_{\lambda^j}). \tag{3.16}$$

As was pointed out before, vector Y^{g^i} indicates that if any component k fails, then it is exactly the minimal path in γ^i which would fail. Vector Y^{g^i} is thus the only minimal path corresponding to the new structure function $\Psi_{g^i}(X)$. The following theorem has been given by O'Brien [26, Chapter 2].

THEOREM 3.1. $\Psi_{g^i}(X)$ *is coherent.*

We give now a representation theorem for a ω-function structure using the above results (see fig. 3.3).

THEOREM 3.2. *A polyfunctional structure could be represented by*

$$\Phi_\omega^*(X) = \bigotimes_{\omega:\,r^j(X)\in\omega} \left\{ \bigotimes_{r^j(X):\lambda\in r^j(X)} \int_{j=1}^{M_\lambda} \prod_{g^i:\,\lambda^j\in g^i} \Psi_{g^i}(X) \right\}, \tag{3.17}$$

where $\otimes$ represents a cartesian product of operational λ functions which lead the polyfunctional system to the operational status given by $r^j(X)$ and to the total structure value, respectively.

Proof. Using Lemma 3.1, we have

$$\Phi_\lambda(X) = \int_{j=1}^{M_\lambda} \prod_{g^i:\,\lambda^j\in g^i} \Psi_{g^i}(X). \tag{3.18}$$

But from (3.18) and from the fact that $\Phi_\lambda(X)$ is a multivalue vector, we can write the bracketed relation from the theorem, and finally $\Phi_\omega^*(X)$ for the case when $\Psi_{g^i}(X)$ is coherent (Theorem 3.1).

As the modular decomposition of a coherent structure is generally unique, it is immediate to ask which to adopt (see fig. 3.3). However, we can choose a modular decomposition which is the most convenient for the maintenance decision making.

3.5. Boolean coherent structures

In this paragraph special attention is given to boolean coherent structures, where each component i ($i = 1, 2, \ldots, n$) can take values $x_i = 0, 1$, and every structure function $\Phi_\lambda(.)$, ($\lambda = 1, 2, \ldots, \omega$), is defined as

$$\Phi_\lambda(X) = \begin{cases} 0 & \text{if the system function } \lambda \text{ is not operational} \\ 1 & \text{if the system function } \lambda \text{ is operational.} \end{cases}$$

The structure function $\Phi_\lambda(X)$ has the same properties as the structure function $\Phi(X)$ defined in ref. [26], where O'Brian gave a detailed account of boolean coherent structures. In order to illustrate such multifunctional systems, we shall dwell on a telephone switching system, which is designed to switch transmission paths and "to provide a variety of signalling and control functions for a large number of telephone customers". In the case when failure of a subsystem or a component occurs, at least a few customers will not receive the intended services. Such a system can be completely or partially operative so that all customers or only some of them receive service. O'Brien noticed that "the design of large complex multifunction systems requires the organization unified entity to meet the performance requirements. If the system requirements permit a level of degraded service, the system designer can make a number of trade-offs in the organization of the equipment to reduce costs and improve the probability of the system performing its intended functions". The following mathematical results will be given without proof. For more details, the reader can refer to [26].

THEOREM 3.3. *In a boolean coherent structure, each component is a member of a minimal path of exactly one module* Ψ_{γ^i}; *the failure of the* Ψ_{γ^i} *is independent, if the failure of the components is independent.*

LEMMA 3.2. *If the structure function* $\Phi_{\lambda^i}(X)$ *is coherent, then it can be represented by*

$$\Phi_{\lambda^i}(X) = \prod_{\gamma^i\,:\,\lambda^i \in \gamma^i} \Psi_{\gamma^i}(X). \tag{3.19}$$

THEOREM 3.4. (A representation theorem). *Any multifunctional structure having* ω *functions and the total structure function* $\Phi_\omega^*(X)$ *can be represented as*

$$\Phi_\omega^*(X) = \sum_{\lambda=1}^{\omega} \left\{ \int_{j=1}^{M_\lambda} \prod_{\gamma^i\,:\,\lambda^j \in \gamma^i} \Psi_{\gamma^i}(X) \right\}. \tag{3.20}$$

For further analysis, we shall consider a few special classes of polyfunctional coherent structures. For each of the models to be considered, we shall assume that each function has only one minimal path, and then $M_\lambda = 1$ for $\lambda = 1, 2, \ldots, \omega$ and $\omega' = \omega$.

Symmetrical function structures correspond to a ω-structure, where each function has a module which is shared in all combinations of the other functions.

We define $\alpha(\gamma^i(X))$ to be the number of functions affected by a particular module Ψ_{γ^i}. α is equal with the number of elements in the set γ_i. For a symmetrical function structure, each distinct λ function has a number of modules equal with $\binom{\omega-1}{\alpha-2}$, which affects α functions, $\alpha = 1, 2, \ldots, \omega$. Each function has a number of $2^{\omega-1}$ modules.

The structure function of a symmetrical multifunctional structure having ω functions is given by relation

$$\Phi_\omega^*(X) = \sum_{\lambda=1}^{\omega'} \left\{ \prod_{\gamma^i\,:\,\lambda \in \gamma^i} \Psi_{\gamma^i}(X) \right\}. \tag{3.21}$$

A block diagram of such a structure is represented in fig. 3.4. In the evaluation of such a structure, it was assumed that the modules corresponding to $\Psi_{\gamma^i}(X)$ exist for all γ^i.

An *independent function structure* with ω functions has only one minimal path for each function and each function is independent of the other functions.

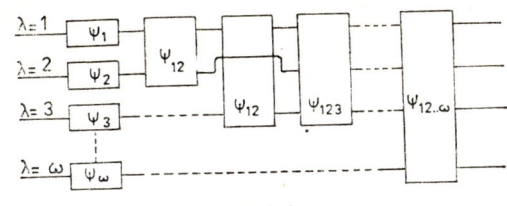

Fig. 3.4

For this kind of multifunctional system the only modules which exist are those corresponding to Ψ_{γ^i} when $\gamma^i = \lambda$. For each function, the coherent structure with ω independent functions consists of one coherent module $\Psi_\lambda(X)$ such that

$$\Phi^*_\omega(X) = \sum_{\lambda=1}^{\omega} \Phi_\lambda(X). \tag{3.22}$$

The block diagram of such a structure is plotted in fig. 3.5.

Hierarchical function structures have a number of functions $\omega = 2^{\chi-1}$, where $\chi \geqslant 1$. These structures are organized in the form of a tree, where the operation of each function depends upon the operation of χ modules. It is assumed that for such structures, each λ function, $(\lambda = 1, 2, \ldots, \omega)$, has only one minimal path. In this case, a hierarchical coherent structure is given by relation

$$\Phi^*_\omega(X) = \sum_{\lambda=1}^{2^{\chi-1}} \left\{ \prod_{\gamma^i : \lambda \in \gamma^i} \Psi_{\gamma^i}(X) \right\}, \tag{3.23}$$

where the only $\Psi_{\gamma^i}(.)$ that exist are those which correspond to a $\gamma^i \in \Gamma^*$ and $\Gamma^* = (1, 2, 3, \ldots, 2^{\chi-1}; 1-2, 3-4, \ldots, 2^{\chi-1}-1, 2^{\chi-1}; 1-2-3-4, \ldots; 1-2-3-4-\ldots-2^{\chi-1})$.

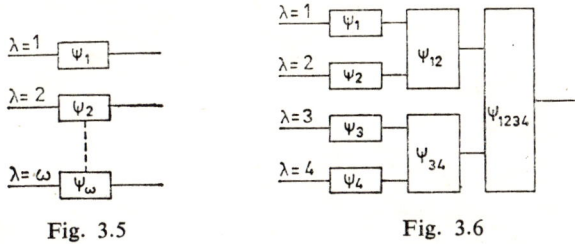

Fig. 3.5   Fig. 3.6

In fig. 3.6 we give a block diagram for a hierarchical multifunctional structure with $\chi = 3$ and $\omega = 2^{\chi-1} = 4$.

As the components of the ω-structure fail, the multifunction system is in one of the 2^n states. When these states are given by the number of functions which are operating, the probability of being in one of these states is $\Pr\{\Phi_\omega^*(X) = a\}$ for $a = 0, 1, \ldots, \omega$, where a is the number of operating functions and $\omega - a = b$ is the number of non-operating functions. Let A_i be the event that the i-th function operates, $(A_i = \Phi_i(X) = 1))$, and let

$$\mathcal{H}_i = \sum_{k_1=1}^{\omega} \cdots \sum_{k_i=k_{i-1}+1}^{\omega} \Pr(A_{k_1} \ldots A_{k_i}). \tag{3.24}$$

The following lemmas can be used to calculate $\Pr\{\Phi_\omega^*(X) = a\}$ and $\Pr\{\Phi_\omega^*(X) \geq a\}$ for structures with identical functions. In this case, each function depends on the same number and similar modules Ψ_{γ^i} so that the functions are not distinguishable (see O'Brien [26]).

The next results will be given without proof.

LEMMA 3.3. *The probability that exactly a functions are operating in a multifunctional system is*

$$\Pr\{\Phi_\omega^*(X) = a\} =$$

$$= \sum_{k=0}^{\omega-a} \binom{\omega}{a}\binom{\omega-a}{l} (-1)^l \Pr\{A_1 \ldots A_{\omega-l}\}. \tag{3.25}$$

LEMMA 3.4. *The probability that at least a functions are operating in a multifunctional system is*

$$\Pr\{\Phi_\omega^*(X) \geq a\} =$$

$$= \sum_{j=0}^{\omega-a} \frac{a}{\omega-j} \binom{\omega-a}{j}\binom{\omega}{a} (-1)^{\omega-a-j} \Pr\{A_1 \ldots A_{\omega-j}\}. \tag{3.26}$$

LEMMA 3.5. *The probability that the first j functions operate in a multifunctional system is*

$$\Pr\{A_1 \ldots A_j\} = \prod_{\gamma^i : 1 \in \gamma^i} \Pr\{\Psi_{\gamma^i}(X) = 1\}. \tag{3.27}$$

$$\prod_{\gamma^i : 2 \in \gamma^i, 1 \notin \gamma^i} \Pr\{\Psi_{\gamma^i}(X) = 1\} \cdots \prod_{\gamma^i : j \in \gamma^j, 1 \ldots j-1 \notin \gamma^i} \Pr\{\Psi_{\gamma^i}(X) = 1\}.$$

Computational relations for symmetrical, independent and hierarchical structures are given in table 3.1. Tables 3.2 and 3.3 show computational relations for the expected number of operating functions and the variance of the number of operating functions.

MULTIFUNCTIONAL SYSTEMS AND EXTENSIONS

Table 3.1

No.	Types of structure	Computational relations	Observations
0	1	2	3
1	Symmetrical dependent structures	$\Pr\{\Phi_\omega^*(X) = a\} =$ $= \sum_{j=0}^{\omega-a} \binom{\omega}{a}\binom{\omega-a}{j}(-1)^{\omega-a-j} \cdot$ $\cdot \prod_{\alpha=1}^{\omega} p_\alpha^{\binom{\omega}{\alpha}-\binom{j}{\alpha}}$	$p_\alpha = \Pr\{\Psi_{\gamma^i}(X) = 1\}$ for all modules Ψ_{γ^i} which have the same α (α corresponds to α ($\gamma^i(X)$)); this variable is defined as the number of elements in the set γ^i.
		$\Pr\{\Phi_\omega^*(X) \geq a\} =$ $= \sum_{j=0}^{\omega-a} \frac{a}{\omega-j}\binom{\omega}{a}\binom{\omega-a}{j} \cdot$ $\cdot (-1)^j \prod_{\alpha=1}^{\omega} p_\alpha^{\binom{\omega}{\alpha}-\binom{j}{\alpha}}$	— ,, —
		$\Pr\{\Phi_\omega^*(X) = a\} =$ $= \sum_{j=0}^{\omega-a} \binom{\omega}{a}\binom{\omega-a}{j}(-1)^{\omega-a-j} \cdot$ $\cdot p^{2^\omega - (2)^j}$	$p_\alpha = p$ for all α
		$\Pr\{\Phi_\omega^*(X) \geq a\} =$ $= \sum_{j=0}^{\omega-a} \frac{a}{\omega-j}\binom{\omega}{a}\binom{\omega-a}{j} \cdot$ $\cdot (-1)^{\omega-a-j} p^{2^\omega - (2)^j}$	— ,, —
2	Independent function structures	$\Pr\{\Phi_\omega^*(X) = a\} =$ $= \binom{\omega}{a} p^a (1-p)^{\omega-a}$	$\alpha = 1$ for all modules Ψ_{γ^i} $\Pr\{\Psi_{\gamma^i}(X) = 1\} = p_1 = p$ for all modules.
		$\Pr\{\Phi_\omega^*(X) \geq a\} =$ $= \sum_{j=a}^{\omega} \binom{\omega}{j} p^j (1-p)^{\omega-j}$	— ,, —

Table 3.1 (continued)

No.	Types of structure	Computational relations	Observations
0	1	2	3
3	Hierarchical dependent structures	$\Pr\{\Phi_\omega^*(X) = a\} =$ $$= \sum_{i=0}^{b/2} \binom{\omega - 2i_1}{a} (1-p_1)^{\omega-a-2i_1} p_1^a \cdot$$ $$\cdot \sum_{i_2=0}^{b/4} \binom{\omega/2 - 2i_2}{\omega/2 - i_1} (1-p_2)^{i_1 - 2i_2} \cdot p_2^{\omega/2 - i_1}$$ $$\vdots$$ $$\sum_{i=0}^{0} \binom{1 - 2i_{\chi-1}}{1 - i_\chi} (1-p_\chi)^{i_{\chi-1} - 2i_\chi} \cdot p_\chi^{1 - i_{\chi-1}}$$	$b = \omega - a$ $\omega = 2^{\chi - 1}$ χ is the number of operations performed per function. It is assumed that the probability of operation for each module in level i is the same.
		$\Pr\{\Phi_\omega^*(X) = a\} =$ $$= \sum_{i_0=0}^{b} \sum_{i_1=0}^{b/2} \binom{\omega - 2i_1}{\omega - i_0} p_1^{\omega - i_0}(1-p_1)^{i_0 - 2i_1} \cdot$$ $$\cdot \sum_{i_2=0}^{b/4} \binom{\omega/2 - 2i_2}{\omega/2 - i_1} p_2^{\omega/2 - i_1}(1-p_2)^{i_1 - 2i_2}$$ $$\vdots$$ $$\sum_{i_\chi=0}^{0} \binom{1 - 2i_{\chi-1}}{1 - i_\chi} p^{1 - i_{\chi-1}}(1-p_\chi)^{i_{\chi-1} - 2i_\chi}$$	
		$\Pr\{\Phi_\omega^*(X) = a\} =$ $$= \sum_{i_1=0}^{b/2} \sum_{i_2=0}^{b/4} \cdots \sum_{i_\chi=0}^{0} \binom{\omega - 2i_1}{a}$$ $$\prod_{j=2}^{\chi} \binom{\omega/2^{j-1} - 2i_j}{\omega/2^{j-1} - i_{j-1}} \cdot$$ $$\cdot (1-p)^{\omega - a - \sum_{l=1}^{\chi-1} i_l} \, p^{a + \omega - 1 - \sum_{l=1}^{\chi-1} i_l}$$	$p_\alpha = a$ for all a
		$\Pr\{\Phi_\omega^*(X) \geq a\} =$ $$= \sum_{i_0=0}^{b} \sum_{i_1=0}^{b/2} \cdots \sum_{i_\chi=0}^{0}$$ $$\prod_{j=1}^{\chi} \binom{\omega/2^{j-1} - 2i_j}{\omega/2^{j-1} - i_{j-1}} \cdot$$ $$\cdot (1-p)^{i_0 - \sum_{l=1}^{\chi-1} i_l} \, p^{2\omega - 1 - \sum_{l=0}^{\chi-1} i_l}$$	— ,, —

Table 3.2

Expected number of operating functions $E\{\cdot\}$

No.	Types of structure	Computational relations	Observations
1	Symmetrical dependent structures	$E\{\Phi_\omega^*(X)\} = \omega p^{2\omega-1}$	$p_\alpha = p$
2	Independent function structures	$E\{\Phi_\omega^*(X)\} = \omega p$	—
3	Hierarchical dependent structures	$E\{\Phi_\omega^*(X)\} = \omega p^\chi$	$p_\alpha = p$ for all α; $\omega = 2^{\chi-1}$

Table 3.3

The variance of the number of operating functions

No.	Types of structure	Computational relations	Observations
0	1	2	3
1	Symmetrical dependent structures	$\operatorname{Var}\{\Phi_\omega^*(X)\} =$ $= \omega p^{2\omega-1}[1+(\omega-1)p^{2\omega-2} - \omega p^{2\omega-1}]$	$p_\alpha = \alpha$ for all α
2	Independent function structures	$\operatorname{Var}\{\Phi_\omega^*(X)\} = \omega p(1-p)$	—
3	Hierarchical dependent structures	$\operatorname{Var}\{\Phi_\omega^*(X)\} =$ $= \omega p^\chi \left[1 - \omega p^\chi + \sum_{i=1}^{\chi-1} 2^{i-1} p^i \right]$	$p_\alpha = \alpha$ for all α $\omega = 2^{\chi-1}$

An example of a 4-function structure*. The theoretical results obtained in the previous sections are used to find the probability that exactly a functions operate in a multifunctional system for each of the three special classes of structure. The probabilities are given in table 3.4 for $\omega = 4$. The probability that at least a functions operate for each of the special classes of structures are shown in table 3.5 for $\omega = 4$. In the case when $p_\alpha = p$ for all α, the probability that at least a functions operate ($a = 1, 2, 3, 4$) was calculated for $0 < p < 1$. The probabilities for each class of structure are given in figs. 3.7—3.10. The mean and variance were found for each class of structure (see table 3.6). The mean and variance for each special structure were plotted for p, $0 < p < 1$ and are given in figs. 3.11 and 3.12.

---

* Each class of structure has a different number of operations per function, a different total number of modules and different structure costs. It is obvious that meaningful comparisons are not possible from the above information.

Table 3.4
Probability that exactly a functions operate

Structure type	$\Pr\{\Phi_4^*(X) = a\}$
Symmetrical	
$a = 0$	$1 - p_1^1 p_2^3 p_3^3 p_4^1 - 6 p_1^2 p_2^5 p_3^4 p_4^1 + 4 p_1^3 p_2^6 p_3^4 p_4^1 - p_1^4 p_2^6 p_3^4 p_4^1$
$a = 1$	$4 p_1^1 p_2^2 p_3^3 p_4^1 - 12 p_1^2 p_2^5 p_3^4 p_4^1 + 12 p_1^3 p_2^6 p_3^4 p_4^1 - 4 p_1^4 p_2^6 p_3^4 p_4^1$
$a = 2$	$6 p_1^2 p_2^5 p_3^4 p_4^1 - 12 p_1^3 p_2^6 p_3^4 p_4^1 + 6 p_1^4 p_2^6 p_3^4 p_4^1$
$a = 3$	$4 p_1^3 p_2^6 p_3^4 p_4^1 - 4 p_1^4 p_2^6 p_3^4 p_4^1$
$a = 4$	$p_1^4 p_2^6 p_3^4 p_4^1$
Independent	
$a = 0$	$1 - 4p + 6p^2 - 4p^3 + p$
$a = 1$	$4p - 12p^2 + 12p^3 - 4p^4$
$a = 2$	$6p^2 - 12p^3 + 6p^4$
$a = 4$	p^4
Hierarchical	
$a = 0$	$1 - 4 p_1^1 p_2^1 p_3^1 + 2 p_1^2 p_2^1 p_3^1 + 4 p_1^2 p_2^2 p_3^1 - 4 p_1^3 p_2^2 p_3^1 + p_1^4 p_2^2 p_3^1$
$a = 1$	$4 p_1^1 p_2^1 p_3^1 - 4 p_1^2 p_2^1 p_3^1 - 8 p_1^2 p_2^2 p_3^1 + 12 p_1^3 p_2^2 p_3^1 - 4 p_1^4 p_2^2 p_3^1$
$a = 2$	$2 p_1^2 p_2^1 p_3^1 + 4 p_1^2 p_2^2 p_3^1 - 12 p_1^3 p_2^2 p_3^1 - 6 p_1^4 p_2^2 p_3^1$
$a = 3$	$4 p_1^3 p_2^2 p_3^1 - 4 p_1^4 p_2^2 p_3^1$
$a = 4$	$p_1^4 p_2^2 p_3^1$

Table 3.5
Probability that at least a functions operate

Structure type	$\Pr\{\Phi_4^*(X) \geqslant a\}$
Symmetrical	
$a = 1$	$h_S(1) = 4 p_1^1 p_2^3 p_3^3 p_4^1 - 6 p_1^2 p_2^5 p_3^4 p_4^1 + 4 p_1^3 p_2^6 p_3^4 p_4^1 - p_1^4 p_2^6 p_3^4 p_4^1$
$a = 2$	$h_S(2) = 6 p_1^2 p_2^5 p_3^4 p_4^1 - 8 p_1^3 p_2^6 p_3^4 p_4^1 + 3 p_1^4 p_2^6 p_3^4 p_4^1$
$a = 3$	$h_S(3) = 4 p_1^3 p_2^6 p_3^4 p_4^1 - 3 p_1^4 p_2^6 p_3^4 p_4^1$
$a = 4$	$h_S(4) = p_1^4 p_2^6 p_3^4 p_4^1$
Independent	
$a = 1$	$h_I(1) = 4p - 6p^2 + 4p^3 - p^4$
$a = 2$	$h_I(2) = 6p^2 - 8p^3 + 3p^4$
$a = 3$	$h_I(3) = 4p^3 - 3p^4$
$a = 4$	$h_I(4) = p^4$
Hierarchical	
$a = 1$	$h_H(1) = 4 p_1^1 p_2^1 p_3^1 - 2 p_1^2 p_2^1 p_3^1 - 4 p_1^2 p_2^2 p_3^1 + 4 p_1^3 p_2^2 p_3^1 - p_1^4 p_2^2 p_3^1$
$a = 2$	$h_H(2) = 2 p_1^2 p_2^1 p_3^1 + 4 p_1^2 p_2^2 p_3^1 - 8 p_1^3 p_2^2 p_3^1 + 3 p_1^4 p_2^2 p_2^1$
$a = 3$	$h_H(3) = 4 p_1^3 p_2^2 p_3^1 - 3 p_1^4 p_2^2 p_3^1$
$a = 4$	$h_H(4) = p_1^4 p_2^2 p_3^1$

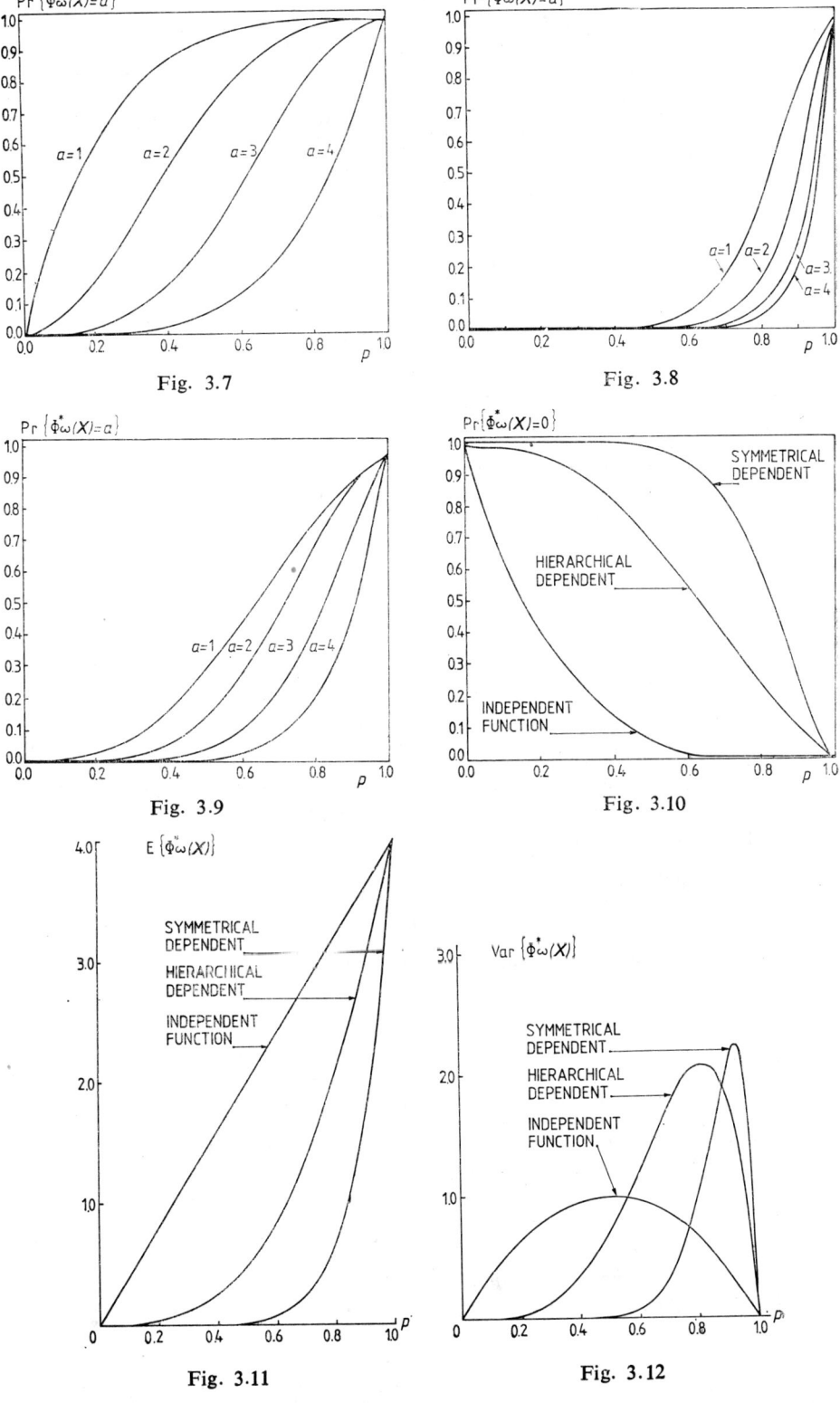

Fig. 3.7

Fig. 3.8

Fig. 3.9

Fig. 3.10

Fig. 3.11

Fig. 3.12

Table 3.6

Mean and variance of $\Phi_4^*(X)$

Structure type	$E\{\Phi_4^*(X)\}=$	$V\{\Phi_4^*(X)\}=$
Symmetrical	$4p^8$	$4p^8[1+3p^4-4p^8]$
Independent	$4p$	$4p(1-p)$
Hierarchical	$4p^3$	$4p^3(1+p+2p^2-4p^3)$

3.6. Structures of fixed time. A general comparison

It is not infrequent in systems engineering that the designer must choose the type of structure he intends to use. In order to make a well-grounded decision, he must first compare the properties of the different structures he wishes to operate with. Let us assume the notation

$$\Pr{}_H\{\Phi_\omega^*(X) \geq a\} = h_H(P, a); \quad \Pr{}_I\{\Phi_\omega^*(X) \geq a\} = h_I(P, a);$$

$$\Pr{}_S\{\Phi_\omega^*(X) \geq a\} = h_S(P, a), \tag{3.28}$$

where a is the number of functions which operate and P is the vector of probabilities such that

$$p_i = \Pr\{\Psi_{\gamma, i}(X) = 1\}. \tag{3.29}$$

DEFINITION 3.9. *Partial Dominance* (a^*). A structure M is defined to partially dominate (a^*) a structure N, if for $a = a^*$, $h_M(P, a^*) \geq h_N(P, a^*)$ for all P.

DEFINITION 3.10. *Complete Dominance* (a^*). A structure M is defined to completely dominate (a^*) a structure N, if it partially dominates (a^*) a structure N for all $a \geq a^*$.

In what follows reliability performances will be structured according to the different kinds of systems described above.

THEOREM 3.5. *For a multifunctional boolean coherent structure* a) *if* $p_\alpha = p$ *for all* α, *then* $h_I(P, a) \geq h_S(P, a)$ *for all* $0 \leq p \leq 1$ *and* $a = 1, 2, \ldots, \omega$; b) *if* $p_\alpha = p$

for all α and $\omega = 2^{\alpha-1}$, then $h_I(\mathbf{P}, a) \geq h_S(\mathbf{P}, a)$ for all $0 \leq p \leq 1$ and $a = 1, 2, \ldots, \omega$.

For other types of structures the reliability of the components can be improved. An immediate question of realistic pertinence is "What should the values of $\mathbf{P}$ be such that dominance can be reversed?" [26]. This question may be formulated as in the

Problem. If $h_A(\mathbf{P}_M, a) \leq h_B(\mathbf{P}_N, a)$ for $p_M = p_N = p$, then find the minimum K where $p_M = p$, $p_N = p^K$, such that $h_M(\mathbf{P}_M, a) \geq h_N(\mathbf{P}_N, a)$.

In case that $p = e^{-\lambda t}$, K is the ratio of the mean time to failure for the components of the structures under consideration, since

$$p_M = e^{-\lambda t} = e^{-\lambda_M t}; \quad p_N = e^{-\lambda K t} = e^{-\lambda_N t}, \quad \text{or} \quad \frac{\lambda_N}{K} = \lambda_M. \tag{3.30}$$

Hence, $K = \dfrac{\lambda_M}{\lambda_N}$.

The solution to this problem is difficult to find. The implicit function theorem and the properties of polynomials were used by O'Brien ([26], p. 82) to show that function $K(p)$ exists and is continuous in the interval $0 < p < 1$. In this case, the expression for $K(p)$ cannot be solved explicitly.

It is important to note that the probability function can be written as

$$h_M(\mathbf{P}, a) = \sum_{i=0}^{m} c_i p^{i+n}; \quad h_N(\mathbf{P}, r) = \sum_{i=0}^{u} d_i p^{i+v}. \tag{3.31}$$

The next results give information on the variation of $K(p)$ in the study of dominance for different structures.

LEMMA 3.6. *For a multifunctional structure, $h_N(\mathbf{P}^K, a)$ is strictly decreasing in K for all $0 < p < 1$ and $K > 0$, where K is not a function of p.*

Proof. (see ref. [26], p. 83). For any value of $p = p_0$,

$$\frac{d}{dK} h_N(\mathbf{P}_0^K, a) = \sum_{i=1}^{u} d_i p_0^{(i+v)K} (i+v) \ln(p_0) = \ln(p_0) \sum_{i=0}^{n} d_i (i+v) p_0^{(i+v)K} =$$

$$= \ln(p^0) p_0^K \sum_{i=1}^{u} d_i (i+v) p_0^{(i+v)} p_0^{(i+v)K-K} = \ln(p_0) p_0^K \sum_{i=1}^{u} d_i (i+v) (p_0^K)^{(i+v)-1}. \tag{3.32}$$

Note that

$$\sum_{i=1}^{u} d_i (i+v)(p_0^K)^{(i+v)-1} = \frac{d}{dp} h_N(\mathbf{P}, a) \bigg|_{p=p_0^K}. \tag{3.33}$$

Since $h_N(\mathbf{P}, a)$ is a cumulative distribution function, it is also an increasing function of p and

$$\frac{d}{dp} h_N(\mathbf{P}, a) \bigg|_{p=p_0^K} \geq 0. \tag{3.34}$$

Since $\ln(p_0) \leq 0$, then

$$\frac{d}{dK} h_B(a, p_0^K) \leq 0. \tag{3.35}$$

LEMMA 3.7. *For a multifunctional structure if* $h_M(P, a) \leq h_N(P, a)$ *and* $h_M(P, a) = h_N(P^K, a)$, *where* $K = f(p)$, *then*

$$K(0) = \frac{n}{v} \quad \text{and} \quad K(1) = \frac{\dfrac{d}{dp} h_M(1, a)}{\dfrac{d}{dp} h_N(1, a)}. \tag{3.36}$$

Proof (see ref. [26], p. 85). Before calculating $K(0)$ and $K(1)$, it is necessary to examine $h_M(P, a)$ and $h_N(P, a)$ within a neighbourhood of $p = 0$ and $p = 1$. Thus, $h_M(P, a)$ can be made arbitrarily close to $h_N(P, a)$ within a neighbourhood of $p = 0$ if the lowest order terms of each polynomial have the same degree. Therefore, $K(0) = \dfrac{n}{v}$. If $(1 - \varepsilon)$ is substituted for p, then $h_M(P, a)$ can be made arbitrarily close to $h_N(P, a)$ within a neighbourhood of $p = 1$ if the lowest-order terms of each polynomial in ε have the same degree.

If we substitute $p = (1 - \varepsilon)$ in

$$h_M(P, a) = \sum_{i=0}^{m} c_i p^{(n+i)} = \sum_{i=0}^{m} c_i (1 - \varepsilon)^{n+i} \tag{3.37}$$

and use the binomial theorem, then

$$h_M(P, a) = \sum_{i=0}^{m} c_i \sum_{j=0}^{n+i} \binom{n+i}{j} (-\varepsilon)^j. \tag{3.38}$$

Since

$$\binom{n+i}{j} = 0 \text{ for } j > n + i,$$

then

$$h_M(P, a) = \sum_{j=0}^{m+n} \sum_{i=0}^{m} c_i \binom{n+i}{j} (-\varepsilon)^j =$$

$$= \sum_{i=0}^{m} c_i + \sum_{j=1}^{m+n} \sum_{i=0}^{m} \binom{n+i}{j} (-\varepsilon^j). \tag{3.39}$$

The quantity $h_M(P, a)$ is a distribution function and then we have

$$h_M(1, a) = \sum_{i=0}^{m} c_i = 1. \tag{3.40}$$

Therefore,
$$h_M(\mathbf{P}, a) = 1 + \sum_{j=1}^{m+n} \sum_{i=0}^{m} c_i \binom{n+i}{j} (-\varepsilon)^j. \tag{3.41}$$

The corresponding value for the degree of the low-order term of ε is equal with $\sum_{i=0}^{m} c_i(n+1)$.

Similarly, for $h_N(\mathbf{P}, a)$ the degree of the low-order term of ε is $\sum_{i=0}^{u} d_i(v+i)$ so that

$$K(1) = \frac{\sum_{i=0}^{m}(n+i)c_i}{\sum_{i=0}^{u}(v+i)d_i}. \tag{3.42}$$

This is equivalent to

$$K(1) = \frac{\dfrac{dh_M(\mathbf{P}, a)}{dp}}{\dfrac{dh_N(\mathbf{P}, a)}{dp}} \bigg|_{p=1}. \tag{3.43}$$

In case that

$$\frac{d}{dp} h_N(\mathbf{P}, a) \bigg|_{p=1} = 0, \tag{3.44}$$

then $K(1)$ does not exist and $K(1)$ must be infinite.

LEMMA 3.8. *For a multifunctional system, if $h_M(\mathbf{P}, a) \leqslant h_N(\mathbf{P}, a)$ and $K(p)$ is such that $h_M(\mathbf{P}, a) = h_N(\mathbf{P}^{K(p)}, a)$, then $K(p) \geqslant 1$ for $0 < p < 1$.*

LEMMA 3.9. *For a multifunctional system, if* a) $K(0) > K(1)$, *then* $\dfrac{dK}{dp}\bigg|_{p=1} < 0$, b) $K(0) < K(1)$, *then* $\dfrac{dK}{dp}\bigg|_{p=1} > 0$, c) *in both cases* $\dfrac{dK}{dp}\bigg|_{p=0} = 0$.

Proof (see [26]).

$$\frac{dh_M(\mathbf{P}, a)}{dp} = \frac{dh_N(\mathbf{P}^K, a)}{dp} = \frac{\partial h_N(\mathbf{P}^K, a)}{p} + \frac{\partial h_N(\mathbf{P}^K, a)}{K} \frac{dK}{dp} \tag{3.45}$$

or

$$\frac{dK}{dp} = \frac{\dfrac{dh_M(a, \mathbf{P})}{dp} - \dfrac{\partial h_N(a, \mathbf{P}^K)}{\partial p}}{\dfrac{\partial h_N(a, \mathbf{P}^K)}{\partial K}}. \tag{3.46}$$

Taking the derivatives,

$$\frac{dK}{dp} = \frac{\sum_{i=0}^{m}(n+i)c_i p^{n+i-1} - \sum_{i=0}^{u}(i+v)K(p)d_i p^{(i+v)K(p)-1}}{\dfrac{\partial h_N(a, \mathbf{P}^K)}{\partial K}}. \qquad (3.47)$$

Substituting $n = vK(0)$, the numerator equals with

$$\sum_{i=0}^{m} vK(0)c_i p^{n+i-1} - \sum_{i=0}^{u} vK(p)d_i p^{(i+v)K(p)-1} + \sum_{i=0}^{m} ic_i p^{n+i-1} - \sum_{i=0}^{u} iK(p)d_i p^{(i+v)K(p)-1}. \qquad (3.48)$$

If $p \to 1$, the numerator equals with

$$vK(0)\sum_{i=0}^{m} c_i - vK(1)\sum_{i=0}^{u} d_i + \sum_{i=0}^{m} ic_i - K(1)\sum_{i=0}^{u} id_i. \qquad (3.49)$$

Because $\sum_{i=0}^{m} c_i = \sum_{i=0}^{u} d_i = 1$ and $h_M(1, a) = h_N(1, a) = 1$ by previous results,

$$\frac{\sum_{i=0}^{m} ic_i}{\sum_{i=0}^{u} id_i} = K(1). \qquad (3.50)$$

By substituting back we can write

$$\frac{dK}{dp}\bigg|_{p=1} = \frac{v(K(0) - K(1))}{\dfrac{\partial h_N(\mathbf{P}^K, a)}{\partial K}}. \qquad (3.51)$$

As it is known that

$$\frac{\partial h_N(\mathbf{P}^K, a)}{\partial K} = \sum_{i=0}^{u} d_i(i+v)p^{(i+v)K}\ln(p) = p\ln(p)\frac{d}{dp}h_N(\mathbf{P}, a), \qquad (3.52)$$

and taking the limit for $p \to 1$, this term becomes negative. Finally, $\dfrac{dK}{dp}\bigg|_{p=1}$ has the sign of $v(K(1) - K(0))$. If $K(1) > K(0)$, $\dfrac{dK}{dp}\bigg|_{p=1} > 0$ and if $K(1) < K(0)$, then $\dfrac{dK}{dp}\bigg|_{p=1} < 0$.

For the second part of the theorem, we can notice that as $p \to 0$,

$$\frac{\partial h_N(\mathbf{P}^K, a)}{\partial K} \to \infty. \qquad (3.53)$$

For the classes of structure considered (symmetrical, independent, hierarchical) "it is more costly to design these classes of structure with an equivalent probability of all functions operating than it is to design them with an equivalent expected number of functions operating" [26].

A numerical example ([26], p. 99). Let us consider a multifunctional structure with $\omega = 4$. $K(p)$ was calculated such that $h_H(\mathbf{P}, a) \geqslant h_I(\mathbf{P}^{K(p)}, a)$, $h_S(\mathbf{P}, a) \geqslant h_I(\mathbf{P}^{K(p)}, a)$, $h_S(\mathbf{P}, a) \geqslant h_H(\mathbf{P}^{K(p)}, a)$.

The variation of $K(p)$'s is shown in figs. 3.13, 3.14 and 3.15. The values of K_1 and K_2 are given in table 3.7.

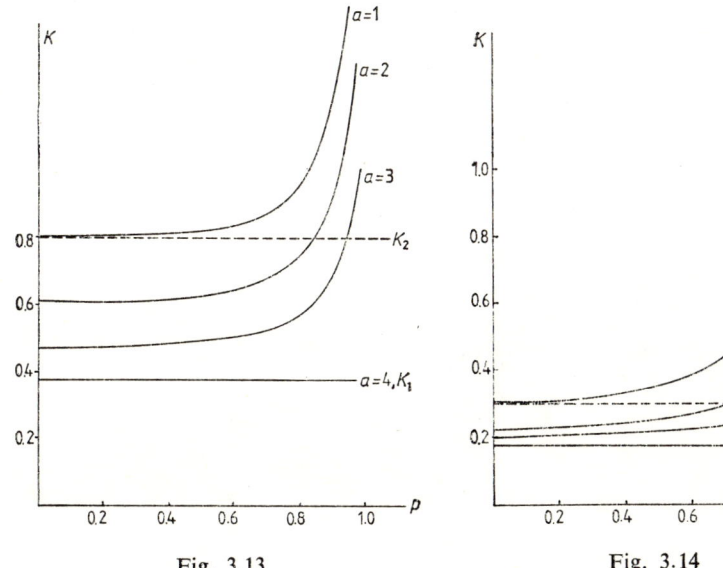

Fig. 3.13   Fig. 3.14

In the practice of systems engineering, designers and systems analysts are often faced with situations where structures are equivalent with respect to reliability performance. In this case, equivalence criteria for multifunctional systems as well as complementary properties to the concept of equivalence must be found.

Criterion 3.1. A complex multifunctional system M is equivalent to system N, if the probability of all functions operating is the same such that

$$h_M(\mathbf{P}, \omega) = h_N(\mathbf{P}, \omega). \quad (3.54)$$

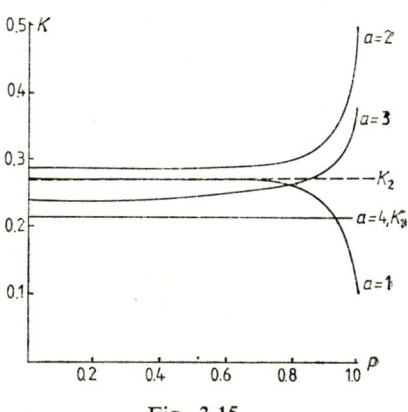

Fig. 3.15

Table 3.7

Types of structure	K_1	K_2
Symmetrical: independent	$\dfrac{15}{4}$	$\dfrac{8}{1}$
Hierarchical: independent	$\dfrac{7}{4}$	$\dfrac{3}{1}$
Hierarchical: symmetrical	$\dfrac{15}{7}$	$\dfrac{8}{3}$

Criterion 3.2. A complex multifunctional system M is equivalent to a system N, if the expected number of operating functions is the same such that

$$E_M\{\Phi_\omega^*(X)\} = E_N\{\Phi_\omega^*(X)\}. \quad (3.55)$$

The following lemmas derive from some of the properties for systems which are equivalent by either of these criteria and which have different properties depending on the type of structures.

LEMMA 3.10. ([26], p. 93). *If two structures are equivalent by criterion* 3.1 *such that*

$$h_M\{\mathbf{P}_N, \omega\} = h_N\{\mathbf{P}^{K_1(p)}, \omega\}, \quad (3.56)$$

then $K_1 = \dfrac{n_M}{n_N}$, *where* n_M, n_N *represent the number of components in the structures* M *and* N, *respectively.*

COROLLARY 3.1. ([26], p. 94).

If $h_S(\mathbf{P}, a) = h_I(\mathbf{P}^{K(p)}, a)$, then $K_1 = \dfrac{2^\omega - 1}{\omega}$; $\quad (3.57)$

If $h_H(\mathbf{P}, a) = h_I(\mathbf{P}^{K(p)}, a)$, then $K_1 = \dfrac{2\omega - 1}{\omega}$; $\quad (3.58)$

If $h_S(\mathbf{P}, a) = h_H(\mathbf{P}^{K(p)}, a)$, then $K_1 = \dfrac{2^\omega - 1}{2\omega - 1}$. $\quad (3.59)$

LEMMA 3.11. *If two structures are equivalent by criterion* 3.2 *such that* $E_M\{\Phi_\omega^*(X)\} = E_N\{\Phi_\omega^*(X)\}$, *where* $p_B = p^{K_2}$, $p_A = p$, *then clearly*

a) *if* $E_S\{\Phi_\omega^*(X)\} = E_I\{\Phi_\omega^*(X)\}$, *then* $K_2 = 2^{\omega-1}$, $\quad (3.60)$

b) *if* $E_H\{\Phi_\omega^*(X)\} = E_I\{\Phi_\omega^*(X)\}$, *then* $K_2 = \chi$, $\quad (3.61)$

c) *if* $E_S\{\Phi_\omega^*(X)\} = E_H\{\Phi_\omega^*(X)\}$, *then* $K_2 = \dfrac{2^{\omega-1}}{\chi}$. $\quad (3.62)$

LEMMA 3.12. $K_2 \geqslant K_1$ *for all three classes of structure.*

Proof. For the comparison of symmetrical to independent structures, $p_S = p$, $p_I = p^K$. If $K_2 \geqslant K_1$, then $2^{\omega-1} \geqslant \dfrac{2^\omega - 1}{\omega}$ or $\omega 2^{\omega-1} \geqslant 2^\omega - 1$. Thus, $\omega \geqslant 2 - \dfrac{1}{2^{\omega-1}}$, which is true. So, $K_2 \geqslant K_1$. For the comparison of symmetrical to hierarchical structures, $p_S = p, p_H = p^K$. If $K_2 \geq K_1$, then $\dfrac{2^\omega - 1}{\chi} \geq \dfrac{2^\omega - 1}{2\omega - 1}$, $(2\omega - 1) 2^{\omega - 1} \geq (2^\omega - 1) \chi$ or $\omega 2^\omega - 2^{\omega - 1} \geq 2^\omega \chi - \chi$. Thus, $2(\omega - \chi) \geq 1 - \dfrac{\chi}{2^{\omega - 1}}$ which is true. So $K_2 \geq K_1$.

3.7. Economic aspects concerning multifunctional systems design

For the components which follow an exponential distribution of failure, the mean time to failure will decrease. For complex technical structures, the reliability of the component decreases as K increases and should therefore cost less.

3.8. General optimization of boolean hierarchical dependent structures

When the systems engineer wants to design a complex multifunction system, he must choose that structure which optimizes the system. The use of redundancy may improve the reliability and safety performance of a system.

According to O'Brien [26], the following criteria for optimization should be considered. a) Select the structure which maximizes the system performance subject to the constraint that the system cost is less than b; b) Select the structure which minimizes the system cost, subject to the constraint that the system performance is greater than $\mathscr{R}$. The measure of system performance to be used in the probability is that more than r functions operate, i.e. $\Pr\{\Phi_\omega^*(X) \geq r\}$ for a particular r. Vector X of the component states may or may not vary with time.

The hierarchical dependent structure with $\omega = 2^{\chi - 1}$ functions has ξ operations per function, where ξ can be greater than χ. A different type of equipment may exist at each structural level in the tree (a hierarchical structure has at most χ different types of equipment, and equipment of type j is shared between 2^{j-1}, $(j = 1, 2, \ldots, \chi)$, functions). For technical reasons, we consider that the equipments for a particular operation are all identical (each operation that uses an equipment of type j takes a total of $\omega/2^{j-1}$ equipments for all functions). The cost of a single piece of equipment of type j depends on the number of functions it serves ($u_j = c + h2^{j-1}$, $h > 0$; $c > 0$). Concerning reliability aspects, the probability that a piece of equipment operates depends on the number of functions it serves ($p_j = p^{e_j}; e_j > 1$).

Finally, it is considered that all equipments of type j are independent. These assumptions appear reasonable for electronic systems which use integrated circuits or even for microprocessors.

Let $\boldsymbol{\theta}$ be a vector, where $\{\theta_j\}$ is the number of operations performed by equipments of type j, $(j = 1, \ldots, \chi)$.

Since all operations must be performed by some equipment types, then

$$\sum_{j=1}^{\chi} \theta_j = \xi. \tag{3.63}$$

Vector $\boldsymbol{\theta}$ is then the variable the designer is trying to determine in order to optimize the structure.

Let $\boldsymbol{C}$ be a vector, where each component c_j is the total cost of the system for using an equipment of type j, such that

$$c_j = \frac{\omega}{2^{j-1}} u_j = \frac{\omega}{2^{j-1}}(c + h2^{j-1}); \quad \Pr\{\Phi_\omega^*(X) \geq a\} = h(\boldsymbol{P}, a); \quad \boldsymbol{C} = \sum_{j=1}^{\chi} c_j \theta_j.$$

The optimization problems can be written as (see O'Brien [26], Chap. 5) below.

Problem I. max: $\Pr\{\Phi_\omega^*(X) \geq a\} = \max_{\boldsymbol{\theta}} h(\boldsymbol{\theta}, a)$, for $0 < p < 1$, $a = \omega - s = 2^{\chi-1} - 2^i - 1$, $(i = 0, 1, 2, \ldots, \chi - 1)$, such that $\theta_i \geq 0$, $(i = 1, 2, \ldots, \chi)$, $\boldsymbol{1}\boldsymbol{\theta}' = \xi$; $\mathscr{C} \geq \boldsymbol{C}\boldsymbol{\theta}'$.

Problem II. min $\boldsymbol{C}\boldsymbol{\theta}' = c$ for $a = \omega - s = 2^{\chi-1} - 2^i - 1$, $(i = 0, 1, 2, \ldots, \chi - 1)$, such that $\theta_i \geq 0$, $(i = 1, 2, \ldots, \chi)$; $\boldsymbol{1}\boldsymbol{\theta}' = \xi$. $\boldsymbol{1}\boldsymbol{\theta}' = h(\boldsymbol{\theta}, a) \geq \mathscr{R}$, for $p = p_0$, $h(\boldsymbol{\theta}, a) \geq h(\hat{\boldsymbol{\theta}}, a)$ for all $\hat{\boldsymbol{\theta}}$ such that $\boldsymbol{C}\hat{\boldsymbol{\theta}}' \leq c$ for $p \neq p_0$.

DEFINITION 3.11. a) *Better with respect to performance:* A vector $\hat{\boldsymbol{\theta}}$ is better with respect to performance than a vector $\boldsymbol{\theta}$ if

$$\boldsymbol{C}\hat{\boldsymbol{\theta}} \leq \boldsymbol{C}\boldsymbol{\theta}', \quad 0 < p_j < 1, \quad h(\hat{\boldsymbol{\theta}}, a) > h(\boldsymbol{\theta}, a). \tag{3.64}$$

b) *Better with respect to cost:* A vector $\hat{\boldsymbol{\theta}}$ is better with respect to cost than a vector $\boldsymbol{\theta}$ if

$$\boldsymbol{C}\hat{\boldsymbol{\theta}} < \boldsymbol{C}\boldsymbol{\theta}', \quad 0 < p_j < 1, \quad h(\hat{\boldsymbol{\theta}}, r) \geq h(\boldsymbol{\theta}, r). \tag{3.65}$$

c) *Inadmissible with respect to performance.* A vector $\boldsymbol{\theta}$ is said to be inadmissible with respect to performance if there is a vector $\hat{\boldsymbol{\theta}}$ which is better with respect to performance than $\hat{\boldsymbol{\theta}}$ for some $0 < p_j < 1$.

d) *Inadmissible with respect to cost:* A vector $\boldsymbol{\theta}$ is said to be inadmissible with respect to cost if there is a vector $\hat{\boldsymbol{\theta}}$ which is better with respect to cost than $\boldsymbol{\theta}$ for some $0 < p_j < 1$.

THEOREM 3.6. *If a vector $\boldsymbol{\theta}$ is inadmissible with respect to performance, then it cannot be an optimal solution to Problem I.*

THEOREM 3.7. *If a vector $\boldsymbol{\theta}$ is inadmissible with respect to cost, then it cannot be an optimal solution to Problem II.*

LEMMA 3.13. *If* $\mathscr{E} = \{e_i^*\}$, $(i = 1, 2, \ldots, \chi)$, *where*

$$e_i^* = e_{\chi+1-i}\left[\frac{\omega}{2^{i-1}} - \left[\frac{s}{2^{i-1}}\right]\right] - e_1\left[1 - \left[\frac{s}{2^{\chi-1}}\right]\right]$$

so

$$h^*(\boldsymbol{\theta}, r) = e_1\chi\left[1 - \left[\frac{s}{2^{\chi-1}}\right]\right]\ln p + \sum_{i=1}^{\chi-1} e_i^*\theta_i \ln p$$

and

$$\ln h(\boldsymbol{\theta}, r) = \text{constant} + \sum_{i=1}^{\chi-1} e_i^*\theta_i \ln p,$$

where the constant is defined as

$$\sum_{i=1}^{\chi} \ln\left(\frac{\frac{\beta}{2^{i-1}} - 2\left[\frac{s}{2^i}\right]}{\frac{\beta}{2^{i-1}} - \left[\frac{s}{2^{i-1}}\right]}\right) + e_1\chi\left(1 - \left[\frac{s}{2^{\chi-1}}\right]\right)\ln p, \qquad (3.66)$$

and the total cost of the system is given by the relation

$$C = \sum_{j=1}^{\chi-1} \{2^{\chi-j}(c+\omega h) - (c+\omega h)\}\theta_j + (c+\omega h)\xi$$

or

$$\boldsymbol{\theta}_\xi: \theta_i = \begin{cases} \xi, & i = l \\ 0, & i \neq l \end{cases}$$

$$\boldsymbol{\theta}_{\xi+1}: \theta_i = \begin{cases} 1, & i = l-1 \\ \xi-1, & i = l \\ 0, & i \neq l, l-1 \end{cases}$$

$$\cdots \cdots \cdots \cdots$$

$$\boldsymbol{\theta}_z: \theta_i = \begin{cases} \xi, & i = 1, \\ 0, & i \neq 1. \end{cases}$$

THEOREM 3.8 [26]. *The cost functions corresponding to the sequence of vectors* $\boldsymbol{\theta}_n$ *is strictly increasing.*

In ref. [26] it was shown that, for any $s = 2^{k-1} - 1$, $(k = 1, \ldots, \chi)$, all $\boldsymbol{0}$ are inadmissible, except those that have only two components greater than zero $(\theta_{l-1} + \theta_l = \xi \text{ or } \theta_\chi + \theta_l = \xi)$.

Let us define $l = \max(j: b_j > 0)$. Hence,

$$\boldsymbol{\theta}_0: \theta_i = \begin{cases} \xi, & i = \chi \\ 0, & i \neq \chi \end{cases}$$

$$\boldsymbol{\theta}_1: \theta_i = \begin{cases} 1, & i = l \\ \xi - 1, & i = \chi \\ 0, & i \neq \chi, l \end{cases}$$

$$\boldsymbol{\theta}_2: \theta_i = \begin{cases} 2, & i = l \\ \xi - 2, & i = \chi \\ 0, & i \neq \chi, l \end{cases}$$

$$\vdots$$

LEMMA 3.14 [26]. *The expression* $\sum_{i=0}^{n} A_i p^i (1-p)^{n-1}$ *can be written as* $\sum_{k=0}^{n} B_k p^k$ *if*

$$B_k = \sum_{i=0}^{k} \binom{n-i}{k-i} (-1)^{k-i} A_i. \tag{3.67}$$

LEMMA 3.15 [26]. *The expression* $\sum_{k=0}^{n} B_k p^k$ *can be written as*

$$\sum_{i=0}^{n} A_i p^i (1-p)^{n-1} \quad \text{if,} \quad A_i = \sum_{j=0}^{i} \binom{n-j}{n-i} B_j. \tag{3.68}$$

LEMMA 3.16 [26]. $(1-p^D)^m$ *can be written as* $\sum_{l=0}^{mD} A_{mD-l} p^{mD-l}$

$(1-p)^l$, *where* $A_{mD-l} = \sum_{j=0}^{m} \binom{(m-j)D}{l} \binom{m}{j} (-1)^j.$ (3.69)

LEMMA 3.17 [26]. $(1-p^D)^m$ *can be expressed as a polynomial of degree mD or as a polynomial of degree mD + E which is defined as*

$$(1-p^D)^m = \sum_{l=0}^{mD+E} \left\{ \sum_{j=0}^{m} \binom{(m-j)D+E}{l} \binom{m}{j} (-1)^j \right\} p^{mD+E-l} (1-p)^l. \tag{3.70}$$

THEOREM 3.9 [26]. $h(\theta_n, a) = \sum_{\delta=0}^{rs_1\theta_{j-1}+2s_2\theta_j} A_\delta p^{\beta\xi} \left(\frac{1-p}{p} \right)^\delta,$

where

$$r = 2^{j-2}, \quad s_1 = \left[\frac{s}{2^{j-2}} \right], \quad s_2 = \left[\frac{s}{2^{j-1}} \right], \quad j \leq \max \{i: b_i > 0\},$$

$$\theta_n: \{\theta_i\} = \begin{cases} k, & i = j \\ \xi - k, & i = j-1 \\ 0, & i \neq j, j-1 \end{cases}$$

and $A_\delta = \sum_{l=0}^{[s_2]} \sum_{n=0}^{[s_2]} \binom{\omega/2a}{l} \binom{l}{n} (-1)^n \cdot \sum_{i=0}^{[s_1]} \sum_{h=0}^{[s_1]} \binom{\omega/r - 2l}{i - 2l} \binom{i-2l}{h} (-1)^n \cdot$

$$\cdot \binom{(i-h)r\theta_{j-1} + (l-n)2r\theta_j}{\delta}.$$

An example for a 4-function structure

Problem I. $\min_\theta C\theta' = \min_\theta C^*\theta' + (c+h4) 3$, subject to the constraints $\theta_i > 0$, $(i = 1, 2, 3)$, $1\theta' = 3$, $h(\theta, r)|_{p=.95} \geq R^* = .75$, where $C^* = (3, 1, 0)$, for $c = 1$, $h = 0$.

MULTIFUNCTIONAL SYSTEMS AND EXTENSIONS

Problem II. max $h(\theta, r)$ subject to the constraints $\theta_i \geq 0$, $(i = 1, 2, 3)$, $\mathbf{1\theta} = 3$, $C\theta' < c^* \overset{\theta}{=} 10$, $C\theta' = C^*\theta' + (c + h4)3$, and $C^* = (3, 1, 0)$, for $c = 1$, $h = 0$.

When $s = 1$ or $\omega - s = a = 3$, $B(s) = (001)$ and the only admissible solutions are $\{\theta_1, \theta_3 : \theta_3 + \theta_1 = 3\}$, $\theta_1 = (0, 0, 3)$; $\theta_2 = (1, 0, 2)$; $\theta_3 = (2, 0, 1)$; $\theta_4 = (3, 0, 0)$.

To calculate A_δ for each θ_n, let $a = 2^{l-2} = 1$; $s_1 = 1$; $s_2 = 0$. From the above results,

$$A_\delta = \sum_{i=0}^{1} \sum_{l=0}^{1} \binom{4}{i} \binom{i}{l} (-1)^l \binom{(i-l)\theta_1}{\delta},$$

which, when evaluated, yields $A_0 = 1$, $A_\delta = \binom{4}{1}\binom{\theta_1}{\delta}$ for $\delta > 0$. Appropriate results are given in table 3.8. When the system performance is evaluated at $p_0 = .95$

Table 3.8

	$h(\theta_n, 3)$
$C^* \theta_1' = 0$	$h(\theta_1, 3) = p^{12}$
$C^* \theta_2' = 3$	$h(\theta_2, 3) = p^{12} + 4p^{11}(1-p)$
$C^* \theta_3' = 6$	$h(\theta_3, 3) = p^{12} + 8p^{11}(1-p) + 4p^{10}(1-p)^2$
$C^* \theta_4' = 9$	$h(\theta_4, 3) = p^{12} + 12p^{11}(1-p) + 12p^{10}(1-p)^2 + 4p^9(1-p)^3$

for each θ_n, only the structures corresponding to θ_n (where $n \geq 3$) satisfy the constraint to the first optimization problem. The optimum structure for this example corresponds to the vector $\theta_3 = (2, 0, 1)$ (see also fig. 3.16) and the minimum cost is $C\theta_3' = C^*\theta_3' + 3 = 9$. The optimum structure for Problem II is also $\theta_3 = (2, 0, 1)$.

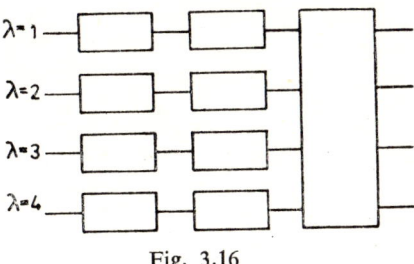

Fig. 3.16

When $s = 3$ or $\omega - s = a = 1$, $B(s) = (001)$ and the only admissible solutions are $\{\theta_3, \theta_2 : \theta_2 + \theta_2 = 3\}$ and $\{\theta_2, \theta_1 : \theta_2 + \theta_1 = 3\}$. $\theta_1 = (0, 0, 3)$, $\theta_2 = (0, 1, 2)$, $\theta_3 = (0, 2, 1)$, $\theta_4 = (0, 3, 0)$, $\theta_5 = (1, 2, 0)$, $\theta_6 = (2, 1, 0)$, $\theta_7 = (3, 0, 0)$.

When $\theta_3 + \theta_2 = 3$, let $a = 2^{l-1} = 2$, $s_1 = 1$, $s_0 = 0$, and finally $A_0 = 1$
$A_\delta = \binom{2}{1}\binom{2\theta_2}{\delta}$ for $\delta > 0$. Other results are given in table 3.9.

Table 3.9

$C^* \theta'_n$	$h(\theta_n, 1)$
$C^* \theta'_2 = 2$	$h(\theta_2, 1) = p^{12} + 4p^{11}(1-p) + 2p^{10}(1-p)^2$
$C^* \theta'_3 = 4$	$h(\theta_3, 1) = p^{12} + 8p^{11}(1-p) + 12p^{10}(1-p)^2$ $+ 8p^9(1-p)^3 + 2p^8(1-p)^4$
$C^* \theta'_4 = 6$	$h(\theta_4, 1) = p^{12} + 12p^{11}(1-p) + 30p^{10}(1-p)^2 + 40p^9(1-p)^3$ $+ 30p^8(1-p)^4 + 12p^7(1-p)^5 + 2p^6(1-p)^6$
$C^* \theta'_5 = 7$	$h(\theta_5, 1) = p^{12} + 12p^{11}(1-p) + 50p^{10}(1-p)^2$ $+ 160p^9(1-p)^3 + 110p^8(1-p)^4 + 72p^7(1-p)^5$ $+ 26p^6(1-p)^6 + 4p^5(1-p)^7$
$C^* \theta'_6 = 8$	$h(\theta_6, 1) = p^{12} + 12p^{11}(1-p) + 62p^{10}(1-p)^2$ $+ 168p^9(1-p)^3 + 246p^8(1-p)^4 + 212p^7(1-p)^5$ $+ 110^6(1-p)^6 + 32p^5(1-p)^7 + 4p^4(1-p)^8$
$C^* \theta'_7 = 9$	$h(\theta_7, 1) = p^{12} + 12p^{11}(1-p) + 66p^{10}(1-p)^2$ $+ 220p^9(1-p)^3 + 414p^8(1-p)^4 + 468p^7(1-p)^5$ $+ 330p^6(1-p)^6 + 144p^5(1-p)^7 + 36p^4(1-p)^8$ $+ 4p^3(1-p)^9.$

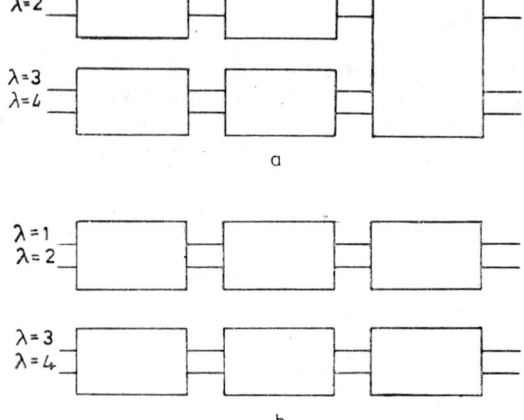

Fig. 3.17

When the system performance is evaluated at $p_0 = .95$ for each θ_n, the optimum structure corresponds to $\theta_3 = (0, 2, 1)$ (plotted in fig. 3.17 a) and the minimum cost is $C\theta'_3 = C^*\theta' + 3 = 7$. The optimum structure for Problem II corresponds to $\theta_4 = (0, 3, 0)$ (plotted in fig. 3.17 b).

General procedure for finding an optimal hierarchical structure

Problem I. For $s = 2^{k-1} - 1$, $(p = 1, 2, \ldots, \chi)$, we have to find $j = \max n$ such that $\mathscr{C}\theta'_n < c^*$, and the feasible solution is the decision vector $\theta_0, \theta_1, \ldots, \theta_j$. To evaluate $\theta'_j s$, we must go through the following steps.

Step 1: Calculate the values of A_δ for all θ_m, $(m = 0, 1, 2, \ldots, j)$.

Step 2: Calculate the vector corresponding to $\max A_\delta$ for all δ and obtain the maximum value of performance for $p \in [0, 1]$.

Problem II. For values of $s = s^{k-1} - 1$, $(p = 1, 2, \ldots, \chi)$, we must calculate the values of A_δ for all θ_n.

Step 1: Select the vector with the minimum n such that $h(\theta_n, a)|_{p=p_0} > \mathscr{R}$.

Step 2: Find A_δ, $\forall\, \delta$ corresponding to all θ_n greater than the A_δ corresponding to θ_j, $j < n$.

Step 3: Find the structure with a minimum cost which meets the performance constraints for a given $p = p_0$, $p \in [0, 1]$.

Finally, we notice that in order to find an optimal structure, it would be necessary to find the value of the cost function and to reduce the performance to $h(\theta, a) = p^{\omega \xi} \sum_{i=0}^{n} B_\delta \left(\frac{1-p}{p}\right)^i$ for all vectors θ which are not inadmissible and then select the optimal structure for each problem by examining the B_δ and $\mathscr{C}\theta'$ for all θ', [26].

3.9. Conception and measurement of system complexity

Complexity is nowadays a property of systems that is largely independent of their specific content. This fact raises an immediate question. Which system is more complex, a large-scale chemical plant, a nuclear power station, a railway system or a microprocessor? On closer examination, it appears that the study of system complexity must no doubt rely on information theoretical considerations, using concepts of the general systems theory. "Modern technology has brought with it an appreciable need for the thorough analysis and understanding of our many man-made systems. This is partly because such systems are becoming larger and, supposedly, more intricate. As a result, systems analysis as a recognizable science is coming into its own.

... In spite of this, the notion of complexity has not been investigated to the extent that it can be given a quantitative measure, though we are apt to make intuitive comparisons about the complexities of different systems" [190].

In a comprehensive attempt to analyse the complexity of systems, Ferdinand [190] uses the concept of entropy which is currently developed in the field of information theory. The basis of his analysis is the principle of maximum entropy. The defect entropy in a complex system is defined by the entropy functional

$$H(\sigma, m) = \ln Z(\sigma, m) - E(\sigma, m) \ln \sigma, \qquad (3.71)$$

where

$$Z(\sigma, m) = (1 - \sigma^{m+1})/(1 - \sigma) \qquad (3.72)$$

and

$$E(\sigma, m) = \sigma \frac{d}{d\sigma} \ln Z(\sigma, m) \qquad (3.73)$$

represents the mean number of defects in a complex system. In the above relations σ denotes the coefficient of complexity for the given system and m is the maximum number of defects which may occur in the system.

LEMMA 3.18. *The defect entropy $H(\sigma, m)$ for a complex system has the following properties:* a) *it is a strictly monotonic increasing function of the coefficient $\sigma(0 \leqslant \sigma \leqslant 1)$ for any given value of m*; b) *it is a strictly monotonic increasing function of m.*

It will be further considered that "a system is described as being perfectly simple when $\sigma = 0$, in which case the defect entropy and the expected number of defects are both equal to zero. A system is perfectly complex when $\sigma \to \infty$. The point $\sigma = 1$ is described as separating simplicity from complexity..." [190].

For the case when $\sigma = 1$, the defect entropy has the maximum value $H(1, m) = \ln(m + 1)$, whereby the mean number of defects in the system is given by $E(1, m) = \frac{1}{2} m$. Let us define now the following parameters of a complex system: r represents the number of defects in the system; m is the maximum number of defects which can occur ($m \geqslant 0$); k denotes the mean number of defects expected in the system ($k \geqslant 0$); $p(r)$ is the probability that r defects occur in the system.

The following relations are immediate:

$$p(r) \geqslant 0; \qquad (3.74)$$

$$\sum_{r=0}^{m} p(r) = 1; \qquad (3.75)$$

$$\sum_{r=0}^{m} rp(r) = k; \qquad (3.76)$$

$$H(p) = - \sum_{r=0}^{m} p(r) \ln p(r). \qquad (3.77)$$

By using a Lagrange functional with β and γ as multipliers, we have

$$\mathcal{H} = - \sum_{r=0}^{m} p(r) \ln p(r) - \beta \sum_{r=0}^{m} p(r) - \gamma \sum_{r=0}^{m} rp(r) \qquad (3.78)$$

$$\frac{\partial \mathcal{H}}{\partial p(r)} = 1 + \beta + \ln p(r) + \gamma r = 0; \quad \forall r. \qquad (3.79)$$

Lagrange multiplies β and γ could be determined by conditions (3.75) and (3.76).

From the above relation, we can write relations

$$e^{-(\beta+1)} \sum_{r=0}^{m} e^{-\gamma r} = 1 \tag{3.80}$$

and

$$k = e^{-(\beta+1)} \sum_{r=0}^{m} r e^{-\gamma r} = \left(\sum_{r=0}^{m} r e^{-\gamma r} \right) \Big/ \left(\sum_{r=0}^{m} e^{-\gamma r} \right). \tag{3.81}$$

Ferdinand [190] defines the coefficient of complexity σ as given by relation

$$\sigma = e^{-\gamma}, \tag{3.82}$$

and then

$$\sigma \frac{d}{d\sigma} \ln Z(\sigma, m) = k > 0, \tag{3.83}$$

where

$$Z(\sigma, m) = \sum_{r=0}^{m} \sigma^r. \tag{3.84}$$

The summation in eqn. (3.80) yields

$$e^{\beta+1} = (1 - \sigma^{m+1})/(1 - \sigma), \quad \sigma < 1$$
$$= (\sigma^{m+1} - 1)/(\sigma - 1); \quad \sigma \geqslant 1,$$

and from eqn. (3.81), we have

$$p(r) = e^{-(1+\beta)} e^{-\gamma r} = \sigma^r (1 - \sigma)/(1 - \sigma^{m+1}); \quad \sigma < 1$$
$$= \sigma^r (\sigma - 1)/(\sigma^{m+1} - 1); \quad \sigma \geqslant 1.$$

THEOREM 3.10. [190]. *The mean number of defects k to be expected in a complex system and the defect entropy associated as a function of σ and m are given by relations*

$$E(\sigma, m) = \sigma \frac{d}{d\sigma} \ln Z(\sigma, m) = \frac{\sigma}{1-\sigma} - \frac{(m+1)\sigma^{m+1}}{1-\sigma^{m+1}}; \quad \sigma < 1$$

$$= m - \left(\frac{a}{1-a} - \frac{(m+1)a^{m+1}}{1-a^{m+1}} \right); \quad a \leqslant 1, \text{ where } a = 1/\sigma,$$

and

$$H(\sigma, m) = \ln[(1-\sigma^{m+1})/(1-\sigma)] - \left\{ \frac{\sigma}{\sigma-1} - \frac{(m+1)\sigma^{m+1}}{1-\sigma^{m+1}} \right\} \ln \sigma;$$

$\sigma < 1$ and $\ln[(1-a^{m+1})/(1-a)] - \left\{ \dfrac{a}{1-a} - \dfrac{(m+1)a^{m+1}}{1-a^{m+1}} \right\} \ln a; \; a \leqslant 1.$

Special values of $E(.,.)$ and $H(.,.)$ for $\sigma = 0, 1, \ldots, \infty$ are given in table 3.10.
When the maximum number of defects m in the system takes very large values ($m \to \infty$), then

$$H(\sigma, \infty) = - \ln(1 - \sigma) - \sigma \ln \sigma/(1 - \sigma); \quad 0 \leqslant \sigma \leqslant 1$$
$$= - \ln(1 - a) - a \ln a/(1 - a); \quad 0 \leqslant a \leqslant 1.$$

Table 3.10

Complexity coefficient σ	Mean number of defects $E(\sigma, m)$	Defect entropy $H(\sigma, m)$
0	0	0
1	$\dfrac{m}{2}$	$\ln(m + 1)$
∞	m	0

According to Ferdinand [190], it would therefore appear that the region of most practical significance is the region around the point $\sigma = 1$ which separates simplicity from complexity. Thus, systems with which one is usually familiar exist between the thresholds of simplicity and complexity.

Let us consider now that a complex system is represented as a general directed graph $\mathscr{G}_n(n; \varkappa n^\mu \lambda^n)$, where n is the number of nodes and $\varkappa n^\mu \lambda^n$ is the number of lines in the graph. Under this assumption, the coefficient of complexity is of an exponential form, $\sigma = \exp(-\alpha/\varkappa n^\mu \lambda^n)$, where $\alpha, \varkappa, \mu$ and λ are the system parameters. The functions H and E associated with the complex system represented as a general directed graph $\mathscr{G}(.)$ are given by relations

$$H(\sigma, \varkappa n^\mu \lambda^n) = n \ln \lambda + \mu \ln n + \ln \varkappa + A(\alpha) + \mathcal{O}(n^{-\mu} \lambda^{-n}) \qquad (3.85)$$

$$E(\sigma, \varkappa n^\mu \lambda^n) = B(\alpha) \varkappa n^\mu \lambda^n - C(\alpha) + \mathcal{O}(n^{-\mu} \lambda^{-n}), \qquad (3.86)$$

where $A(\alpha) = \alpha B(\alpha) + \ln \dfrac{1}{\alpha}(1 - e^{-\alpha})$, $B(\alpha) = 1/\alpha - 1/(e^\alpha - 1)$,

$$C(\alpha) = \frac{1}{2} + 1/(e^\alpha - 1) - \alpha e^\alpha (e^\alpha - 1)^2.$$

The model parameters λ and μ are given by the structure of the complex system.

DEFINITION 3.12. The critical size n_c of a complex system represents the value of n for which the term linear in n becomes dominant in eqn. (3.85) of the defect entropy $H(.)$.

Remark 3.1. For $n \geqslant n_c$, the expected number of defects in the system becomes dominantly exponential in n in its behaviour.

Systems were not born complex. As Ferdinand righteously [190] observed, "In the initial formation of an organization, the relations between the constituents are usually simple ones. As the organization grows, relationships within the system are also expanded. These new relationships tend to be rather contrived or strained, compared with the primary relationships which formerly existed".

In a comprehensive attempt to develop a formal theory of system complexity, Ferdinand observed that "in many instances, complexity may very well prove to be a function of m", and in this case it is clear that "the coefficient of complexity σ is a function of m, where m is large" [190].

The region of most practical interest for the complexity coefficient σ is around $\sigma = 1$. Let us define

$$\sigma = 1 - \varepsilon = 1 - \frac{\alpha}{m}, \tag{3.87}$$

where $\varepsilon \to 0$ such that α remains finite.

THEOREM 3.11. *For a system with* $\sigma = 1$,

$$E(\sigma, m) \cong \frac{m}{2}\left(1 - \frac{\alpha}{3}\right) \bigg/ \left(1 - \frac{\alpha}{2} + \frac{\alpha^2}{6}\right)^2, \tag{3.88}$$

$$H(\sigma, m) \cong \ln(m + 1). \tag{3.89}$$

Proof. The proof is direct by applying Taylor's theorem for $E(\cdot)$ and $H(\cdot)$ at the point $\sigma = 1$.

The next theorems will be given without proofs; for details, see ref. [190].

THEOREM 3.12. *For a simple system in which the coefficient of complexity has the functional form* $\sigma = e^{-\alpha/m}$ *($\alpha > 0$), the following expressions are true.*

$$Z(e^{-\alpha/m}, m) = \frac{m}{\alpha}(1 - e^{-\alpha}) + \frac{(1 + e^{-\alpha})}{2} + \mathcal{O}\left(\frac{\alpha}{m}\right) \tag{3.90}$$

$$E(e^{-\alpha/m}, m) = mX(\alpha) - Y(\alpha) + \mathcal{O}\left(\frac{\alpha}{m}\right) \tag{3.91}$$

$$H(e^{-\alpha/m}, m) = \ln m + U(\alpha) + \mathcal{O}\left(\frac{\alpha}{m}\right), \tag{3.92}$$

where

$$X(\alpha) = \frac{1}{\alpha} - \frac{1}{e^\alpha - 1} \tag{3.93}$$

$$Y(\alpha) = \frac{1}{2} + \frac{1}{e^\alpha - 1} - \alpha \frac{e^\alpha}{(e^\alpha - 1)^2} \tag{3.94}$$

$$U(\alpha) = \alpha X(\alpha) + \ln \frac{1}{\alpha}(1 - e^{-\alpha}). \tag{3.95}$$

THEOREM 3.13. *For a complex system in which the coefficient of complexity has the functional form* $\sigma = e^{\alpha/m}$, $(\alpha \geq 0)$, *the following expressions are true.*

$$E(e^{\alpha/m}, m) = [1 - X(\alpha)] m + Y(\alpha) + \mathcal{O}\left(\frac{\alpha}{m}\right) \tag{3.96}$$

$$H(e^{\alpha/m}, m) = \ln m + C(\alpha) + \mathcal{O}\left(\frac{\alpha}{m}\right). \tag{3.97}$$

LEMMA 3.19. *In the region of the critical point* $\sigma = 1$, *the mean number of faults is proportional to* $\ln m$; *for large values of* m,

$$E(e^{-\alpha/m}, m) \cong \begin{cases} mX(\alpha) & \text{for } \alpha > 0 \text{ — a simple system} \\ m[1 - X(|\alpha|)] & \text{for } \alpha \leq 0 \text{ — a complex system.} \end{cases} \tag{3.98}$$

For practical considerations, it is clear that m is also related to system size. A complex sytem $\mathscr{S}$ described by the graph $\mathscr{G}$ $(n; s)$ is of size n, where n is the number of nodes and s is the sum of the number of nodes and edges. It is clear that the number of maximum possible defects m for the system is equal to s.

THEOREM 3.14 [190]. *Given a complex system* $\mathscr{S}$, *defined by the general graph* $\mathscr{G}$ $(n; \varkappa n^\mu \lambda^n)$, *and by the complexity coefficient* $\sigma = \exp(-\alpha/\varkappa\, n^\mu \lambda^n)$, *the measures of complexity for* $\mathscr{S}$ *are given by*

$$H(\sigma, \varkappa n^\mu \lambda^n) = n \ln \lambda + \mu \ln n + \ln \varkappa + U(|\alpha|) + \mathcal{O}(|\alpha|\, n^{-\lambda} \mu^{-n}). \tag{3.99}$$

$$E(\sigma, \varkappa n^\mu \lambda^n) = \begin{cases} X(\alpha)\, \varkappa n^\mu \lambda^n - Y(\alpha) + \mathcal{O}(\alpha n^{-\lambda} \mu^{-n}); & \sigma < 1 \text{ — simple system} \\ [1 - X(|\alpha|)]\, \varkappa n^\mu \lambda^n + Y(|\alpha|) + \mathcal{O}(|\alpha|\, n^{-\lambda} \mu^{-n}); & \sigma \geq 1 \text{ — complex system} \end{cases} \tag{3.100}$$

where $\alpha, \varkappa, \mu$ and λ are parameters of the system and $X(\alpha), Y(\alpha)$ and $U(\alpha)$ are given by equations (3.93), (3.94) and (3.95), respectively.

The critical size n_c of a complex system is the solution to the equation

$$n \ln \lambda = \mu \ln n. \tag{3.101}$$

The *critical inequality of systems* is of the form (see ref. [190])

$$\frac{1}{\mu} \ln \lambda \leq \frac{1}{e}, \tag{3.102}$$

which "in physical terms,... implies that unless the relation given [above] is satisfied, the critical size of the system is given by $n_c = 1$, implying that the expected number of faults in the system is from the outset dominantly dependent on the size of the system".

As was previously emphasized, the modularity in complex system structure has a quantitative effect on the value of the number of expected defects in the system.

THEOREM 3.15. *Given a system defined by the directed graph* $\mathscr{G}(.,.)$ *and by the coefficient* σ *that obeys an exponential law, the expected number of faults has* the following functional form.

$$E \cong \begin{cases} X = X(\alpha) & \text{for } \sigma < 1 \\ X = 1 - X(\alpha) & \text{for } \sigma \geq 1 \\ X = 1 & \text{for } \sigma \gg 1. \end{cases} \quad (3.103)$$

In the case when we consider t independent systems whose coefficients of complexities obey an exponential law and the sizes of the systems are $\eta_j n$, $j = 1, 2, \ldots, t$, $\left(\sum_{j=1}^{t} \eta_j = 1 \right)$, and if the system is defined by $\mathcal{G}(\eta_j n, \varkappa(\eta_j n)^\mu \lambda^{(\eta_j n)})$, $(j = 1, 2, \ldots, t)$, then the expected number of faults in the j-th system is

$$E_j \cong X\varkappa(\eta_j^\mu) \, n^\mu (\lambda^{\eta_j})^n, \quad (3.104)$$

and the total number of expected faults in the r systems is

$$E^* = \sum_{j=1}^{t} E_j \cong X\varkappa n^\mu \sum_{j=1}^{t} \eta_j^\mu (\lambda^{\eta_j})^n. \quad (3.105)$$

COROLLARY 3.2 [190]. *The critical size of the system comprising t independent components is greater than or equal to that of the original system. Modularity reduces the effective λ of a system.*

In real cases, those t systems are not independent. They form a system of t dependent subsystems. If the system of size t is characterized by the parameters $\mu_1, \lambda_1, \varkappa_1, X_1$, then the number of expected faults is

$$E' \cong X_1 \varkappa_1 t^{\mu_1} \lambda_1^t. \quad (3.106)$$

The total number of faults for the system with t subsystems is

$$\hat{E} = \hat{E}' + E^*. \quad (3.107)$$

DEFINITION 3.13. The ratio of faults $\rho = \dfrac{\hat{E}}{E}$ is given by

$$\rho = X_1 \varkappa_1 t^{\mu_1} \lambda_1^t / X \lambda^n \varkappa n^\mu + \sum_{j=1}^{t} \eta_j^\mu (\lambda^{\eta_j - 1})^n. \quad (3.108)$$

For $\eta_j = 1/r$, $j = 1, 2, \ldots, t$, the ratio of faults becomes

$$\rho_1 = (X_1 \varkappa_1 / X\varkappa)(\lambda_1^t / \lambda^n)(t^{\mu_1} / n^\mu) + (1/t)^{\mu-1}(1 - \lambda)^{n-n/t}. \quad (3.109)$$

It has been observed "that the impact of modularity becomes more significant with increasing n, λ and μ" and that "small deviations of component size from optimal size result in very little change in the ratio [of faults]" [190].

In the design of engineering systems, complexity could also be viewed as a dynamic process falling under four classes, i.e. organized, unorganized, long-term and short-term. Throughout the evolution of engineering systems, "1) Complexity tends to be higher in the long-run than in the short-run. This characteristic is consistent with the argument that in the evolution of engineering systems easier problems are solved first and more complex, costly problems afterwards. 2) The role of random elements (chance factors) is of major significance in the evolution of engineering systems. 3) In the short-run, the total complexity is almost solely composed of unorganized complexity.

... in the short-run, the determinants of complexity are many and the effect of each quite small".

Sahal considers that the development of a theory of system complexity must rely on information theory and that it requires the highly restrictive assumptions of the weak-time invariance of the complex engineering system under study. The original variables of the system are transformed into functions representing the aspects of invariant phenomena in the real system under study.

As Gottinger emphasized, "great significance has been attached to the concept of complexity in the science of the artificial, i.e. in computer or system science, where it has been considered as a key scientific problem warranted by the fact that complexity pertains to almost every aspect of model building, development and prediction".

The concept of complexity is not disentangled from such attributes as largeness, size, multi-dimensionality. Hierarchy and complexity are intimately related.

In a basic work on the "architecture of complexity", Simon argues that "The fact then, that many complex systems have a nearly decomposable, hierarchic structure is a major facilitating factor enabling us to understand, to describe, and even to see such systems and their parts... If there are important systems in the world which are complex without being hierarchic, they may to a considerable extent escape our observation and our understanding. Analysis of their behaviour would involve such detailed knowledge and calculation of the interactions of the elementary parts that it would be beyond our capacities of memory or computation".

It was already mentioned in the informational approach to complexity that the system size and the maximum number of defects m are interconnected. In this analysis, Gottinger argues in this respect that "A system which develops beyond its complexity bound, e.g. that is required to perform more complex tasks than is inhibited in its design, is not going to survive, but to break down. Complexity can be thought of as a property of design, but it could also be thought of as the level of understanding of the design-users: the people, groups, institutions involved who inform, act, compute..., etc. We would call the relationship between the level of contact (or control complexity) and the design complexity simple *evolution-complexity relation*. The smaller this relationship, the more balanced (stable) the system tends to be; on the other hand, the larger this relationship, the more unbalanced (unstable) the system will become.

... the control complexity is significantly lower than the design complexity. Of course, a collapse need not occur if corrections of the rules are made in time. This is why *learning, adaptation* or even *control* receive a prominent place in achieving survival of such systems".

3.10. Diagnostic questionnaires for complex systems

System diagnostics plays an important role in the field of systems engineering where it has a large domain of applications.

A *questionnaire* is considered to be a collection of questions that finally locate a unique fixed element e in a set E by resolving E into the singleton sets $\{e_1\}$, $\{e_2\},\ldots,\{e_m\}$ (see refs. [263] and [264]).

As Duncan [263] observed, "Questionnaires appear under a variety of names which make important contributions to many fields. A medical doctor makes use of a diagnostic schedule, an electronics technician follows a trouble-shooting routine, a biologist designs a taxonomic key, the medical technologist economizes with a group testing program, the statistician presents a weighing design, the operations researcher devises a search scheme... Intuitively, a questionnaire can be thought of as a sequential pattern of questions, perhaps heterogeneous in that questions vary in resolution".

Questionnaire theory operates with the concept of informational entropy [263], [264] as well as with an alternative informational measure, called informational energy, which was developed by Onicescu et al. [265].

Let us define a question set $A = \{q\}$ and a state space $E = \{e_\alpha\}$ containing subsets Λ ($\alpha \in$ set $\{A\}$). According to Duncan [263], we shall use the following

DEFINITION 3.14. a) The question q on Λ is a collection of non-empty subsets of Λ which cover Λ; b) Let $\mathscr{L}$ be an arbitrary collection of subsets of E and let Q be a function which assigns a question $Q\Lambda$ to each $\Lambda \in \mathscr{L}$; c) A subset Λ may be reached by Q after k questions if Λ is an answer to $Q\Lambda'$ (Λ' is reached by Q after $k-1$ questions); d) The function Q is a questionnaire if $E \in \mathscr{L}$ and if it assigns a question $Q\Lambda$ to exactly those Λ reached by Q; e) A questionnaire is valid if $\{e_\alpha\}$ may be reached by Q for each $e_\alpha \in E$.

In what follows, we shall assume that when the answer to $Q\Lambda$ is $\{e_\alpha\}$, a charge $c_\alpha(Q)$ is assessed. The appropriate charging scheme is denoted by Γ. Γ is called question-based if Λ_α and Λ_β (α and β are labels for the states in E), derived by the same finite sequence of questions, imply that $c(\Lambda_\alpha) = c(\Lambda_\beta)$, where $c(\Lambda)$ is the charge assessed when Λ is reached. The average charge $E_p[C(Q)] = \int c_\alpha(Q) \, dp(e_\alpha)$.

A particular class of questionnaires is obtained by using directed trees and they are defined as arborescence questionnaires [263].

DEFINITION 3.15. a) A questionnaire Q is said to be an arborescence questionnaire if the sets in $Q\Lambda$ are disjoint for all Λ reached by Q; it is a lattice questionnaire if it is not an arborescence questionnaire; b) A questionnaire is called *homogeneous* if each question used has the same resolution; it is called *heterogeneous* if the resolutions may be different.

By using the Shannon entropy $(H(p) = -\sum_{i=1}^{m} p_i \log p_i)$, a lower bound on the average charge becomes available.

LEMMA 3.20. *Let $\{c_i\}$ be any set of constants. Then there exists a unique D satisfying*

$$\sum_{i=1}^{m} \left(\frac{c_i}{2}\right)^{-1/D} = 1 \qquad (3.110)$$

and

$$DH(p) \leq \sum_{i=1}^{m} p_i c_i, \qquad (3.111)$$

for all probability vectors p, with equality holding if and only if $p_i = (2^{c_i})^{-1/D}$ for $i = 1, 2, \ldots, m$.

The state space for E is finite such that $|E| = m < \infty$. The valid questionnaire Q accepts terminal nodes γ_i, $(i = 1, 2, \ldots, r)$. The terminal nodes of Q are identified either with $\{e_i\}$, $(i = 1, 2, \ldots, m)$, or some arbitrary symbols λ_i, $(i = m+1, \ldots, r)$. The quantity n_{id} denotes that a count can be made of the number of questions of each resolution d required to reach λ_i.

DEFINITION 3.16 [263]. If k_i is the smallest integer such that

$$Q^{-k_i}\lambda^i = E, \qquad (3.112)$$

then

$$n_{id} = \sum_{j=1}^{k_i} \sigma_{\{d\}}(|Q\,Q^{-j}\lambda_i|), \qquad (3.113)$$

where

$$\sigma_{\{d\}}(x) = \begin{cases} 1, & d = x \\ 0, & \text{otherwise,} \end{cases} \qquad (3.114)$$

and the average charge for Q is given by relation

$$E_p[C(Q)] = \sum_{i=1}^{m} \sum_{d=1}^{\infty} p_i n_{id} \log d. \qquad (3.115)$$

LEMMA 3.21. *If each terminal node of a tree can be reached after exactly w_d questions of resolution d, $(d = 2, 3, \ldots)$, then the tree has Πd^{w_d} terminal nodes.*

THEOREM 3.16. *If a questionnaire Q is valid and uses precisely n_{id} resolution-d-questions to determine e_i, $(i = 1, 2, \ldots, m)$, then*

$$\sum_{i=1}^{m} \prod_{d=1}^{\infty} d^{-n_{id}} \leq 1. \qquad (3.116)$$

COROLLARY 3.3. *If Q is a questionnaire, then it is adapted to E if and only if*

$$\sum_{i=1}^{m} \sum_{d=1}^{\infty} d^{-n_{id}} = 1. \qquad (3.117)$$

Certain relationships between the Shannon entropy and the average charge could be established (see Duncan, ref. [263]).

THEOREM 3.17. *For the case when Q is a valid heterogeneous questionnaire and $C(Q)$ is the random charge based on resolution d question, if $E = \{e_1, e_2, \ldots, e_m\}$ is a finite state space and $p = [p_1, p_2, \ldots, p_m]$ is a probability vector, then*

$$H(p) \leqslant E_p[C(Q)] \tag{3.118}$$

only for $n_{id} = 0$, $d > m$ and

$$p_i = \prod_{d=2}^{m} d^{-n_{id}}, \quad (i = 1, 2, \ldots, m). \tag{3.119}$$

Assuming that the charge for a resolution-d-question is $c(d)$, we have

$$\sum_{i=1}^{m} \sum_{d=2}^{m} p_i n_{id} c(d) \geqslant H(p), \tag{3.120}$$

with equality if the relation $p_i = \prod_{d=2}^{m} d^{-n_{id}}$ is equivalent to the relation $c(d) = \log d$.

Considering that a resolution-d-question is Shannon efficient if it partitions E into d sets of equal probability, we obtain the following (see Duncan, ref. [263])

THEOREM 3.18. *The Shannon lower bound is attained by a valid questionnaire Q if and only if each question is Shannon efficient.*

THEOREM 3.19. *Suppose that the state space E is countably infinite. The average charge for a valid questionnaire is never less than the countable Shannon entropy. In case that the entropy has a finite value, then there exists a valid questionnaire with average charge*

$$H(p) \leqslant \inf_Q E_p[C(Q)] \leqslant H(p) + 1, \tag{3.121}$$

where $H(p) = -\sum_{i=1}^{\infty} p_i \log p_i$.

Proof. If

$$H^A(p) = -\sum_{i=1}^{A} p_i \log p_i - \delta_A \log \delta_A, \tag{3.122}$$

where $\delta_A = p_{A+1} + p_{A+2} + \ldots$, in the sense of Duncan ([263], p. 69), then

$$H^A(p) \leqslant \inf_{Q_A} E_p[C(Q_A)] < H^A(p) + 1, \tag{3.123}$$

where Q_A denotes a questionnaire which determines the true state among the e_j's. It is clear that

$$\lim_{A \to \infty} \sum_{i=1}^{A} p_i \log p_i \to H(p) \tag{3.124}$$

and
$$\lim_{A \to \infty} \sum \delta_A \log \delta_A \to 0, \tag{3.125}$$

such that $\lim_{A \to \infty} H^A(p) \to H(p)$. \hfill (3.126)

In case that the value of $H(p)$ is finite for $A \to \infty$, we have

$$H(p) \leqslant \gamma \leqslant H(p) + 1, \tag{3.127}$$

where
$$\gamma = \liminf_{A \to \infty} E_p[C(Q_A)]. \tag{3.128}$$

Let us define the value

$$\delta_A^{(k)} = \left(\frac{1}{2}\right)^{k-1} \delta_A, \quad (k = 1, 2, \ldots), \tag{3.129}$$

and choose positive integers $A^{(k)}$, $k \geqslant 1$ such that a) $A = A^{(1)}$; b) $A^{(k)} < A^{(k+1)}$; c) $p_{A^{(k)}+1} + \ldots + p_{A^{(k+1)}} \leqslant \delta_A^{(k)}$; and d) $p_{A^{(k)}+1} + \ldots + p_{A^{(k+1)}+1} > \delta_A^{(k)}$.

Let $Q_A^*, Q_{A^{(1)}, A^{(2)}}^*, Q_{A^{(2)}, A^{(3)}}^*, \ldots$ be the best questionnaires for determining the true state among $\{e_1, e_2, \ldots, e_{A^{(1)}}\}$, $\{e_{A^{(1)}+1}, \ldots, e_{A^{(2)}}\}$ and $\{e_{A^{(2)}+1}, \ldots, e_{A^{(3)}}\}$, ... respectively. The best questionnaires Q_A^* can be extended to determine the elements of the set E. If $Q_{(A)}^*$ is an extended questionnaire, then we can prove [263] that

$$\sum_{k=1}^{\infty} q_k(A)\{k + E_p[C(Q_{A^{(k)}, A^{(k+1)}}^*)]\} \leqslant \sum_{k=1}^{\infty} k\delta_A^{(k)} +$$

$$+ \sum_{k=1}^{\infty} q_k(A) \left(-\sum_{i=M^{(k)}+1}^{M^{(k+1)}} \frac{p_i}{q_k(A)} \cdot \log \frac{p_i}{q_k(A)}\right) + \sum_{k=1}^{\infty} q_k(A), \tag{3.130}$$

where $q_k(A) = (p_{A^{(k)}+1} + \ldots + p_{A^{(k+1)}})$, $k \geqslant 1$. \hfill (3.131)

Following relation (3.127), the right-hand side of the above relation becomes

$$5\delta_A + \left(-\sum_{i=A}^{\infty} p_i \log p_i\right) + \sum_{k=1}^{\infty} \sum_{i=A^{(k)}+1}^{A^{(k+1)}} p_i \log q_k(A). \tag{3.132}$$

It can be demonstrated that

$$\lim_{A \to \infty} \sum_{k=1}^{\infty} k\delta_A^{(k)} = 0 \tag{3.133}$$

$$\lim_{A \to \infty} \left[\sum_{k=1}^{\infty} q_k(A) \left(-\sum_{i=A^{(k)}+1}^{A^{(k+1)}} \frac{p_i}{q_k(A)} \log \frac{p_i}{q_k(A)}\right)\right] \to 0$$

and that

$$\sum_{k=1}^{\infty} \sum_{i=A^{(k)}+1}^{A^{(k+1)}} \frac{p_i}{\delta_A} \log \frac{q_k(A)}{\delta_A} \leqslant \sum_{k=1}^{\infty} \sum_{i=A^{(k)}+1}^{A^{(k+1)}} \frac{p_i}{\delta_A} \log \frac{p_i}{\delta_A}. \quad (3.134)$$

Clearly,

$$\lim_{A \to \infty} \sum_{k=1}^{\infty} \sum_{i=A^{(k)}+1}^{A^{(k+1)}} p_i \log q_k(A) = 0. \quad (3.135)$$

Thus, we have shown that there exists a positive integer A so that the questionnaire $Q_{(A)}$ has an average charge close to γ and so that

$$\lim_{A \to \infty} \{\inf_{Q_A} E_p[C(Q_A)]\} = \gamma. \quad (3.136)$$

For the case when $H(p) \to \infty$,

$$H^{A(p)} \leqslant \inf_{Q_A} E_p[C(Q_A)] < \inf E_p[C(Q)], \quad (3.137)$$

where Q is a questionnaire to determine that $e \in E$. But

$$H^A(p) H(p) = +\infty \text{ and then } \inf E_p[C(Q)] = +\infty.$$

3.11. Resilience of complex systems

The concept of resilience has become very popular for the last five years, with special application to systems analysis and systems engineering. It was first introduced by Holling for ecological systems. Later on, it was used in the study of the "New societal equations", where an alternative expression for resilience was employed and computed explicitly.

In the original definition, Holling introduced resilience as "a measure of the ability of a system to absorb changes in state variable plus parameters and still persist".

In the recent approach to the concept of resilience, Grümm distinguished between (a) resilience *in* state space, which corresponds to changes in the state variables and (b) resilience *of* state space, corresponding to changes in the parameters.

However, the phase portrait (see fig. 3.18) is paramount in both definitions; the state space of a given system is subdivided into a finite number of basins, each with a particular attractor. Generally, an attractor is defined as an equilibrium point, a stable orbit, or a more complicated structure. Concerning the system resilience in state space (see fig. 3.19), a perturbation of the state variables can move the system to a different basin where it will have a different long-term behaviour. Unlike resilience in state space, the resilience of state space (see fig. 3.24) indicates that a change in parameter or state variables brings about a structural modification of basins and attractors. Therefore one point may now lie in a different basin and the long-term behaviour of the system changes accordingly.

For a proper definition of resilience, in what follows we shall use several results due to Grümm. So far, the concept of resilience was used mainly in connection with the class of deterministic systems, where the theory of differential dynamical systems was widely exploited for description purposes.

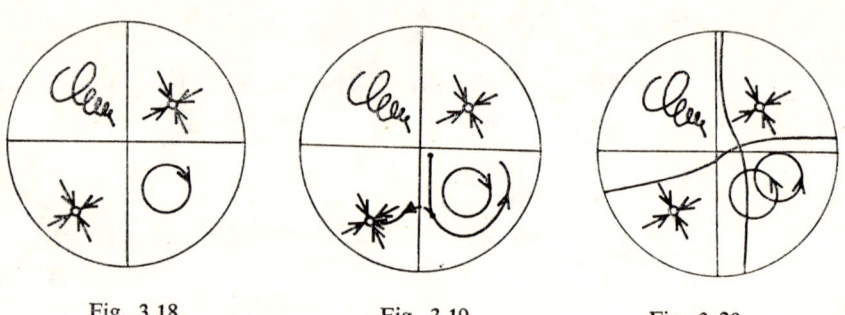

Fig. 3.18   Fig. 3.19   Fig. 3.20

Any given engineering system can be associated with an abstract mathematical model of the form $X(t) = \Phi_t(x)$; $x \in M$; $t \geqslant 1$ or $t \in \mathscr{R}^+$, where M represents the state space of the system and Φ_t indicates the total dynamic evolution of the system over time.

In a general sense, the concept of resilience may be defined as the "system" which "can absorb changes". These changes are supposed to be sudden and external to the system and are not to be included in structure Φ_t. It is evident that for practical reasons engineering systems cannot accept any kind of changes so as to persist. A pertinent question is then "How large can those changes be?"

A cascade approach concerning the above measure for an engineering system should consist of the following steps. "Is a system resilient?" If YES, then: "How resilient is it?" In this case we need a resilience measure. Let us assume that in the state space M, the system has a metric d which represents the distance to the next point of catastrophic behaviour. In discussing the two aspects of resilience, Grümm observed that "Changing the state of the system at one particular time changes one particular orbit, while changing the functional form of the flow or map through a change in parameters, for instance, involves the whole phase portrait".

Concerning the behaviour of a complex system in a dynamic environment, we can identify a finite number of attractors A_i located in basins B_i and separated by separatrices S_j.

A system may be characterized by a non-resilient behaviour in two ways:

R1: The sudden jump of the state variables moves the point describing the system across a separatrix into another basin;

R2: The phase portrait changes such that the system now lies in a basin whose attractor is in a different region of the state space.

Three formal definitions for R1 and R2 will be given below.

DEFINITION 3.17. A dynamic engineering system $\{\Phi_t\}$ in M, which has a finite number of attractors A_i and associated basins B_i, is called resilient in the sense of R1 if $M - \bigcup_i B_i$ has measure zero.

If we define $r(x) = d(x, \mathcal{S})$ for each point in the state space $\mathcal{S}$, then we distinguish different forms for resilience measures such that a) the mean resilience of the basin B_1 is $R_M = \int_{B_1} r(x) \, d\mu(x)$, where μ is a probability measure; b) the trajectory average resilience $R_{av} = \dfrac{1}{\int_0^t \dfrac{dt}{|\dot{x}(t)| \, r(\Phi_t \, x)}}$; c) the volume resilience $R_v = v(B_1)$, where v is the volume of state space.

DEFINITION 3.18. A system has a certain property of structural stability if its equivalence class under the given equivalence notion is open in some C^r-topology *.

We shall use now the Hausdorff distance in defining R2.

For two compact sets A and B, the Hausdorff distance is defined by $d^*(A, B) = \max\{\sup_{x \in A} \inf_{y \in B} d(x, y), \sup_{y \in B} \inf_{x \in A} d(x, y)\}$.

DEFINITION 3.19. Given a system Φ on M and a manifold P such that $\Phi \in P \leqslant \text{Diff}(M)$, the system $\{\Phi\}$ is called resilient in the sense of R2, if there exists a neighborhood U of Φ in the C^1-topology on $\text{Diff}(M)$ such that all systems $\Phi' \in U \subset P$ have the same finite number of attractors.

The appropriate resilience measure is a) the volume sensitivity resilience $\bar{R}_v = \lim_{h \to 0} \dfrac{1}{h} \sup_{\Phi' \in B_h} |v(B_1) - v(B_1')|$, where B_1 is the desired basin, and b) the speed resilience

$$\bar{R}_s = \dfrac{1}{\lim_{h \to 0} \dfrac{1}{h} \sup_{\Phi' \in B_h} \sup_i \{d(A_i, A_i'), d(\bar{B}_i, \bar{B}_i')\}},$$

where B_h denotes a ball of radius h around the system Φ, d is the Hausdorff metric of closed sets in state space and A_i and B_i are defined as attractors and basins respectively.

The concept of resilience is used in system safety engineering where two major concepts in the design of engineering systems — fail safe *vs.* safe fail — appear. "Our traditions generally lead us to attempt to minimize the probability of... failures or unexpecteds. There are many examples of these fail-safety designs in nuclear power plants, the setting of and adherence to fixed environmental or health standards and the design of dams for flood control... The goal, then, is to design systems

---

* A system Φ is C^r-structurally stable, if, for all Φ', C^r-close enough to Φ, there exists a homeomorphism of M transforming orbits of Φ into orbits of Φ'. In this case, Φ' looks exactly like Φ up to a topological deformation.

with broad operational limits and second to confine the operation of the system to a limited region well away from these limits of catastrophe.

... then, there might be a place in environmental institutional or societal management for disaster design — periodic 'mini-disasters' — that prevent the evolution of inflexibility. That, combined with traditional fail-safe design for those parts that are more surely known, monitored and controlled lead away from the hypotheticality trap to systems with rich options for experimentation, mistakes and hence learning".

3.12. Systems engineering and systems resilience

A rough view of the mathematical methods presented is sufficient to realize that the design philosophy behind these aspects is intimately related to the concept of "fail safe" and is appropriate to levels I and II in fig. 3.21.

However, in view of future systems planning (e.g. energy systems), we should not be indifferent to the way in which the technological design and its associated philosophy will evolve. This is not to say that the systems analyst should improve only the reliability models for a particular subsystem of the (energy) system. It

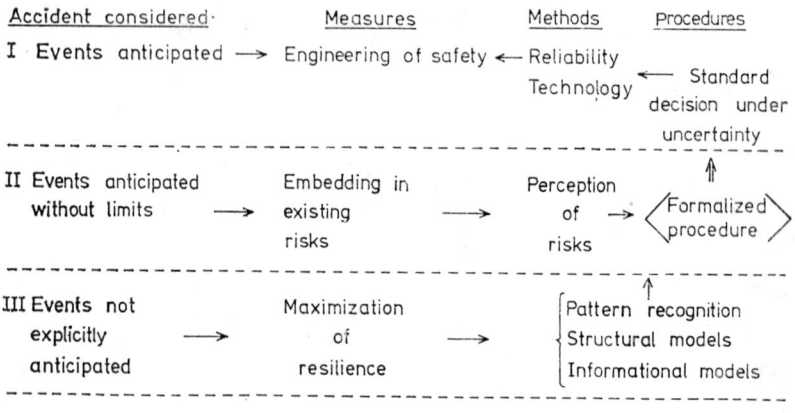

Fig. 3.21

appears equally important to look for systems capable of absorbing events that have not been explicitly anticipated (level III in fig. 3.21). Such systems are called resilient and in our opinion it is important to identify methods and procedures to protect the (energy) system or part of it against eventual catastrophe. Following Holling's advice, we should look for a "safe fail" approach in order to maximize system resilience.

3.12.1. What is a resilient design?

Resilience is known to be the capacity of the system to absorb changes that are assumed to be sudden and external. As was stated earlier, engineering systems are designed to perform one or more functions. Reliability, availability and safety of complex technical systems are performance indices which describe the operational status of the system in time.

For technical, economical and design reasons we cannot afford highly resilient subsystems without looking at the same time into the resilience of the overall system. Thus, we may consider the environment to be a "change generator" and, hence, must find an appropriate design philosophy to protect the system against the large variations in the state space behaviour (resilience in the state space) of the system or changes in the system parameters (resilience of state space).

As in reliability technology, we have to look specifically on the problem of having a system with highly resilient components (subsystems), and see how resilient the overall system is. Resilience of multifunctional systems depends on the structural design of the system.

3.12.2. Safety engineering — risk embedding — resilience

Measures to protect a system against the unknown are either direct physical methods (given by the techniques of safety engineering) or the embedding of existing risks (measured by risk perception methods) and their "translation" into standards and finally the improved ability of the system to protect itself against states of discontinuity, generally known as catastrophic states. This loop is given in fig. 3.21 by events 1—10.

A measure against the unknown for the class of events in stratum III (see fig. 3.21) may be to maximize the system resilience. The hardware of the system must be able to overcome difficult situations. We refer here to resilience as a design methodology for systems engineering. With regard to resilience, the following should be borne in mind.

a) The methodology will be able to assist the systems designer in the initial phase of system definition with designing prescriptions for the hardware and software associated with the system operation (e.g. computer programs to protect the system against disasters, etc.);

b) If some general qualitative models could describe the most probable behaviour of the system, then we can learn appropriate design measures to protect the system. No one can say that we are going to have in the future only brand new (energy) systems. We shall build such futures on the present ones. Much attention must be given to the coupling aspects of (energy) systems from different technological ages. Design strategies must be chosen so that flexible and reliable systems may be obtained;

c) Catastrophes are not necessarily bad; they can be the cause of the system maintenance.

Resilience-aided approach enforced the use of models of thought (e.g. structural models). This approach is qualitative as soon as we are able to describe properly the system behaviour by means of dynamical models and the relationships

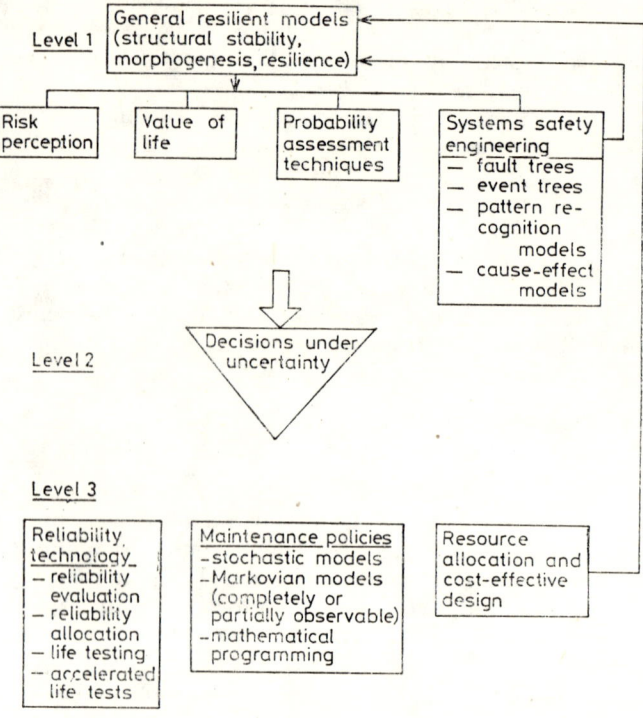

Fig. 3.22

between the design components of the system. A formal approach for incorporating the concept of resilience in the overall design problem of a complex (energy) system is given in fig. 3.22.

Chapter 4

A probabilistic safety modelling methodology for large-scale technical systems

One of the most complex questions for large technical systems is connected with safety analysis. A large variety of applications exists for military equipment, nuclear reactors, satellite systems, etc. For large, complex and costly systems, safety analysis has recently become a very important matter, and much attention has been given to nuclear reactor systems and their socio-economical implications in the case of unsafe functioning. Previous analytical methodologies on safety analysis managed to handle safety questions in connection with the hardware operation and to produce figures specific to components, modules, subsystems, etc. The results were either qualitative (based on empirical data) or relevant to theoretical models using techniques such as "Fault Tree Analysis" or "Synthetic Tree Model" (see Vesely [16] and Fussell [46]). A comprehensive investigation of safety analysis applied to nuclear reactors is given in refs. [104, 105, 106]. From the system definition (see Chapter 2), we can see that, generally speaking, technical systems are open systems. Simultaneous incidents (accidents) could happen such that the overall system could become unsafe in operation. Fault Tree techniques could not answer safety questions under earthquakes, intentional damage, human errors, or a combination of previous events with a spontaneous failure of the equipment hardware. It is precisely the purpose of this chapter to provide a methodology to answer questions in case of uncertainty about the operational states of the system. This methodology which will be given in the form of a Synthetic Probabilistic Tree Model (SPTM), takes good account of the concepts from systems analysis [109]. The results and conclusions derived after an extensive application of safety analysis to a nuclear power station (BWR) are given in Chapter 9.

4.1. Basic definitions

In order to achieve the safety goals of a complex technical system, we shall adopt the following definitions [99, pp. 106].

DEFINITION 4.1. *Safety:* Freedom from those conditions that can cause injury or death to personnel, damage to or loss of equipment or property.

DEFINITION 4.2. *System safety:* The optimum degree of safety within the constraints of operational effectiveness, time and cost, attained by a specific observance of the principles of safety management and engineering throughout the phases of a system life-cycle.

DEFINITION 4.3. *System safety engineering:* An element of systems engineering involving the use of scientific and engineering principles for the timely identification of hazards within the system. This draws upon professional knowledge and specialized skills in mathematics, physics, and the related scientific disciplines, together with the principles and methods of engineering design and analysis to specify, predict and evaluate the safety of the system.

An important concept for SPTM is that of *state variable*, which describes the operational status of the system. For a complex safety analysis we must recall the definition of a system (see Chapter 1) and several concepts from polyfunctional coherent structures. Clearly, we must correlate the internal performance index of the system and its connection with the environmental factors. Also, we must have a specific view of the system topology and its functional and operational characteristics. Standby components (eventually parallel redundancy) are to be considered for the realistic safety figures of the system, whenever improper operation of the basic components is apparent. The most difficult step in SPTM is a good understanding of the operation of the system for further differentiation between primary (initiating) events and the path description through the system. System diagnosis with a Fault Tree Analysis (FTA) is used for calculating certain probabilities in SPTM. The step in the methodology could be represented as shown in fig. 4.1. A hierarchy tree associated with the above figure for the total safety analysis (plotted in fig. 4.2) is used in nuclear reactor safety analysis. In building SPTM, we must bear in mind the boundary conditions for extending a particular section in the tree. Two are the types of boundary to be considered: functional and probabilistic. The functional boundary is concerned with the mutually exclusive and collectively exhaustive properties of the tree events, whereas the second boundary is concerned with trimming branches in the tree when the probability value of a sequence of events is too small (i.e. 10^{-8}) in comparison with a prior reference probability for the most undesirable event.

The *safety transfer function* for components, subsystems or even for a complex system itself can be thought of as a path description for collateral damage. It may also represent the manner in which the functional design of the system responds to the input events (see fig. 4.3). The calculation of the safety transfer function could be supported by the FTA. A *safety discriminator* is defined as an output event due to improper operation of system components. In the case of a nuclear reactor, we can identify as discriminators core melting, design basis accident, etc. It is important to note that the discriminator could be used as a marginal boundary.

In what follows, we shall explain the criteria used for trimming in PTM.

A sequence B can be trimmed whereas sequence A has larger probability and is a "worse" occurrence, its consequence distribution being stochastically dominant over that for sequence B (see fig. 4.4).

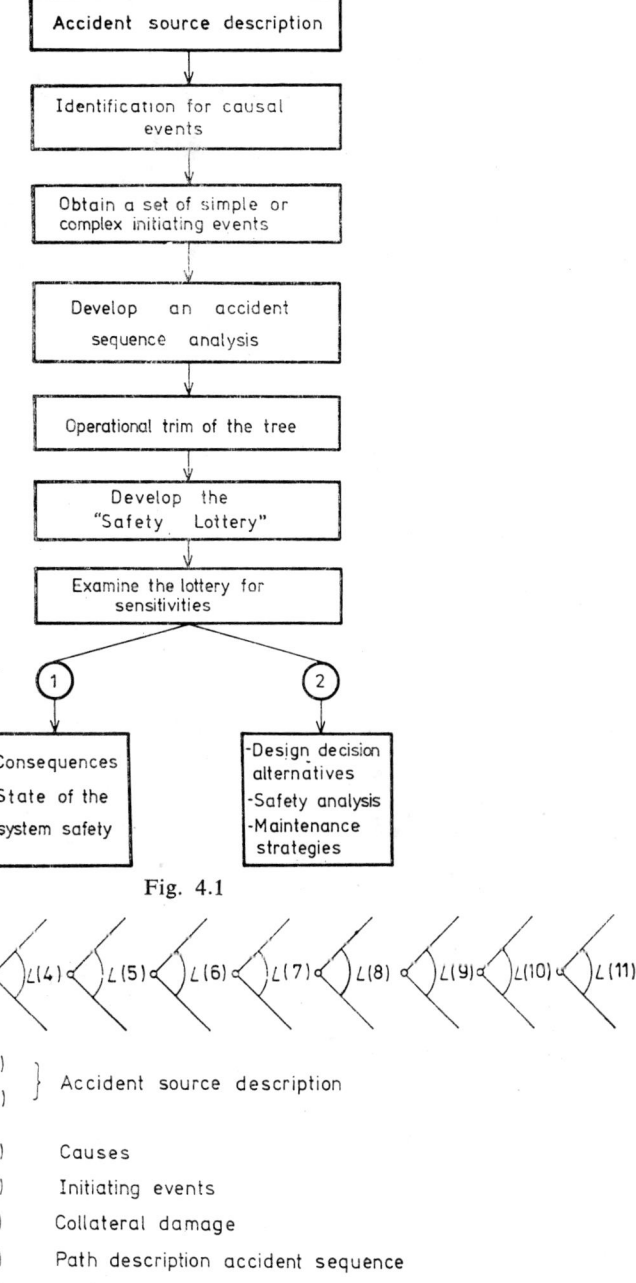

Fig. 4.1

Fig. 4.2

$L(1)$, $L(2)$ } Accident source description

$L(3)$ Causes
$L(4)$ Initiating events
$L(5)$ Collateral damage
$L(6)$ Path description accident sequence
$L(7)$ Accident summary
$L(8)$ Releases to the environment
$L(9)$ Dispersion
$L(10)$ Population exposure } Consequences
$L(11)$ Social cost

Safety Consequences and Utility Model. After the safety problem has been structured and the accident summary for the complex technical system has been decided, we have to construct a value model of consequences for the unsafe system.

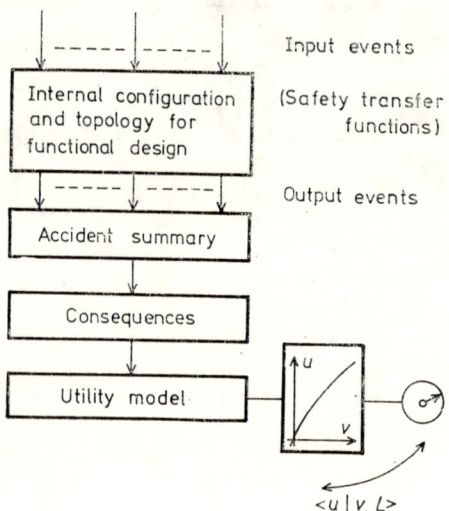

Fig. 4.3

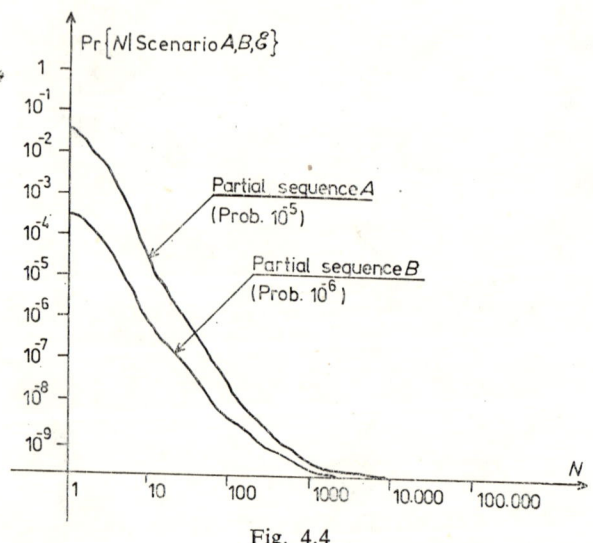

Fig. 4.4

If we can assign a utility measure to the value of consequences, then this could be used in a risk-sensitivity analysis.

Probability Encoding for Safety Analysis. Probability encoding is one of the most important processes in the phase of quantitative safety analysis. We can use the safety figures produced by a FTA, subject to different degrees (i.e. component, subsystem or system level) of investigation.

Sometimes it is quite impossible to produce by a reasonable model probability data which could be eventually used in SPTM. In this case the systems analyst can use the method of probability encoding (by consulting experts opinion) in system operational status. Assessing probabilities for rare events requires a special treatment. For a complete discussion of this matter, see Selvidge [100]. The method of probability encoding provides a clear means for communication about uncertainty in the state of components, subsystems, the overall system and environmental factors *vis-à-vis* the safety goal(s). A complete discussion about the general problem of probability encoding can be found in ref. [101]. Modelling and encoding processes can be viewed together in the safety analysis systems.

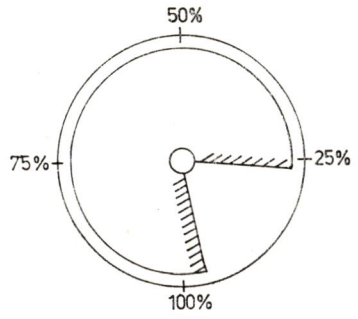

Fig. 4.5

For other kinds of analysis, that are different from such rare events as those occurring in a nuclear reactor, the *"probabilistic wheel"* may be used for encoding subjective probabilities. The idea of using the probability wheel was introduced by Savage [108]. He proposed that the decision-maker should estimate probabilities by considering what odds he would give the events X and $S - X$ if he were an odds-maker. Raiffa [109] suggested the use of an intermediate reference device so that instead of specifying numbers directly, the decision-maker should choose between a bet on the event X and a bet on some reference events. (A probability wheel is shown in fig. 4.5.) The decision-maker must choose between two bets, i.e. a bet b_x on the event X, and another on the event that a randomly-spun pointer on the probability wheel will stop in the unshaded portion. If the shaded portion can be adjusted to a unique fraction of the wheel, denoted by f, so that the decision-maker is indifferent between the two bets, then $P(X)$ is defined to be f. There is little doubt at present that the use of such reference devices can be instrumental to safety analysis. However, they should produce more representative estimates since the decision-maker is allowed to consider the probability of the event autonomously, i.e. without the possible confusion and emotion that may be associated with the complete original encoding and decision problem. (For other details regarding the use of such devices, see ref. [107].)

Bayes' theorem for safety figures. A complete safety analysis could be made using Bayes' theorem (see fig. 4.6). By rolling back the probabilistic tree, we can obtain probabilities of events occurring, given that a particular discriminator has happened. Hence, the probability value of the safety transfer function can be calculated. Let us illustrate this statement by a very simple example from nuclear safety analysis. Pr {Occurrence of Initiating Events/Core Melting, $\mathscr{E}$} =
$$= \frac{\Pr\{E(i)/\text{Core Melt}, \mathscr{E}\}}{\Pr\{\text{Core Melt}/\mathscr{E}\}}.$$

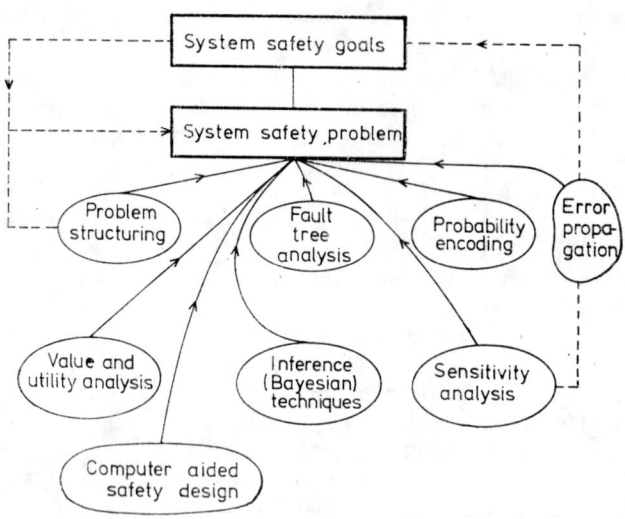

Fig. 4.6

4.2. A mathematical approach to "How safe is safe enough?"

Large and complex technical systems (e.g. nuclear reactors), for which safety is very important, are equipped with special safety devices. There is always a trade-off between the increasing size, degree of sophistication and finally the design philosophy for the safety subsystems, and the risk taken that the whole system will get into an unsafe operational state. These observations raise the well-known question of "How safe is safe enough?" Out of several theoretical and practical considerations, the size of the emergency safety devices is restricted. In this section, we shall give a mathematical formulation for the analysis of marginal safety of systems with large socio-economic side-effects, assuming that the system experiences a most undesirable state of operation (i.e. core melting for a nuclear reactor when the safety system (ECCS) failed to operate properly).

Let us consider that a decision-maker concerned with the design of safety subsystems for a technical system wishes to participate in a normally fatal game, where risk in operation is an inherent characteristic. A constant value A derives from participating in the game. Assume that in the most undesirable operational state of the system, the life normally lost is worth a constant W. However, after participating in this game for many times (i.e. after designing and testing the safety system a few times and keeping a record of its operational data), the decision-maker realizes that he can improve the game by reducing the probability of death p, consequent to the most undesired accident. In practice, this is done by paying a fixed sum $C(p)$ to the game operator (the safety system). The variation of $C(p)$ in pounds is illustrated in fig. 4.7. But in doing this, the decision-maker is confronted with the following questions: "How much should he pay for the safety equipment to reduce the probability of death from an unsafe operational system?" and "How

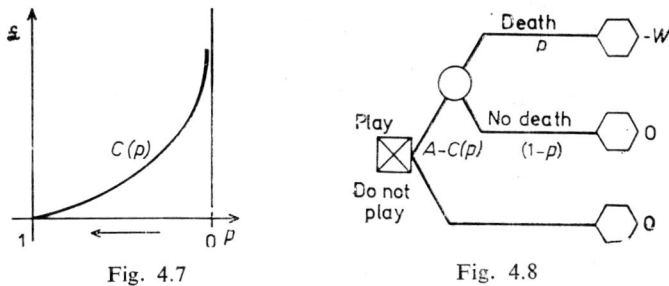

Fig. 4.7 Fig. 4.8

safe is safe enough?" This problem is represented in the form of a decision tree in fig. 4.8. Naturally, the decision-maker will try to maximize his expected profit and so he will optimise the following reward functional: $EP = \max[0, (A - C(p) - pW)]$. In order to find a maximum value for the above functional, let us write

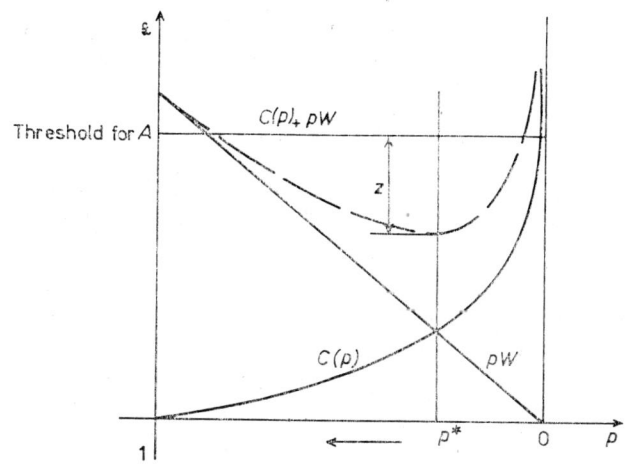

Fig. 4.9

$$\frac{d(EP)}{dp} = \frac{d(A - C(p) - pW)}{dp} = 0,$$ or, after an elementary calculus, $-C'(p) - W = 0$,

$\frac{dC(p^*)}{dp} = -W$, where p^* is an "optimal" probability of death. A pictorial representation is given in fig. 4.9. As is obvious from the above figure, there exists an economic threshold for A such that $Z = A - (C(p) + pW) > 0$.

Z is a function of p subsequent to the design alternative chosen for the safety system. In order to find out the variation for p when realistic technological and economical assumptions are made on the safety design of the system a sensitivity analysis of the input data may be performed. The above results may be also used in a step forward, when the decision-maker is able to buy sequential information or "observables" (see Miller [80]). However, the possibility of buying information sequentially (i.e. by performing different tests on the equipment behaviour or by improving the design) provides the decision-maker with a set of secondary decisions: "which observables should he buy, in which order should he buy them?". In the above analysis we can extend the consideration of utility and risk preference methodologies.

4.3. How safe is "too" safe?

As was emphasized, it is impossible to reach the absolute safety for complex engineering systems. Consequently it is only wise for the systems engineer to decide which risks are acceptable and to what extent risks have to be reduced. The analytical approach to the safety problem assumes either comparison of risks and benefits, or cost-effectiveness procedures for risk reduction. For the first case, the methodology relies on the implicit fact that for a higher expected benefit, a higher level of risk/hazards should be accepted by the decision-maker (e.g. the systems engineer). By performing such an analysis, no clear indication exists of whether these levels of safety are adequate or not or whether the risk must be reduced to levels as low as reasonably achievable or not. To bridge this gap, a cost-effectiveness methodology may be used.

In compliance with this procedure, it is already accepted that (a) the marginal expenses of risk increase with the achieved level for the safety indicator and (b) for any achieved safety level of a given system it is possible to reduce the risk even further, but it is impossible to reach absolute safety.

Figure 4.10 renders a representation for the accepted fact that a practical limit to risk reduction does exist. The conclusion from this diagram is applied to the system element. The detrimental effects associated with any given technology may be employed in a methodology for measuring.

The problems of risk and safety engineering are wider as far as practice is concerned. It is implicit in the production of goods, equipments and services in general.

In what follows, we shall present an input-output model for calculating the total risk involved in the production of goods or services. This model was adapted from the general procedures used in energy analysis. The structure of the input-output coefficient matrix is given by relation (4.1), where an element A_{ij} gives the percentage of the total output of sector j which was achieved as preprocessed goods from sector i.

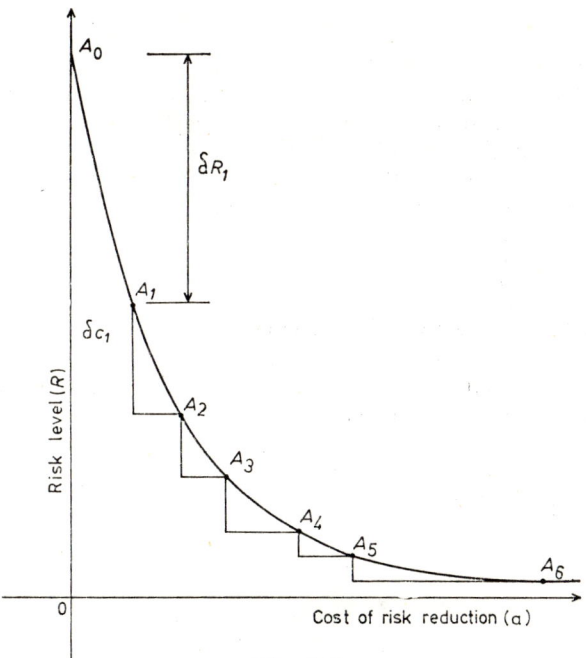

Fig. 4.10

sector	1	2	...	j	...	n	y	Σ
1	A_{11}	A_{12}		A_{1j}		A_{1n}	y_1	X_1
2	A_{21}	A_{22}					y_2	X_2
$\vdots$			to	$\uparrow$				
i $\rightarrow$	from			A_{ij}		A_{in}	y_i	X_i
$\vdots$								
n	A_{n1}	A_{n2}		A_{nj}	...	A_{nn}	y_n	X_n

(4.1)

The total production of any given sector i ($i = 1, 2, \ldots, n$), is denoted by X_i and

$$X_i = \sum_{j=1}^{n} A_{ij} X_j + y_i, \quad (i = 1, 2, \ldots, n), \tag{4.2}$$

where y_i represents the total final consumption of goods from sector i.

In a matrix form, the above relation becomes

$$X = AX + y \tag{4.3}$$

or

$$X = [I - A]^{-1} y, \tag{4.4}$$

where I is the unit matrix and $[I - A]^{-1}$ is defined as the inverse Leontief matrix (if this matrix is developed into a Taylor series, then $[I - A]^{-1} = (I + A + A^2 + \ldots)$).

By introducing the vector for final production V_0 and n processing steps, $V_0 = Iy; V_1 = Ay; \ldots, V_n = A^n y$.

In this case, the total production becomes

$$X = \sum_{n=0}^{\infty} V_n = [I + A + A^2 + \ldots] y = [I - A]^{-1} y. \tag{4.5}$$

A "specific risk matrix" S is defined as $(S = [s_{ij}], s_{ij} = R_{ij}/X_j)^*$, when element R_{ij} denotes the total health effects of type i per year from sector j. The health effect i, which occurs in sector k to enable production of one value unit of final products from sector 1, is given by

$$\theta_{ik1} = S_{ik} [I - A]_{k1}^{-1}. \tag{4.6}$$

If $H_{i1} = \sum_k \theta_{ik1}$, then

$$H_{i1} = \sum_k S_{ik} [I - A]_{k1}^{-1}, \tag{4.7}$$

or, in a matrix formulation, by

$$H = S[I - A]^{-1}. \tag{4.8}$$

The above relation gives the total risk i associated with the processing steps. The total occupational risk of producing DM 1 million safety equipment is

Table 4.1

Total working hours	17,700
Lost working hours	225
Accidental occupational deaths Driving fatalities Chronic occupational deaths	3.93×10^{-3} 2.06×10^{-3} 0.153×10^{-3}
Total deaths	6.14×10^{-3}
$\int$ equivalent death (1 death = 6,000 men-days) $\int$ equivalent lost working days	10.8×10^{-3} 65

* The matrix S consists of a number of lines equal to the number of health effects considered, and n columns defined by sectors.

given in table 4.1. The general cost-effectiveness relationship in risk reduction is outlined in fig. 4.11, where it includes the health effects in the total economic system, as was suggested by Black et al. "Any achievement in technological safety through additional equipment has to be paid for not only by additional costs but also by the occupational and public health effects caused by the production of this safety

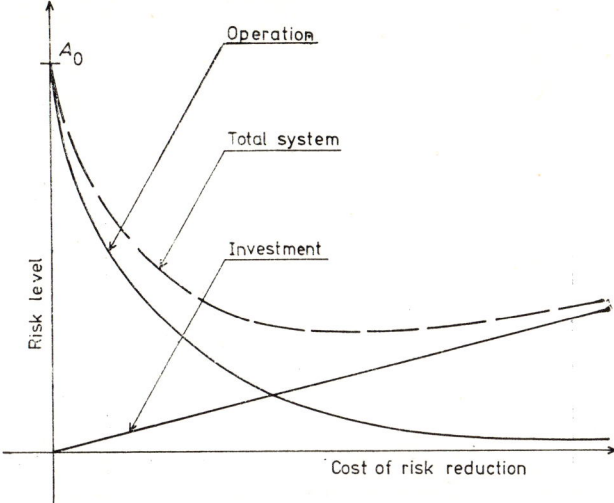

Fig. 4.11

equipment". We shall consider now that this risk is proportional to safety investments. As is obvious from fig. 4.11, "beyond a certain limit the risk increases again with increasing expenditure on safety equipment. The minimum of the risk-cost relationship is given when the marginal cost of risk reduction... is equal to the slope of the linear relationship for investments".

In case that the systems engineer accepts an initial design without appropriate safety measures, a risk A_0 is given by the intrinsic factors of the system and the attached technological process. Additional expenditure on safety devices reduces the expected number of health effects. If the decision-maker is too much risk-averse and the safety investments are too big, then the health impact will be even larger than the reduction expected in future effects. Marginal costs of risk reduction for a few examples are given in table 4.2.

The total risk-cost relationship of the system can be described by the equation

$$R = f(c) = R_0 \exp(-c/c_0) + r_p c, \qquad (4.9)$$

where R represents the risk level for a given system, c denotes the cost of safety equipment, R_0 is the risk of initial design, c_0 is constant for a particular technology or system, and r_p is the specific risk of producing safety equipment *.

---

* Black et al. consider that $r_p = 1$ death/$ 30×10^6.

Table 4.2

No	Area of application	$ 10⁶ per life saved	Lives saved per $ 10⁶
1	Automobile seat belts	0.3	3
2	High-rise flats fire control	40	0.025
3	Food poisoning control	0.03	33
4	Nuclear plants — recombiners based on 1 fatal effect per 10^4 man-rem	17	0.06
	— 6 charcoal beds	43	0.024
	— 12 charcoal beds (proposed)	300	0.003
	— iodine treatment	1,000	0.001
	— remote siting	10,000	0.0001

The minimum for the risk level is

$$\frac{dR}{dc} = R' = \left(-\frac{R_0}{c_0}\right) \exp(-c/c_0) + r_p \tag{4.10}$$

and when the costs are

$$c_{min} = c_0 \ln(R_0/r_p c_0), \tag{4.11}$$

the minimum risk level R_{min} is given by relation

$$R_{min} = r_p c_0 [1 + \ln(R_0/r_p c_0)]. \tag{4.12}$$

When comparing two technologies, these values for the minimum risk level indicate the minimum expected risk achievable for a specific design.

The impact of the specific risk in producing a safety equipment r_p on the value of R' is indicated in fig. 4.12. Clearly, the value of c_{min} is sensitive to the value of r_p. The marginal costs of risk decreasing are defined by relation

$$-\frac{dc}{dR} = -(R')^{-1} = f(c) = \frac{1}{\frac{R_0}{c_0}\exp(-c/c_0) - r_p}. \tag{4.13}$$

A graphical image of this function is given in fig. 4.13. On closer examination, we can observe that the marginal costs of risk reduction yield a very large value at $c = c_{min}$.

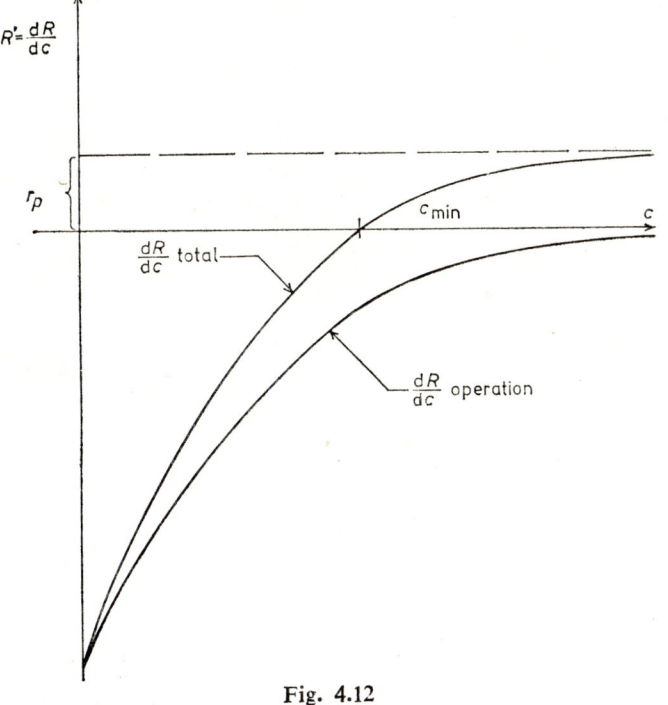

Fig. 4.12

Fig. 4.13

The above approach gives an analytical answer to a new problem "how safe is too safe?". We may conclude that the risk-cost relationship actually shows a minimum beyond which additional expenditures intended to reduce a risk will actually increase it. For a proper use of the cost-effectiveness methodology, the systems engineer must define in an appropriate way the notion of risk in compliance with his specific goals. Additionally, he has to compare health effects, costs and labour requirements and an appropriate discounting measure. Generally, a simple solution to the problem of risk-assessment and safety evolution for complex technologies may be accepted. On the other hand, "further investigation into individual and group attitudes towards risk and risk reduction" must be carried out.

4.4. The model of consequences and utilities

After the safety problem of a technical system has been organized in accordance with the user's goal, it is necessary to assess a value model of consequences for an unsafe system. When a corresponding utility measure can be associated with the value of consequences, this can be used for a risk-sensitivity analysis in the decision and management of the given system.

We shall now give a revised example illustrating the evaluation of a safety methodology for a pilot model (see Howard et al. [115]). A logical presentation for the problem of assessing nuclear reactor safety is shown in fig. 4.14. The system

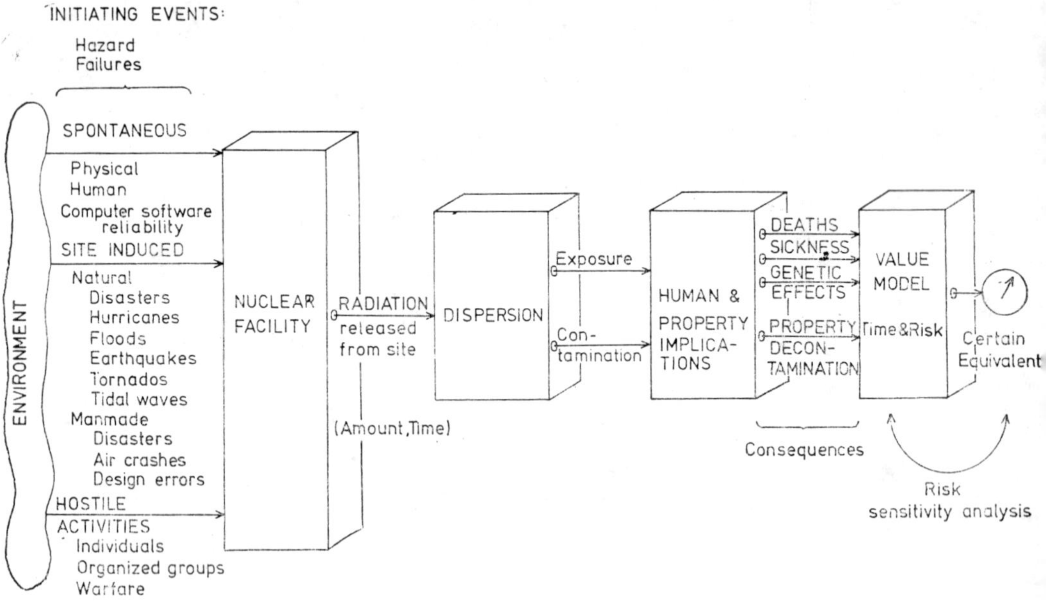

Fig. 4.14

is exposed to various hazards and stochastic events. These hazards could be spontaneous failures (e.g. physical failure of a piece of equipment) or a human error caused by incompetence, inattention, monotony or incapacity. Failures may be caused by the fact that the reactor is at a particular site. Site-induced failures, in their turn, could be caused by natural disasters (e.g. airplane crashes, human errors, computer software failure). The reactor might also be the target of hostile activities.

It is possible that as a result of these hazards, the reactor facility may suffer damage. Of particular interest in this study is the question of whether there is a radioactive release from the site, for such a release could be the cause of public loss. The release from the site can be described by the amount of radioactive element emitted in each future time interval (mr/ci). For a given radioactive release, the magnitude of the loss will be governed by its subsequent spatial and temporal distribution, by the spatial and temporal distribution of population in the surrounding area, and by the land patterns in the area. The dispersion model thus outlines meteorological, geographic and demographic information to predict the human exposure and property contamination produced by a given release (an extensive analysis of this aspect may be found in the Rasmussen report). A human implication model will assess the deaths, sickness and genetic damage which are likely to occur as a consequence of a given exposure. Likewise, the cost of decontaminating or burying property infested with radioactivity is assessed in a property-implication model. A value model reduces to a single value measure (perhaps but not necessarily monetary units) the combined societal evaluation of both human and property losses. Any model of nuclear safety (or even for offshore engineering safety) can be viewed within this logical structure. The question is how much detail to include in each sub-model and how to assess the great number of inputs required. We shall present now a pilot model in the same way as a wind tunnel model is a pilot model for airplane design. The formal structure of the model is simple and very important for making design decisions regarding the full-scale version. The pilot model has the form of a probability tree which reflects the major feature of the safety problem, i.e. sequential uncertainty.

4.5. Basic structure of a pilot model for systems safety evaluation

In the pilot model represented by the probability tree in fig. 4.15, the circular (probability) nodes denote the resolution of uncertainty and the hexagonal figures depict terminal nodes in the tree. Proceeding from left to right through the tree, we obtain a sequence of events which characterize an accident in the nuclear power reactor. If drawn completely, the pilot probability tree has 253 nodes and 252 branches.

As described before, a model of reactor accidents can be broken into a finite number of sub-models, e.g. the reactor site model, the site release model, the dispersion model, the property and human loss model, the value of outcome model. These sub-models were created to represent a "reference" 1000 MWe reactor.

No choice has been made between BWR, PWR or other type. Geographic location and population proximities have been explicitly (probabilistically) modelled in the tree. The reactor is assumed to have been in steady-state, full-capacity operation for one year (including normal maintenance).

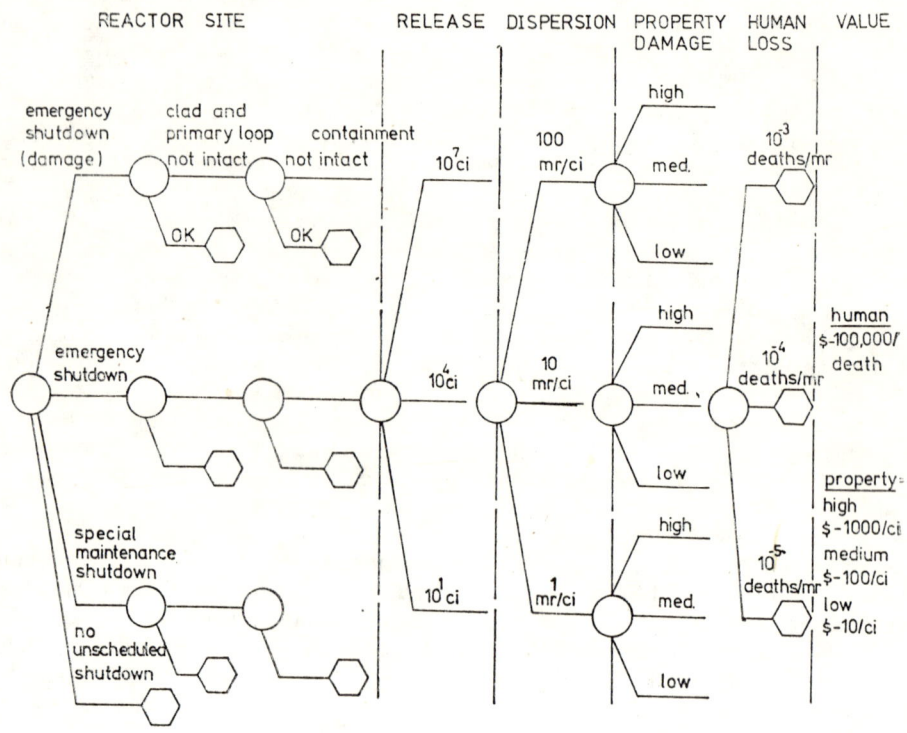

Fig. 4.15

The starting node in the tree represents four mutually exclusive and collectively exhaustive events related to reactor shutdown. Three of these events are chosen to describe the state of the reactor and site immediately after an unscheduled shutdown. The fourth and complementary event is "no surprise trips" for any given year. These events are intended to characterize states of nature after shutdown, rather than to provide scenarios for initiating events which require a reactor trip.

Concerning the evaluation of a pilot safety model using decision analysis, Howard et al. [115] noticed that "It should be clear that presentation of a mutually exclusive and collectively exhaustive set of states after trip is by itself a formidable task. Compounding this list with all possible combinations of initiating events would defeat the purpose of a pilot level analysis. That purpose, again, is to capture only the most "significant" events in each sub-model, or in this sub-model to answer the following: What one fact would I ask, if allowed only one question about the shutdown?"

For the pilot model, the following shutdown events were chosen.

1. Emergency reactor shutdown with mechanical or structural damage to the reactor or site done by the initiating event.

2. Emergency reactor shutdown with no mechanical or structural damage to the reactor site caused solely by the initiating event.

3. Unscheduled reactor shutdown for special maintenance.

4. No abnormal reactor shutdown during the given year.

A detailed description of the causes of event number 1 could include earthquake, structural failure, unpredicted chemical explosion, primary loop rupture, intentional human damage, etc. Event number 2 might have been initiated by reactor excursion, sudden incapacity of a reactor operator, detection of an imminent intentional damage attempt, etc. Event number 3 would represent discovery of crucial reactor components with abnormal "wear" or performance degradation. Event number 4 is a complement to the other three: "no unscheduled reactor shutdown in a given year".

The second set of nodes describes "loss of integrity of the fuel cladding and primary loop" [115]. It is assumed that loss of integrity represents a condition with sufficient clad temperature and fracturing to allow escape of the fission products, including noble gasses. The binary "yes-no" model is chosen for simplicity. The "not intact" branch "guarantees" that radioactivity will be released at least within secondary containment vessel of the reactor.

The third node set reveals whether radioactivity has or has not breached in secondary containment and has also a binary structure. Given "yes" answers to the shutdown and both primary and secondary containment release questions, a state exists in which radioactive material is released from the reactor site. This release and the ensuing results are modelled in subsequent tree sectors. A single parameter, i.e. the "magnitude of the radioactivity released in curies", models the release from the reactor site. Since the matter released could be in the form of particulate or gaseous radioactive material, uncertainty in composition and state is modelled by the single discrete random variable. As with other variables in the probability tree, this variable is naturally continuous, although, for the simplicity of the model, it is represented by three discrete states.

Dispersion is a function of many uncertain variables including weather conditions, point of release, terrain features, etc. Since radioactivity is most harmful to humans and animals, the model must also include consideration of the population exposed and exposure time. The units used are man-rems/curie (mr/ci) and the three branches chosen represent wide, intermediate and narrow dispersion. A few characteristics of the wide dispersion could be prolonged exposure of a large number of people to a relatively homogeneous cloud (i.e. many people absorbed a large and similar dose). "Narrow dispersion" might represent a few people absorbing a very high dose.

The parameters used to model property loss can best be described as clean-up costs. For the pilot model, a dependence assumption was made for the "dispersion-property loss" relationship. Wide dispersion was assumed to yield higher clean-up costs than intermediate or narrow distributions. For simplicity, we have chosen a single variable, i.e. deaths per man-rem, in the analysis of human loss damage.

Uncertainty here represents probabilistic effects of the different types of radioactive material on the human body, the available medical treatment, body resistance and prior accumulated dose, etc. Probabilistic independence is assumed to be arbitrary.

In his study of the value analysis phase in the pilot model for nuclear reactor accident safety, Howard et al. [115] assumed that "The value to society of the loss or injury of one of its citizens is measured and considered by many public decision makers, though almost always implicitly. Airlines safety standards, public highway design, law enforcement, and decisions of state all depend implicitly and importantly on the value of human life. In an emergency, how much would society pay to save one of its nameless members? $ 1.00? An amount equal to the total annual federal budget? The amount clearly lies between these extremes. The dollar value may be a function of an individual's wealth, position, background, etc. and of the emergency situation in which he or she is found. The analysis could also be done in terms of death, injuries, farmland irradiated or lost political merit points. In any case, the measure must be understood and accepted by the decision-making individual or group."

4.6. A quantitative model for safety analysis

The data used in this section are arbitrary and are given for the sake of illustration only.

Three items concerning modelling philosophy deserve attention before the numbers are described. Prior to the assessment of the probability of occurrence of any event, the event must be precisely defined. Combined with a probability assignment of 0.01, the first event described becomes "an emergency reactor shutdown with material damage that would occur only once every one hundred years." As the probability is lowered, the same event definition connotes a more severe though less frequent event. 'Incredible' events can sometimes be defined in this indirect manner. The "one thousand year flood" may have more meaning to the casual observer than a precise definition of water volume, velocity, damage, etc. along with its probability assignment. When this tree is updated using current information, precise definition of events must be developed to obtain the associated probability assignments.

Another observation on modelling techniques is that as event definitions are revised, and probability assignments consequently vary, the conditional probabilities attached to subsequent events must also be refined or at least re-evaluated. Failure to do so invalidates the model.

As was clearly emphasized by Howard et al. [115], probabilistic independence of events in decision analysis is a strong assumption and is quite often not justifiable in modelling physical processes. We shall adopt this assumption in three cases: the quantity of radioactivity released is assumed to be independent of the initiating event; dispersion patterns are assumed to be independent of the quantity of radioactivity released; finally, human loss levels per man-rem are also assumed to be independent of the release magnitude. However, probabilistic dependence is represented in the conditional probability assignments of the property loss model and elsewhere throughout the tree.

Consider the following scenario to illustrate the process of assigning probabilities. An honest clairvoyant, who has professional and civic interest in reactor safety, has employed his powers of prediction to produce this forecast:

> Reactor A, of specifications B, C, D, etc., in year E will experience an emergency shutdown with damage to structure or mechanical systems.

From fig. 4.16, it is clear that the likelihood of this event is assigned as .01/year. He goes even as far as to state that

> Fuel cladding and primary loop integrity will be lost and a hole greater than one square foot will exist in the final containment system.

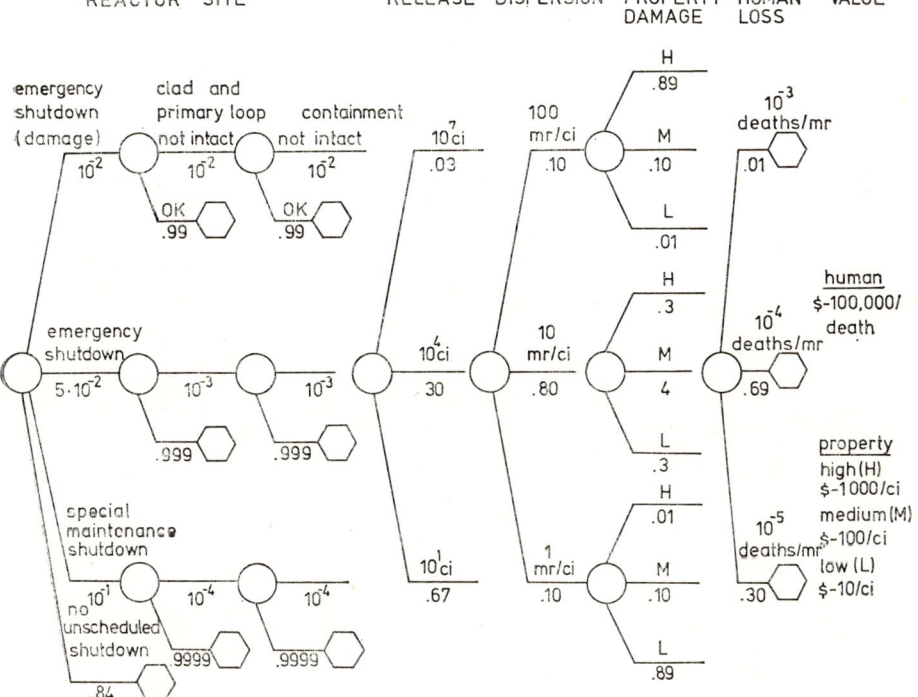

Fig. 4.16

The marginal probability of each of these events is .01, and their joint probability is 10^{-4}. It is suggested that save for a loss of electrical power, the community would suffer relatively little physical damage if events following a reactor trip terminate with no release to the environment.

The clairvoyant reads on:

The thousand curies were released...

The probability on this branch was assigned to be 0.3, which is ten times that of the more severe 10^7 release, while the most likely release of 10 curies was set at .67 (these values are assumed to be mutually exclusive and collectively exhaustive; thus, the probabilities sum to 1.0). Several trade-offs are associated with the selection of the number of branches for the node. A choice of three branches (again, selected for illustration only) gives the range of possibilities, yet it keeps the size of the remainder of the tree within reasonable bounds for computation. An overriding consideration, though, is whether the "accuracy" of this portion of the model as the number of branches is increased, (1) exceeds the modelled detail

Table 4.3

LOSS LOTTERY

LOSS IN DOLLARS	PROBABILITY	CUMULATIVE
−.11000000E+12	.28061700E−10	.28061700E−10
−.10100000E+12	.31530000E−11	.31214700E−10
−.10010000E+12	.31530000E−12	.31530000E−10
−.20000000E+11	.20119293E−08	.20434593E−08
−.11000000E+11	.63819873E−08	.84254466E−08
−.10100000E+11	.23893434E−08	.10814790E−07
−.10010000E+11	.94590000E−11	.10824249E−07
−.20000000E+10	.70595670E−08	.17883816E−07
−.11000000E+10	.85033257E−08	.26387142E−07
−.10100000E+10	.94590000E−10	.26481732E−07
−.20000000E+09	.42064173E−08	.30688149E−07
−.11000000E+09	.11224680E−08	.31810617E−07
−.10100000E+09 ←	.31530000E−10 ←	.31842147E−07
−.10010000E+09	.31530000E−11	.31845300E−07
−.20000000E+08	.20119293E−07	.51964593E−07
−.11000000E+08	.63819873E−07	.11578447E−06
−.10100000E+08	.23893434E−07	.13967790E−06
−.10010000E+08	.94590000E−10	.13977249E−06
−.20000000E+07	.70595670E−07	.21036816E−06
−.11000000E+07	.85033257E−07	.29540142E−06
−.10100000E+07	.94590000E−09	.29634732E−06
−.20000000E+06	.42064173E−07	.33841149E−06
−.11000000E+06	.90452213E−08	.34745671E−06
−.10100000E+06	.70417000E−10	.34752713E−06
−.10010000E+06	.70417000E−11	.34753417E−06
−.20000000E+05	.44933088E−07	.39246726E−06
−.11000000E+05	.14253105E−06	.53499831E−06
−.10100000E+05	.53362003E−07	.58836031E−06
−.10010000E+05	.21125100E−09	.58857156E−06
−.20000000E+04	.15766366E−06	.74623522E−06
−.11000000E+04	.18990761E−06	.93614283E−06
−.10100000E+04	.21125100E−08	.93825534E−06
−.20000000E+03	.93943320E−07	.10321987E−05
−.11000000E+03	.18801339E−07	.10510000E−05
.00000000E+01	.99999895E+00	.10000000E+01

of another part and (2) causes the precision of the model to surpass the "probability assignment capabilities" of the reactor safety community.

The clairvoyant returns with more revelations...

The weather conditions will be such that 100 man-rems/curie will be absorbed by the population, and property loss will be on the "medium" branch in the model.

The dispersion of 100 man-rems/curie (total dose = 10^6 man-rems) tells us that a large number of people over a wide land area absorbed a fairly uniform dose. Although the model indicates that, given this dispersion, the likeliest property damage level would be high, our friend reveals that damage was moderate.

The joint probability of all the events in this scenario is the product of probabilities along each branch traversed or $(10^{-2}) (10^{-2}) (10^{-2}) (.3) (.1) (.1) (.01) =$ $= 3 \cdot 10^{-11}$.

The probability tree and the uncertain losses it represents can be termed a loss lottery. The loss lottery represented by this probability tree is shown in table 4.3 which lists losses in the order of their descending absolute values. By definition, the tree is the model of a complex and uncertain process. By making the decision to build nuclear power generating stations, society decides that in order to obtain a cheaper energy source it will exist with the potential perils of major reactor accident, quantitatively represented as a loss lottery. A hypothetical alternative to existing with the uncertainties of the reactor accident lottery is the development of a new nuclear power source for communities, which presents no risk of reactor accident. This "ideal risk-free" reactor and its site are identical to the reference reactor except that it is completely safe with no chance of harm to the adjacent community due to accidental radioactivity release. What price would the community pay for this guaranteed system? It is well known that technology today cannot economically produce such a reactor. However, the concept is useful since it places an upper bound on the amount to be spent by any rational decision maker for a safer reactor, and is a guide to the value of additional research and development for reactor accident prevention (see fig. 4.17).

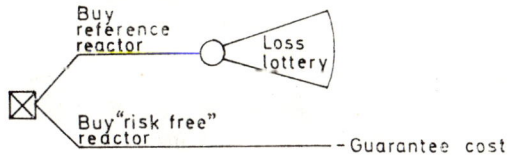

Fig. 4.17

The first alternative is the purchase of a reference design reactor for any given community at the current cost and with the current assessment of the reactor accident loss lottery. The second alternative buys this same system for the same cost plus the cost of a 100% guarantee of safety. For a guaranteed cost of zero risk, any rational decision-maker would take the second alternative. Again, what is the highest guarantee cost the community would tolerate while still choosing the second path? The maximum guarantee price is termed the certain equivalent for the loss lottery in alternative 1. More precisely, the certain equivalent is the cash amount (profit or loss) such that the decision maker is indifferent about keeping a lottery with its inherent uncertainty and accepting the cash (profit or loss) for sure. Utility theory provides the foundation for the use of this concept to the probability tree model in this memo.

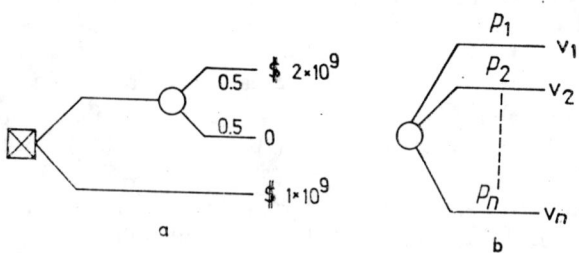

Fig. 4.18

Assuming natural aversion to risk on the part of society itself, there is a limiting case of utility theory that is of interest here. If the decision-maker (e.g. government community) is indifferent to risk, then it can be described as an expected value decision-maker. This represents the case in which society would be indifferent between the following two free alternatives: (A) A fifty-fifty chance at winning two billion or winning nothing; (B) One billion dollars for sure.

The decision problem is represented graphically in fig. 4.18 a. For the node above (see fig. 4.18 b), the expected value $\bar{v}$ is $\sum_{i=1}^{n} p_i \cdot v_i$.

By a recursive application of this rule from terminal nodes on the right back to the starting nodes on the left we obtain the expected or mean value of the loss lottery (probability tree). For a risk-neutral individual or group, the expected loss is the certain equivalent loss, or, in the reactor accident case, the maximum guarantee price. Figure 4.19 shows the expected value at each node in the tree. As was already mentioned, the use of the arbitrary numbers assigned to the tree places an upper bound of only $ 164 per year on additional reactor safety expenditures for any site similar to the reference design.

Since communities are usually not risk-neutral, this expected loss understates the negative social utility of the reactor accident loss lottery. As the decision-making body moves from a risk-neutral to a risk-averse attitude, the certain equivalent for the lottery decreases. Figure 4.20 shows how the certain equivalent of the pilot probability tree falls with increased risk aversion for an exponential utility function. As gamma becomes larger, the certain equivalent decreases rapidly. This rapid decrease indicates extreme certain equivalent sensitivity to risk attitude. With $\gamma = 10^{-8}$, the certain equivalent approximates the geatest loss, i.e. $1.1 \cdot 10^{11}$ monetary units as shown in table 4.3. To eliminate this limited choice of an exponential utility function, the next step could be the assessment of the appropriate utility function for the community or the "decision maker" and the corresponding gamma.

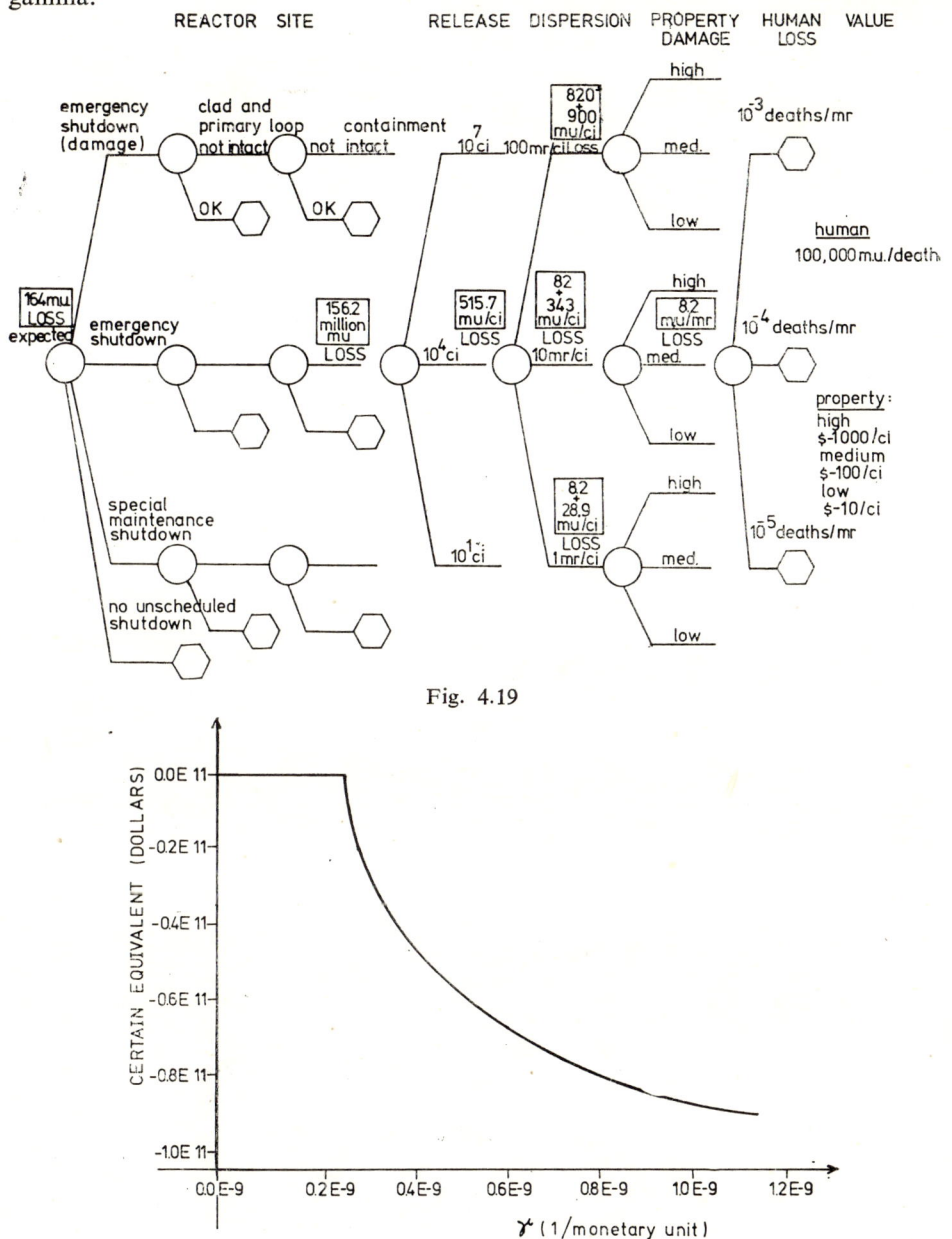

Fig. 4.19

Fig. 4.20

Part II

Models, algorithms and optimization techniques in systems engineering

Chapter 5

Modelling aspects of reliability, maintenance and safety strategies using probabilistic systems analysis

Complex systems are too costly to be immediately replaced after some degradation occurs in the performance level of any structure function which is under observation. Large production and monetary losses could result from an unsatisfactory operation of the system due to lack of maintenance. On the other hand, system complexity and its probabilistic behaviour in time imply that maintenance policies cannot rely on experience and common sense alone. However, when analysing such systems, we must design appropriate realistic models that should meet the characteristics of the real world (i.e. complexity, uncertainty, risk) [15]. These models need not give a faithful description of the real world but at least should handle most of its practical aspects, and then, by using mathematical procedures and logical decisions, should provide the decision-maker with a set of possibly better management solutions (see also ref. [15]). As Bonder righteously noticed, "Modelling is the set of activities that extends from the statement of management's problem through to the abstraction and rationalistic procedures that produce the quantitative formal premises of the process dynamics" [98].

In this chapter, we shall introduce maintenance management models using Markovian processes. This class of models has practical limitations. Under a dichotomic behaviour of any component, there exist 2^n states of the system for a Markov model ($2^{10} > 10^3$, $2^{20} > 10^6$). Thus, for systems of practical interest, "dimensionality" limits a complex analysis of the process. So, it is necessary to take advantage of the structure functions $\Phi_\lambda(\underline{X})$, ($\lambda = 1, 2, \ldots, \omega$) which partition the state space of the components in order to reduce the dimensionality of the Markov model (see fig. 5.1). In a complex systems engineering approach to technical systems, the five-dimensional treatment of the maintenance of a complex system could be also used (see fig. 5.2). By a rough examination of diagrams 5.1 and 5.2, the following observations are immediate.

1°. The dynamics of the system can be described by a Markov or a semi-Markov process, where some appropriate probabilistic measures for the system behaviour can be evaluated.

2°. The degree of observability of the system will be introduced. We can say that the system is completely observable; an observer who is outside the dynamic mechanism of the system, can identify precisely the state of the system at any point in time. To have more information about the state of the process in the case when a Markovian model is used, an independent observation space could

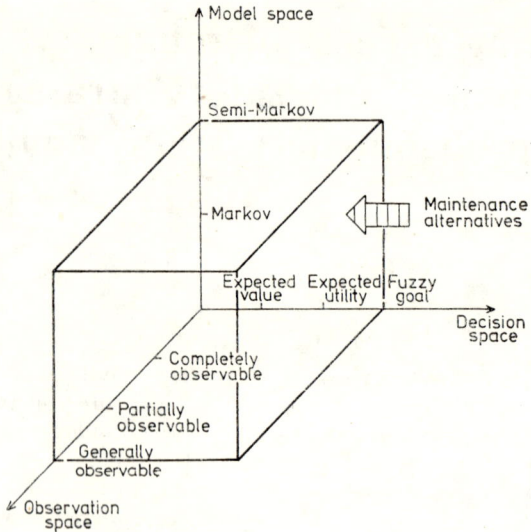

Fig. 5.1

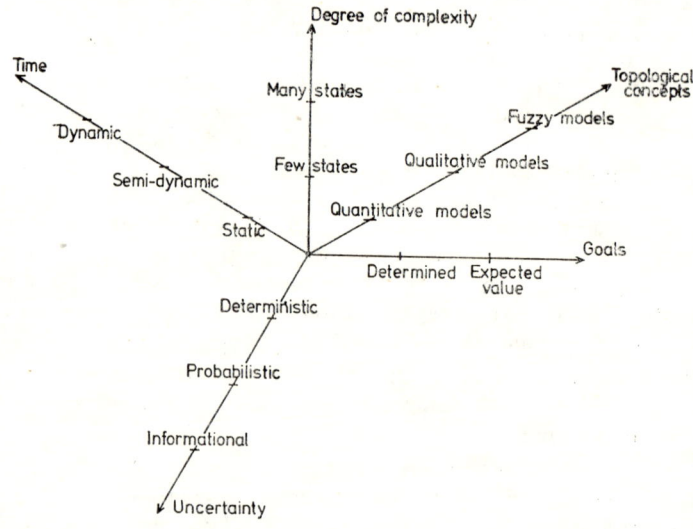

Fig. 5.2

be attached to the overall analysis and then we obtain a partially observable process. Furthermore, if these observations are considered as causes in a cause-effect model, then the process is called generally observable (see fig. 5.3).

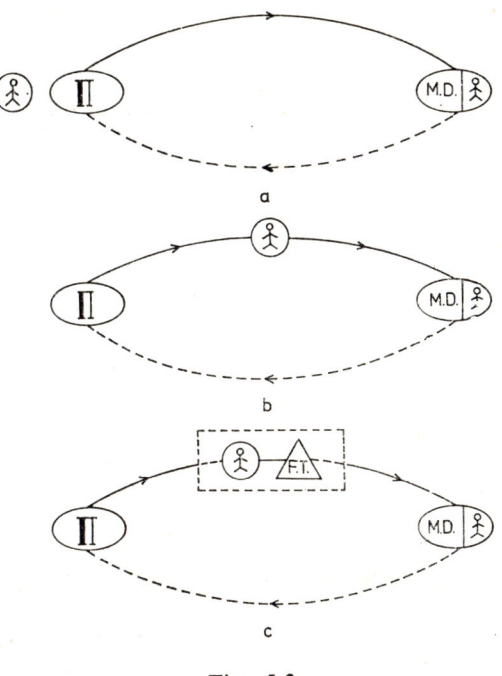

Fig. 5.3

3°. Finally, we have to take decisions for maintenance; even a "do nothing" alternative will be included in the decision space of the MDP. Following the theory of economic behaviour (see Neumann and Morgenstern [93]), we can define an expected value or an expected utility decision-maker. In the most general case, we can identify a decision-maker the goals of which are fuzzy defined (see Bellman and Zadeh [94]).

5.1. Markovian decision processes. Basic definitions

We shall give basic definitions of MDP which will be used throughout the remaining sections of the book. An MDP is a stochastic sequential process [7, 8, 10]. We have to consider that the system can be described by a discrete-time Markov chain. Decisions of each epoch and costs (rewards) are associated with each observable state. A Markov process describes the behaviour of a system by saying that it is a certain state at a given time. "The probability law of its future state of existence depends only upon the state it is in and not on how the system arrived in that state" [11]. The occurrence of a particular system state can be regarded as the occurrence of an event. For the case when the times of interest are discrete, suppose that the possible outcomes of a trial are the events $E_1, E_2, \ldots, E_N$. Define $p_{ij} = \Pr\{E_j$ occurs at the next trial$/E_i$ occurred on the preceding trial $\mathscr{E}\}$, $\pi_i = \Pr\{E_i$ is the outcome at the initial trial$/\mathscr{E}\}$, where $\mathscr{E}$ represents *a priori* knowledge about the process evolution for probability assessment. In the above statement, p_{ij} represents the transition rate probability for the Markov process and π_i is known as the probability state vector. A more precise definition on a Markov process will be given in the sequel.

Let $X(t)$ be a random variable whose value indicates the state of the system at time t, when t takes discrete values: $\{X_t : t = 0, 1, 2\ldots\}$. The process is said to be a Markov process if, for any set of n time points $t_1 < t_2 < \ldots < t_n$ in the index set of the process, the conditional distribution of $X(t_n)$, given the values for $X(t_1), X(t_2), \ldots, X(t_{n-1})$, depends only on $X(t_{n-1})$. More precisely, for any real number $q_1, q_2, \ldots, q_n$, $\Pr\{X(t_n) \leq q_n/X(t_1) = q_1, \ldots, X(t_{n-1}) = q_{n-1}\} = \Pr\{X(t_n) \leq q_n/X(t_{n-1}) = q_{n-1}\}$. The above probability measure is known as p_{ij} defined above, with the additional property that $\sum_{j=1}^{N} p_{ij} = 1$, $(i = 1, 2, \ldots, N)$.

For the use of semi-Markov processes, we shall accept the definition given by Pyke [39].

"A semi-Markov process is a stochastic process which moves from one to another of a countable number of states, with the successive states visited forming a Markov chain,..., the process stays in a given state a random length of time, the distribution function (d.f.) of which may depend on this state as well as on the one to be visited next. It is thus a Markov chain for which the time scale has been randomly transformed."

We may conclude that a semi-Markov process is characterized by a probability transition matrix $P = [p_{ij}]$ and also by a holding-time mass function that gives information on the length of time during which that process is in a particular state $H(m) = [h_{ij}(m)]$, (see [8, Chap. 15]).

As previously mentioned, in many applications of systems engineering (i.e. machine replacement, medical diagnosis), the use of partially observable Markovian processes for which the underlying Markovian process is a discrete-time finite state Markov (semi-Markov) process is profitable. In addition, we have a finite number of possible outputs. The policy space for the maintenance process using MDP is denoted by the cartesian product of all actions of state maintenance. A generalized MDP is represented in fig. 5.4; we can identify the core process, the observation process (the stochastic inspection) and the decision process. We shall give an example from systems engineering in order to illustrate the practical value of such models.

Let us consider a nuclear reactor system of the BWR type (see fig. 5.5). The components belonging to any system inside the nuclear reactor (i.e. Emergency Core Cooling System), undergo natural deterioration; the deterioration law can be represented by a Markov (semi-Markov) process. To obtain information about the state of the components, we must perform (from the control room or appropriate safety places) observations (a stochastic inspection), which for practical reasons can be only finite in number. With such information about the probable state of the components, the decision-maker (the maintenance engineer) has to take repair/replacement decisions. However, for a more complicated (realistic) case,

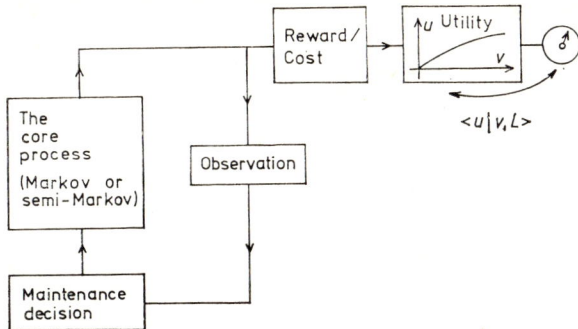

Fig. 5.4

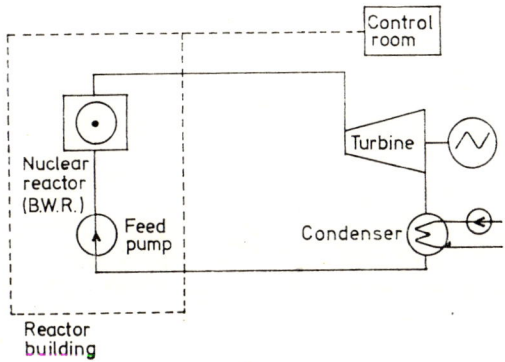

Fig. 5.5

the decision-maker can be a risk-sensitive person. Now, a review of possible optimization solutions for the above MDP will be given (see fig. 5.6).

Completely observable MDP have been widely investigated. The policy iteration algorithm given by Howard for Markov [7] and semi-Markov [8] processes develops in two phases; the value iteration phase (VIP) and the policy improvement phase (PIP). The VIP consists in solving a set of finite number of linear equations, whereas PIP is concerned with performing the optimization test. The algorithm is iterative in essence. Manne [74], and later Mine and Osaki [10] presented for the first time a linear programming (LP) solution to completely observable MDP.

Though less investigated, partially observable MDP have a more complicated structure. Well-known dynamic programming solutions are limited to processes with a small number of states and observations. Sondik [13], [31] developed a new algorithm for the finite horizon case of partially observable Markov processes using the piecewise linear property of the above processes. For the case of infinite horizon [13], it was shown that Howard's policy iteration algorithm works for certain stationary policies called "finitely transient policies". Independently, Satia [34] developed an implicit enumeration algorithm for the case of partially observ-

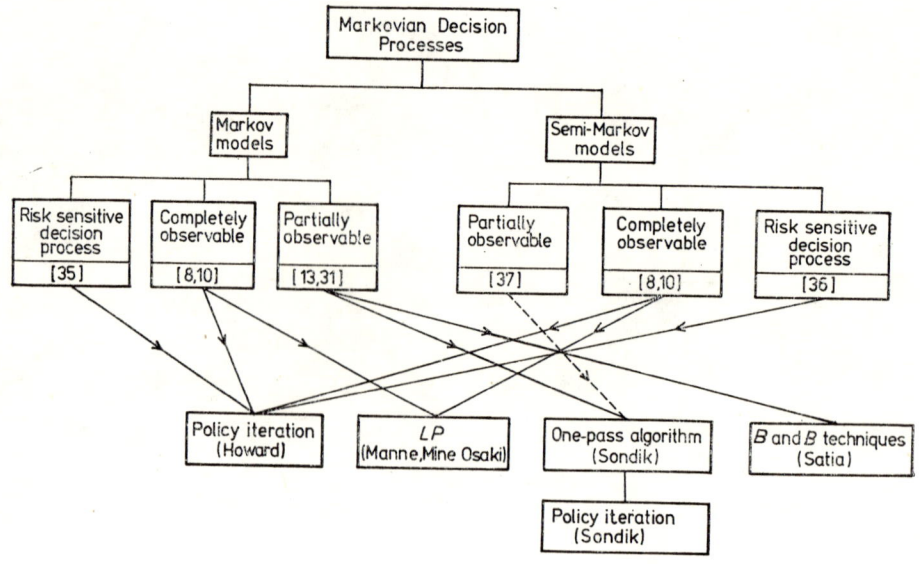

Fig. 5.6

able processes. His algorithm is in fact an approach by a Branch and Bound method. Both algorithms have practical limitations for large scale MDP (see Sondik [13]). Other successive approximation techniques and state space partitions have been suggested [95] for large-scale MDP.

5.2. Basic definitions in operational engineering technical systems

The aim of this section is to introduce basic definitions from systems engineering with regard to the operational status of technical systems in a repair environment and effectiveness criteria for the support problem. Before the definition of a system, we can identify two distinct processes: one refers to the system operational state, on-line process, and the other is about the system's "connection" with the environment (i.e. man-machine interface concerning operational and maintenance process, logistics and supply effectiveness). The latter will be called the off-line process (see fig. 5.7). Definitions and comments for the components of the on-line process will be given in the sequel.

Reliability: Generally speaking, system reliability is defined as a probability and it relates to the frequency of the system failures. We shall accept the following [72, p. 6].

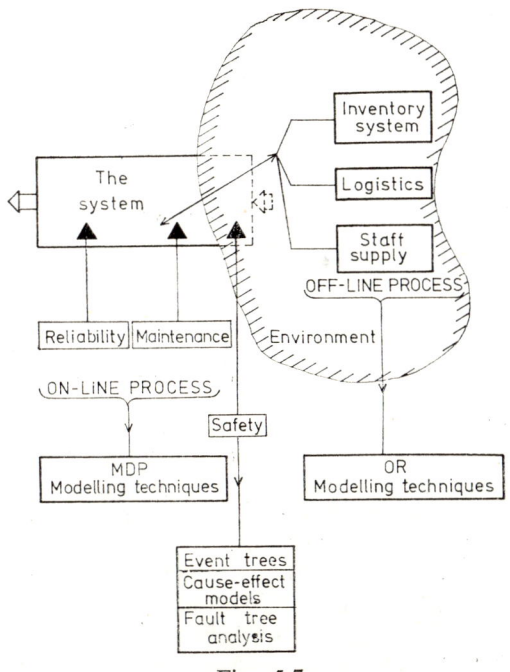

Fig. 5.7

DEFINITION 5.1. *Reliability* is the probability that a system will perform satisfactorily for at least a given period of time when used under stated conditions.

The system reliability depends on the reliability designed in each independent component of the system, its topology and the amount of "built in" redundancy. Reliability of technical systems can be increased by redundancy techniques; various approaches to optimal system redundancy have been intensively investigated in the literature (for a general survey, see Gheorghe [55]). An alternative technique for improving system reliability is by preventive maintenance such as inspection, repair, replacement and overhaul of equipment and its components.

Maintainability: Following ref. [72], we shall accept the following

DEFINITION 5.2. *Maintainability* is defined as the probability that a failed system will be restored to operable condition in a specified down time.

In other words, system maintainability is the probability that, by maintenance decision actions under stated conditions, a failed system (or some of its components)

will be restored to normal operable condition within a specified total down time. For the on-line process modelling, we shall adopt the following reasonable assumptions: (1) There is only a finite number of decision alternatives of system maintenance for each operational state (2) There is a finite number of alternative hardware designs for certain equipment components of the system (3) Additional "value of information" for modelling and probability encoding is obtained by performing a detailed system inspection or from expert judgment on the operational status of the system.

The off-line support problem is concerned with the inventory management and allocation of spares and repair parts. However, this problem is considered after an on-line support strategy has been established.

5.3. Hybrid OR methods for maintenance and safety modelling

For large and complex systems, owing to limitations of the "course of dimensionality", it is theoretically difficult and practically impossible to use MDP for an overall approach to problems of systems maintenance. On the other hand, the above processes (MDP) cannot incorporate a wide range of technical and managerial constraints because of their quite fixed structure. Probability assessment is also a limitation in an extensive use of MDPs for the maintenance problem. In order to answer a larger variety of practical questions about systems maintenance and safety, we must use some other relative OR techniques, such as mathematical programming. The well-known modelling problem of real world *vs.* model world for the maintenance problem is graphically represented in fig. 5.8. A similar approach to that presented in the above figure is given in Chapter 8.

For the problem of maintenance management of polyfunctional systems, we can identify four distinct steps in the model phase: (1) a decomposition rule for large scale systems modelling, (2) a dynamic probabilistic model for the analysis of the stochastic behaviour in accordance with the decomposition rule, (3) formulation of technological and maintenance constraints for the system and the use of a mathematical programming technique, (4) conclusions concerning practical aspects of optimal maintenance strategies for the complex system. To analyse the optimal maintenance strategies for polyfunctional technical systems, hybrid OR techniques have been used. However, a hierarchical decomposition technique has been implicitly introduced (see fig. 5.9), when a general co-ordinator (maintenance cost) was available as a goal for the system modelling. A pictorial representation of the process of maintenance modelling, data acquisition and implementation for real systems is presented in fig. 5.10. The reader could find a mapping between the above "practical" approach to the maintenance problem and the mathematical modelling process. However, we deal with stochastic processes of deterioration, inspection and the risk-sensitive character of the maintenance decision process which raise still more difficulties. It is the author's experience that the safety question for large-scale

MODELLING OF RELIABILITY, MAINTENANCE AND SAFETY

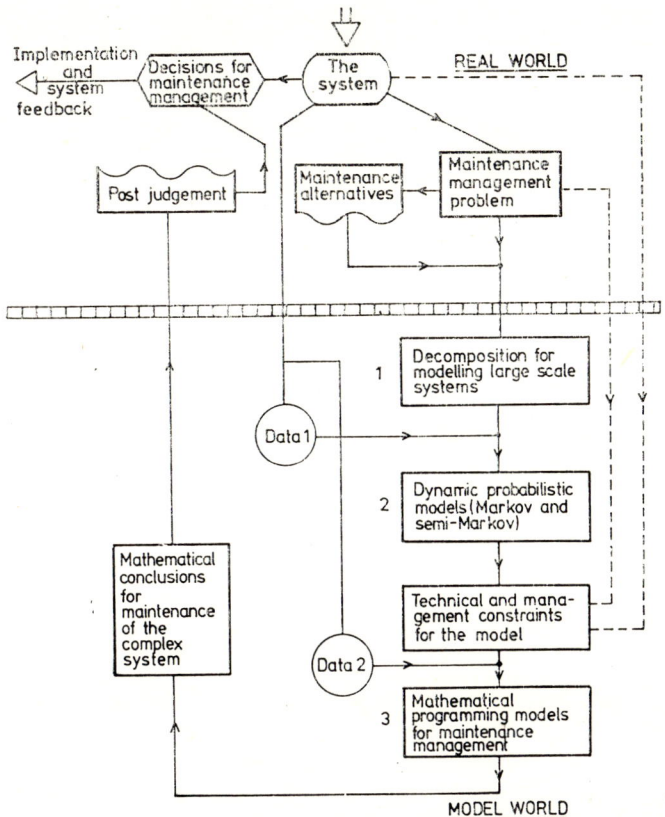

Fig. 5.8

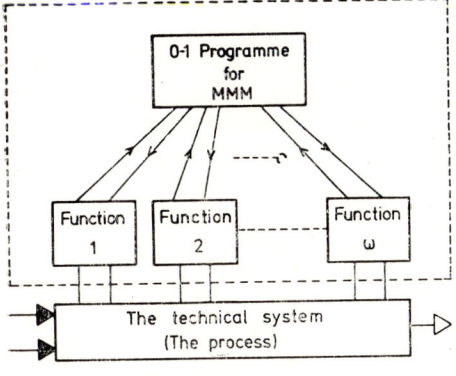

Fig. 5.9

engineering systems cannot be handled by the standard techniques of stochastic processes. However, we have to use systems analysis to structure the safety problem, and inference techniques to find the combined effects of the safety figures of the system.

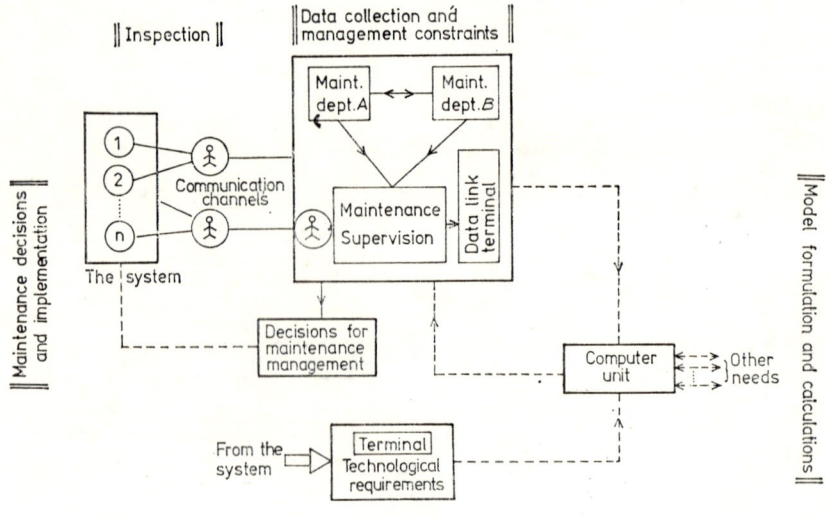

Fig. 5.10

5.4. Reliability prediction of complex systems with Markov behaviour having many failed states

Complex systems operate in a repair environment. Their sub-systems have known or measurable constant failure and repair rates. Such systems have acceptable states A_i, $(i = 1, 2, \ldots, r)$, and failed states F_j, $(j = 1, 2, \ldots, m)$. We develop now a reliability model using a stationary Markov model [2]. We can define a discrete finite-dimensional state space given by $A \cup F$, where $A = $ set $\{A_i; i = 1, 2, \ldots, r\}$, $F = $ set $\{F_j; j = 1, 2, \ldots, m\}$ and the discrete time set is T. The state probability functions are defined by

$$s_i(n) = \Pr\{S(n) = A_i\}, \quad n \in T, \tag{5.1}$$

$$s_j(n) = \Pr\{S(n) = F_j\}, \quad n \in T. \tag{5.2}$$

The set of states in A defines a transient class where the states in set F are absorbing states. We can define a one-step transition matrix M for the Markov process with the state space $A \cup F$,

$$M = \left[\begin{array}{c|c} A & B \\ \hline 0 & 1 \end{array} \right],$$

where

$$A = [a_{ik}] = [\Pr\{S(n+1) = A_k \mid S(n) = A_i\}] \quad (5.3)$$

is an $(r \times r)$ matrix;

$$B = [b_{ij}] = [\Pr\{S(n+1) = F_j \mid S(n) = A_i\}] \quad (5.4)$$

is an $(r \times m)$ matrix;

$$1 = [\delta_{ju}] = [\Pr\{S(n+1) = F_u \mid S(n) = F_j\}] \quad (5.5)$$

is an $(m \times m)$ unit matrix

$$\delta_{ju} = \begin{cases} 1 & \text{for } j = u \\ 0 & \text{for } j \neq u \end{cases} ; \quad \text{and} \quad (5.6)$$

$$0 = [0] = [\Pr\{S(n+1) = A_i \mid S(n) = F_j\}] \quad (5.7)$$

is an $(m \times r)$ null matrix.

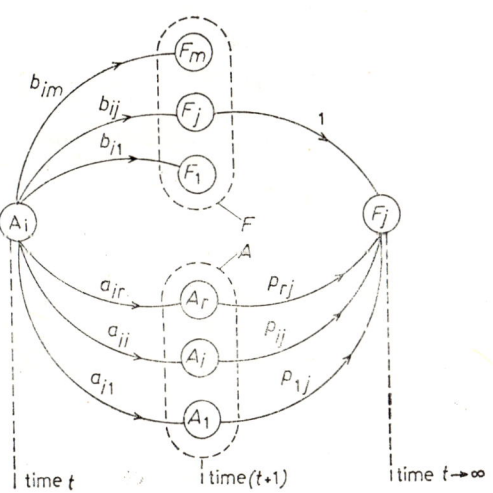

Fig. 5.11

A possible evolution of a complex system having many failed states is plotted in fig. 5.11. In the present analysis, the transient state A_i is called an *acceptable state* if it characterizes the satisfactory operation of the complex system. By contrast, "the absorbing state F_j is called a failed state if it characterizes some unsatisfactory mode of operation of the complex system" [2].

If $s(n)$ is a $(1 \times (r + m))$-state probability vector with components $s_i(n)$ and $s_j(n)$, then

$$s(n) = s(0) M^n, \qquad (5.8)$$

where $s(0)$ is the vector of state probabilities for time $n = 0$.

The reliability function of a complex system with a Markov deterioration is given by the probability that at time n the complex system is in some acceptable state:

$$\mathscr{R}(n) = \Pr\{S(n) \in A\} = \sum_{i=1}^{r} s_i(n). \qquad (5.9)$$

If we define p_{ij} to be the steady transition probability that a complex system that starts in a state A will eventually end up in a specific failed state F_j, then

$$p_{ij} = \Pr\{S(\infty) = F_j | S(n) = A_i\}. \qquad (5.10)$$

In a matrix formulation, $P = [p_{ij}]$ with $(r \times m)$ as matrix dimensions. As was already proved by De Mercado [2],

THEOREM 5.1. *For a technical system with r acceptable states and m failed states operating in a repair environment with a known matrix M, the $(r \times m)$ steady transition probability matrix P is given by*

$$P = [1 - A]^{-1} B. \qquad (5.11)$$

We can further define $p_{ij}(n)$ to be the probability that at time n the complex system is in failed state F_j, given that at time $n = 0$ it was in the acceptable state A_i:

$$p_{ij}(n) = \Pr\{S(t + n) = F_j | S(t) = A_i\}. \qquad (5.12)$$

In a matrix formulation, $P(n) = [p_{ij}(n)]$ with $(r \times m)$ dimensions.

THEOREM 5.2. *For a technical system with r acceptable states and m failed states with a known matrix M, the $(r \times m)$-matrix $P(n)$ satisfies the relation*

$$P(n) = A^{n-1} B + P(n - 1). \qquad (5.13)$$

We shall investigate now moments of the first time to failed state for complex technical systems with a Markov behaviour. The generating function $g_{ij}(z)$ is defined by using the z-transform such that

$$g_{ij}(z) = \sum_{n=1}^{\infty} z^n p_{ij}(n). \qquad (5.14)$$

The moments $\tau_{ij}(h)$ of the first time to a failed state are defined as the moments when the complex system jumps for the first time from state A_i to state F_j,

$$\tau_{ij}(h) = \frac{d^h}{dz^h}(g_{ij}(z))\bigg|_{z=1}. \tag{5.15}$$

If we define

$$G(z) = [g_{ij}(z)] \tag{5.16}$$

and

$$\tau(h) = [\tau_{ij}(h)], \tag{5.17}$$

then we can prove the following

THEOREM 5.3. $\tau(h) = \dfrac{d^h}{dz^h} G(z)\bigg|_{z=1},$ (5.18)

where

$$G(z) = \frac{z}{1-z}[1-zA]^{-1}B. \tag{5.19}$$

The exit probability function $w_i(n)$ is defined as the probability that a complex technical system will pass from the acceptable state A_i into the set F in a number of n units of time (see fig. 5.12).

$$w_i(n) = \Pr\{S(n+t) \in F \mid S(t) = A_i\} =$$

$$= \sum_{j=1}^{m} \Pr\{S(n+t) = F_j \mid S(t) = A_i\} =$$

$$= \sum_{j=1}^{m} p_{ij}(n). \tag{5.20}$$

THEOREM 5.4. If $W(n)$ is an $(r \times 1)$-column vector of the exit probability functions, then

$$w_i(1) = \sum_{j=1}^{m} b_{ij} \tag{5.21}$$

$$W(n) = AW(n-1). \tag{5.22}$$

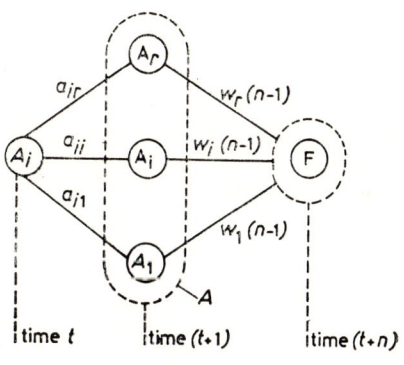

Fig. 5.12

Extend the problem of the exit probability functions; the moments of the first exit time from the acceptable class A will be investigated in the sequel.

Moments of the first time for the case when the system jumps from state A_i into a class F, will be described by $\tau_i(h)$. In terms of the z-transform, the generating function $c_i(z)$ becomes

$$c_i(z) = \sum_{n=1}^{\infty} z^n w_i(n). \tag{5.23}$$

The moments of the first time to failure are

$$\tau_i(h) = \frac{d^h}{dz^h}(c_i(z))\bigg|_{z=1}. \tag{5.24}$$

THEOREM 5.5. *The generating functions $g_{ij}(z)$ and $c_i(z)$ are related as*

$$c_i(z) = \sum_{j=1}^{m} g_{ij}(z) \tag{5.25}$$

or, in an equivalent form, as

$$\tau_i(h) = \sum_{j=1}^{m} \tau_{ij}(h). \tag{5.26}$$

THEOREM 5.6 [2]. *If $C(z)$ is an $(r \times 1)$-column vector of the generating functions $c_i(z)$ and if $\tau(h)$ is an $(r \times 1)$-vector of the moments $\tau_i(h)$ for a complex system operating in a repair environment, then*

$$\tau(h) = \frac{d^h}{dz^h}(C(z))\bigg|_{z=1} \tag{5.27}$$

with

$$C(z) = \frac{z}{1-z}[1-zA]^{-1}B', \tag{5.28}$$

where

$$B' = [b'_1 \ldots b'_r]^{\text{tr}} \tag{5.29}$$

and

$$b'_i = \sum_{j=1}^{m} b_{ij}. \tag{5.30}$$

Example 1 [2]. A telecommunication network consists of two channels as shown in fig. 5.13. Table 5.1 gives the state assignment for describing the operational behaviour of the network. The associated transition graph is given in fig. 5.14. The failure repair and delay-repair rates are $\lambda = 0.002/h$, $\mu = 0.004/h$ and $\rho = 0.2/h$, respectively. The associated partitioned matrix M of the one-step transition for the marked process is

$$M = \begin{array}{c} \\ A_1 \\ A_2 \\ A_3 \\ F_1 \\ F_2 \end{array} \begin{array}{c} A_1 \\ \left[\begin{array}{ccccc} 0.998 & 0.002 & 0 & 0 & 0 \\ 0 & 0.798 & 0.2 & 0.002 & 0 \\ 0.004 & 0 & 0.994 & 0 & 0.002 \\ \hline 0 & 0 & 0 & 1 & 0 \\ 0 & 0 & 0 & 0 & 1 \end{array}\right] \end{array}.$$

MODELLING OF RELIABILITY, MAINTENANCE AND SAFETY

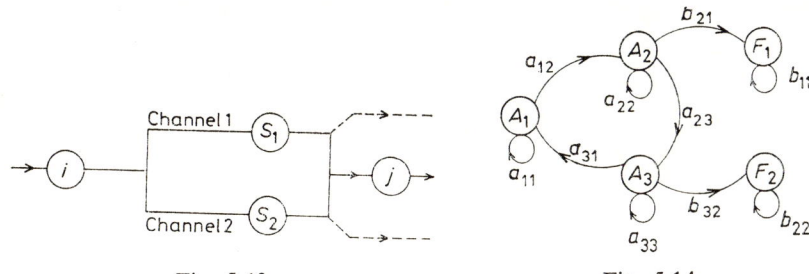

Fig. 5.13   Fig. 5.14

Table 5.1

No.	state	The word description for the state
1	A_1	S_1 and S_2 provide a path from i to j.
2	A_2	S_1 fails and S_2 provides the only path. Repairs to S_1 are not yet started.
3	A_3	Repairs to S_1 start. S_2 provides the connection between i and j.
4	F_1	S_2 fails before repairs to S_1 have begun.
5	F_2	S_2 fails before repairs to S_1 are completed.

The computer simulation results for $P(n)$, $n = 10$ and 50, as well as the reliability performance function $R(n)$ for different initial $s(.)$ vectors are shown in fig. 5.15.

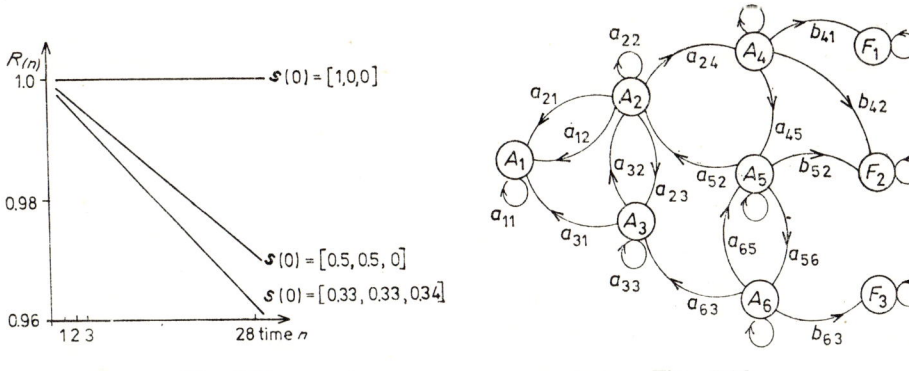

Fig. 5.15   Fig. 5.16

Example 2 [2]. Let us consider that a complex system in a repair environment yielded the transition graph represented in fig. 5.16. The associated transition matrices A and B are included into the structure of the matrix M.

$$M = \begin{array}{c|cccccc|ccc} & A_1 & A_2 & A_3 & A_4 & A_5 & A_6 & F_1 & F_2 & F_3 \\ \hline A_1 & 0.90 & 0.045 & 0 & 0.055 & 0 & 0 & 0 & 0 & 0 \\ A_2 & 0.20 & 0.45 & 0.15 & 0.1 & 0 & 0 & 0 & 0 & 0 \\ A_3 & 0.6 & 0.1 & 0.3 & 0 & 0 & 0 & 0 & 0 & 0 \\ A_4 & 0 & 0 & 0 & 0.55 & 0.20 & 0 & 0 & 0.15 & 0 \\ A_5 & 0 & 0.4 & 0 & 0 & 0.20 & 0.20 & 0.10 & 0 & 0.10 \\ A_6 & 0 & 0 & 0.35 & 0 & 0.10 & 0.10 & 0.25 & 0 & 0.20 \\ \hline F_1 & 0 & 0 & 0 & 0 & 0 & 0 & 1 & 0 & 0 \\ F_2 & 0 & 0 & 0 & 0 & 0 & 0 & 0 & 1 & 0 \\ F_3 & 0 & 0 & 0 & 0 & 0 & 0 & 0 & 0 & 1 \end{array}$$

Computer results for systems having many failed states are presented in fig. 5.17.

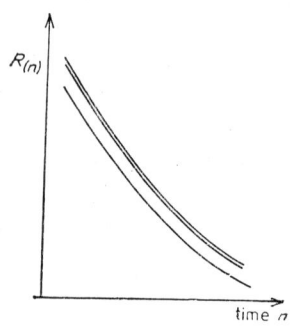

Fig. 5.17

5.5. Reliability and availability prediction of polyfunctional technical systems with semi-Markov structure

As was mentioned, most of the analysis of complex engineering systems is concerned with reliability prediction. For the normal operation of power systems, short-term reliability calculations have been made. In most of the research carried out so far, it was assumed that a system has Markov behaviour, and that the system performs only one function. However, a semi-Markov model is much more realistic; it is characterized by a probability transition matrix, and also by a holding-time mass function that gives information about the time-span the process will occupy a particular state [1, 6, 8, 27].

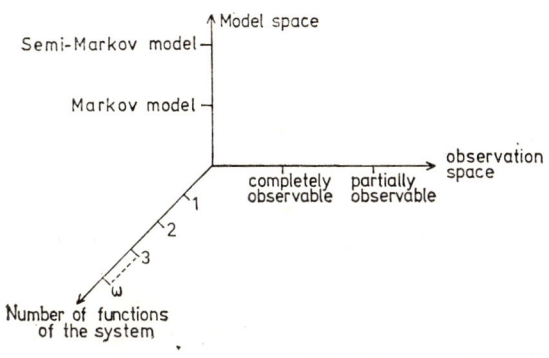

Fig. 5.18

In this paragraph we analyse the reliability and availability of a polyfunctional technical system which performs a finite number of functions and whose behaviour is described by a Markov and semi-Markov model. The system is considered to be either completely or partially observable. Practical examples are included. Reliability prediction in polyfunctional systems can be represented as in fig. 5.18. When the system is polyfunctional, we can use either Markov or semi-Markov models which may be either completely [8] or partially observable [13, 15].

In the following sections, we shall also consider the calculation of reliability and availability predictions when the system is completely or partially observable, respectively, under a Markov or a semi-Markov behaviour.

5.5.1. Completely observable complex systems

The performance index of the total structure function $\Phi_\omega^*(X)$ represents the operational state of the system (e.g. $\Phi_\omega^*(X) = 0, 1, \ldots, \omega$ and $N = \omega + 1$). We illustrate this statement by considering an electric power system for which demand is met at time t, when at least g out of ω generators are operating; however, for the underlying semi-Markov model, state 1 is for $g = 0$ and state N is for $g = \omega$. The state of operation for the system given by $\Phi_\omega^*(X)$ can be divided into three major classes: the system capable of performing complete operation (A), undergoing repair (F) and incapable of operation (C). The class A states are fully operational; A_i, $(i = 1, 2, \ldots, r)$; $A_i \in A$ (for the above example $r = \omega - g$). The class F is for system undergoing repair; F_j, $(j = 1, 2, \ldots, s)$; $F_j \in F$ corresponding to intermediate performance levels for $\Phi_\omega^*(X)$. The class C is for non-operational states; C_k, $(k = 1, 2, \ldots, c)$; $C_k \in C$. Under the deterioration level of $\Phi_\omega^*(X)$ given by $c = (\omega - r - s + 1)$, the polyfunctional system does not perform its designed requirements.

The system state space is the union $S = A \cup F \cup C$. The reliability function is given by the probability that the system will be in a fully operational state

$$\mathcal{R}(n) = \Pr\{S(n) \in A\}, \tag{5.31}$$

at time n.

5.5.2. The exit probability matrix

Consider a completely observable polyfunctional system with a semi-Markov behaviour. At each transition of the system, we can identify the system state. Generally, we may consider that two kinds of states exist for a system: the operational states $Q_k \in Q$, $(k = 1, 2, \ldots, L)$, where $Q = A \cup F$ and the trapping state, $d \in C$.

The probability of going from state $k \in Q$ to class C in t time periods is defined as an exit probability [2] and

$$\gamma_k(n) = \Pr\{S(n + u) \in C \mid S(u) = k\}. \tag{5.32}$$

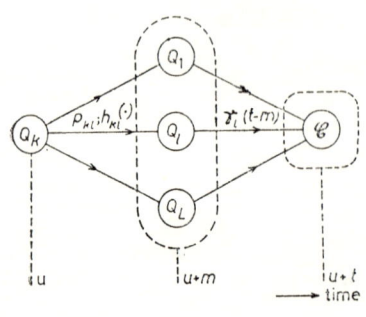

Fig. 5.19

Figure 5.19 shows an evolution diagram for such a behaviour. As was mentioned earlier in this chapter, a semi-Markov process used in the study of systems operational behaviour is given by the probability rate transition p_{ij} for the system to make a transition from state i, $i \in Q \cup C$, to state j, $j = Q \cup C$ and by the waiting time probabilities $h_{ij}(m) = \Pr\{\tau_{ij} = m\}$, $(m = 0, 1, 2, \ldots)$ for which $h_{ij}(0) = 0$ (see ref. [8], p. 578).

Using basic concepts from the theory of semi-Markov processes ([8], Chap. 15), the probability exit function (5.32) satisfies the relation

$$\gamma_k(n) = \sum_{m=0}^{n} \sum_{l \notin C} p_{kl} h_{kl}(m) \gamma_l(n - m) + \sum_{m=0}^{n} \sum_{l \in C} p_{kl} h_{kl}(m), \tag{5.33}$$

where $k = 1, 2, \ldots, L$ and $n = 0, 1, 2, \ldots$. For the case when the system accepts only one trapping state d, the above equation becomes

$$\gamma_k(n) = \sum_{l=1}^{L} p_{kl} \sum_{m=0}^{n} h_{kl}(m) \gamma_l(n - m), \tag{5.34}$$

$k = 1, 2, \ldots, L; n = 0, 1, 2, \ldots, \gamma_k(0) = 0$.

In matrix form, the exit probability of the system from an operational state Q is $\Gamma(n)$, i.e. a diagonal matrix with the dimensions $(L \times L)$ and $\gamma_k(n)$ as matrix components.

$$\Gamma(n) = \sum_{m=0}^{n} (P \,\square\, H(m))\, \Gamma(n-m), \quad (n = 0, 1, 2, \ldots), \tag{5.35}$$

where $\square$ is the symbol of the congruent multiplication of two given matrices. By particularization to a Markov process [8], we can have results similar to those given by De Mercado ([2], Theorem 3).

$$H(m) = U\,\delta(m-1),\ (m = 0, 1, 2, \ldots),\ \Gamma(n) = \sum_{m=0}^{n} (P \,\square\, U\,\delta(m-1))\, \Gamma(n-m) =$$

$$= \sum_{m=0}^{n} P\,\delta(m-1)\, \Gamma(n-m) = P\,\Gamma(n-1),$$

where U is the unity matrix (a square matrix with all elements equal with one) and $\delta(m-1)$ is a unit step function defined by

$$\delta(m-1) = \begin{cases} 1 & \text{if } m = 1 \\ 0 & \text{otherwise.} \end{cases}$$

The process described above can also be represented by using flow graph theory which, as was already mentioned, has many applications in the field of systems engineering.

Using a geometric transformation applied to equation (5.35), we can write

$$\Gamma^g(z) = [I - P \,\square\, H^g(z)]^{-1}. \tag{5.36}$$

We give now computational relations for determining the value of the reliability prediction function. Additional computations will be introduced for the failure interval transition probability.

Consider a polyfunctional system described only by operational kinds of states A and F. States in class F are lexicographically ordered — the system experiences states $F_1, F_2, \ldots, F_l, \ldots, F_s$ in the increasing order of the index l, such that $F_j \succeq F_l$. The system is degraded (see fig. 5.20); for reliability prediction we use the concept of "evolution diagram" given by Girault [3].

The partition matrix is

$$P = \left[\begin{array}{c|c} B & C \\ \hline 0 & D \end{array}\right], \tag{5.37}$$

where D has the special form (5.38) including information about the lexicographical order of states in class F. The elements of P are p_{ij} $(p_{ij} > 0,\ \sum_{j=1}^{N} p_{ij} = 1,$

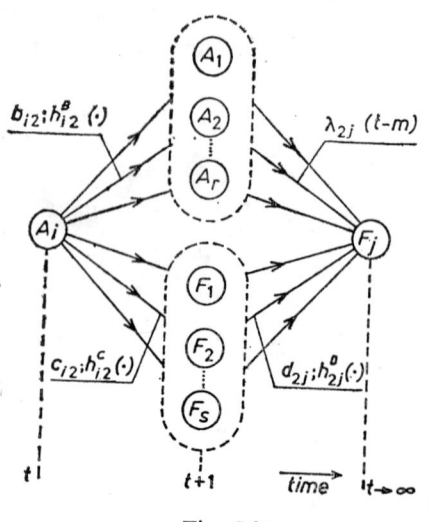

Fig. 5.20

$i = 1, \ldots, N$), i.e. the probability that the system moves from state i, ($i = 1, 2, \ldots, N$), to state j, ($j = 1, 2, \ldots, N$), in the next transition. The holding time function is $H(m)$.

$$D = \begin{bmatrix} 0 & d_{12} & \cdots & d_{1r} \\ & 0 & \cdots & d_{2r} \\ & & 0 & \vdots \\ & & & \ddots & \vdots \\ & & & & 1 \end{bmatrix}. \quad (5.38)$$

Let $\alpha(t)$ be the $(1 \times (r+s))$ vector of state probabilities defined by

$$\alpha_k(n) = \begin{cases} \alpha_i(t) = \Pr\{S(n) = A_i | \mathcal{E}(t)\}, (i=1,2,\ldots,r) \\ \alpha_j(t) = \Pr\{S(n) = F_j | \mathcal{E}(t)\}, (j=1,2,\ldots,s) \end{cases} \quad (5.39)$$

In order to calculate system reliability, we have to calculate the interval transition probabilities for the underlaying semi-Markov process and following [8, pp. 585], the interval transition probability matrix is

$$\Omega(t) = {}^{>}W(t) + \sum_{m=0}^{t} (P \square H(m)) \Omega(t-m), \quad (t = 0, 1, 2, \ldots), \quad (5.40)$$

where ${}^{>}W(t)$ is the diagonal matrix of the complementary cumulative probability distribution for the waiting time $\left({}^{>}w_i(t) = \sum_{m=t+1}^{\infty} w_i(m)\right)$, and $w_i(.)$ is the mass probability function for τ_i (waiting time in any of the states $i \in S$), such that

$$w_i(.) = \sum_{j=1}^{N} p_{ij} h_{ij}(.). \quad (5.41)$$

For a semi-Markov process

$$\alpha(t) = \alpha(0) \Omega(t-m), \quad (t = 0, 1, 2, \ldots), \quad (5.42)$$

where $\alpha(0)$ is the vector of the initial (time $t = 0$) state probabilities. By virtue of (5.31), the system reliability is given by

$$\mathcal{R}(t) = \sum_{i: i \in A, j \in A} \alpha_i(t), \quad (t = 0, 1, 2, \ldots). \quad (5.43)$$

The failure interval transition probabilities [8, 2] (the probability that the system will be in a state F at time t if at time 0 it was in a state A, $\lambda_{ij}(n) =$

$$= \Pr\{S(n) = F_j \mid S(0) = A_i\}) \quad \text{are (see also fig. 5.20)}$$

$$\lambda_{ij}(n) = \sum_{k=1}^{r} b_{ik} \sum_{m=0}^{t} h_{ik}(m) \lambda_{kj}(n-m) + \left(\sum_{l=1}^{s} c_{il} \sum_{m=0}^{t} h_{il}(m)\right) \cdot \left(\sum_{l=1}^{s} d_{lj} \sum_{m=0}^{t} h_{lj}(n-m)\right), \quad (5.44)$$

$$(i = 1, 2, \ldots, r; \; j = 1, 2, \ldots, s; \; t = 0, 1, 2, \ldots),$$

$\lambda_{ij}(0) = 0$.

By changing the summation order in relation (5.44) and by introducing the congruent multiplication notation for two matrices, relation (5.44) can be written in an $(r \times s)$-matrix formulation:

$$\Lambda(n) = \sum_{m=0}^{t} (B \boxdot H^B(m)) \Lambda(n-m) + \sum_{m=0}^{t} (C \boxdot H^C(m))(D \boxdot H^D(m)); \quad (5.45)$$

$\Lambda(0) = 0, \; \Lambda(1) = C$,

where B, C, D are the matrix representations or transition probabilities of system states using the concept of "evolution diagram" (see fig. 5.20). $H^B(.), H^C(.)$ and $H^D(.)$ are matrix representations for the mass function of waiting time probabilities in accordance with the given partition for the matrix P. From the above model, by particularization to a Markov process and using assumptions from ref. [2] for $D(D=1)$ — the system is not allowed lexicographic deterioration —

$$\Lambda(n) = \sum_{m=0}^{t} (B \boxdot U\delta(m-1)) \Lambda(n-m) + \sum_{m=0}^{t} (C \boxdot U\delta(m-1)) =$$

$$= \sum_{m=0}^{t} B\delta(m-1) \Lambda(n-m) + \sum_{m=0}^{t} C\delta(m-1) = B\Lambda(t-1) + C, \quad (5.46)$$

which is similar to [3, (14)] for $n \to \infty$ ($\Lambda = \lim \Lambda(n)$) and every element of Λ is given by $\lambda_{ij} = \Pr\{S(\infty) = F_j \mid S(n) = A_i\}$.

Applying a geometric transformation to (5.45), we have

$$\Lambda^g(z) = [I - B \boxdot H^{B_g}(z)]^{-1} \cdot [C \boxdot H^{C_g}(z)] \cdot [D \boxdot H^{D_g}]. \quad (5.47)$$

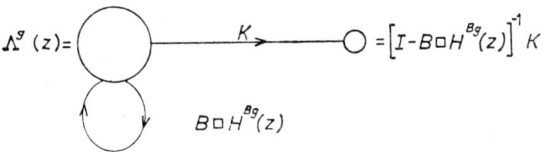

Fig. 5.21

For the above case, it is easy to see that computation of the failure interval transition probabilities rests only on the matrix B, called the "core matrix" [8, pp. 586] for our semi-Markov reliability model. In the matrix flowgraph (see fig. 5.21), K is the transmission output matrix of the process

$$K = [C \boxdot H^{C_g}(z)] \cdot [D \boxdot H^{D_g}(z)]. \quad (5.48)$$

5.5.3. Partially observable systems and availability prediction functions

In general, as was already mentioned, systems are only seldom entirely observed. For most systems (engineering, biological, etc.) we can have information about the system dynamics only when inspecting the system (making observations) [13, 15, 37].

As was mentioned earlier, a partially observable semi-Markov process is important for modelling a nuclear reactor. A BWR system is a good example, where no direct information about the state of the reactor or some subsystems is available, unless the block reactor-turbine stopped and a pertinent waiting-time is necessary before inspection and maintenance procedures (the constant of radioactivity). By defining a set of appropriate independent observations (using, for example, the control instrumentation system from the control room) and by previous experience about the reactor dynamics we can build more powerful models for reliability prediction and system availability.

In this section, the system availability will be calculated when the process deterioration experiences a semi-Markov behaviour and a finite set of observations are available. Following Rau [11, p. 239], the point availability is "defined to be the probability that the system is in an up state (i.e. either operating or operable) at a specified time" such that

$$\mathcal{A}(t) = \Pr\{S(t) \in A \mid S(0) \in \mathcal{S}\}, \quad (t = 0, 1, \ldots). \tag{5.49}$$

Let the system dynamics be described as previously, and let

$$Z_{ij} = \Pr\{S(t) = j \mid S(t-1) = i, \mathcal{E}(t)\}, \quad (i = 1, 2, \ldots, N; \, j = 1, 2, \ldots, N), \tag{5.50}$$

and $h_{ij}(m) = \Pr\{\tau_{ij} = m\}$.

An *observer* can see that a transition has happened in the performance level of the structure function (this leads to $Z_{ii} = 0$, $i = 1, 2, \ldots, N-1$), and $Z_{NN} = 1$, but he cannot identify the system state exactly. Information about the system state is available only by inspecting it, after which we can have all possible outputs from the process in a sequence of R independent observations, e.g. $\theta = 1, 2, \ldots, R$ (see fig. 5.22).

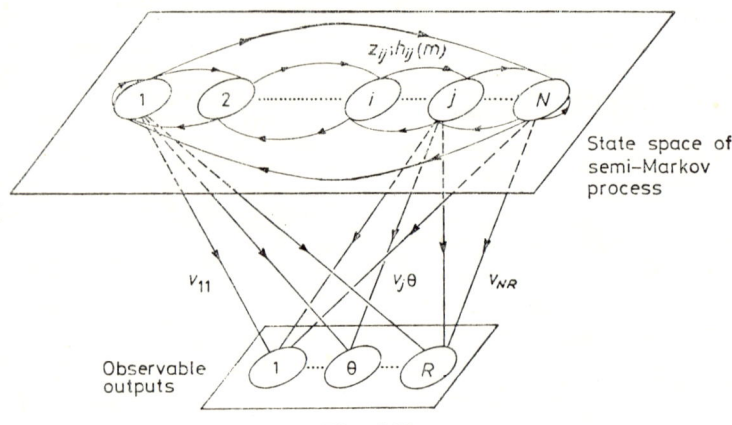

Fig. 5.22

The inspection process, which is stochastic, takes place after a transition in the system state occurred; we can define a random variable $V_{j\theta} = \Pr\{O(t) = \theta \mid S(t) = j, S(.), \tau_i = m, \mathscr{E}(.)\}$, $(\theta = 1, 2, \ldots, R)$, where $(\cdot)$ indicates a random starting point for the system before time t and before any real transition out of state i has happened. From a previous inspection, we can have information about the prior distribution of the system states as given below:

$$\alpha(.) = [\alpha_1(.), \alpha_2(.), \ldots, \alpha_N(.)]. \tag{5.51}$$

The partition in the system state space is similar to that for completely observable systems, i.e. classes A, F and C. Define

$$\pi_j^t(\alpha(.) \mid \theta, m) = \Pr\{S(t) = j_j \mid O(t) = \theta, S(.) = i, \tau_i = m, \mathscr{E}(.)\} \tag{5.52}$$

$$(j = 1, 2, \ldots, N; \; t = 0, 1, 2, \ldots).$$

Using Bayes' theorem for the statistical inference of the inspection process, we can write

$$\pi_j^t(\alpha(.) \mid \theta, m) = \frac{\sum_{i=1}^{N} \sum_{m=0}^{t} \Pr\{O(t) = \theta \mid S(t) = j, \tau_i = m, \mathscr{E}(t)\}}{\Pr\{O(t) = \theta, \tau_i = m \mid \mathscr{E}(.)\}}$$

$$\cdot \Pr\{S(t) = j, S(t-1) = i, \tau_i = m \mid \mathscr{E}(.)\}. \tag{5.53}$$

Using Bayes' theorem for the last term in (5.53), we can write

$$\Pr\{S(t) = j, S(t-1) = i, \tau_i = m \mid \mathscr{E}(.)\} = \Pr\{S(t) = j, \tau_i = m \mid S(t-1) = i,$$

$$\mathscr{E}(.)\} \Pr\{S(t-1) = i \mid \mathscr{E}(.)\}. \tag{5.54}$$

From (5.51), we can write

$$\alpha_i(0) = \Pr\{S(0) = i \mid \mathscr{E}(.)\}, \quad (i = 1, 2, \ldots, N). \tag{5.55}$$

Using basic definition of semi-Markov processes [8, Chap. 15],

$$\Pr\{S(t) = j, \tau_i = m \mid S(t-1) = i, \mathscr{E}(.)\} = \delta_{ij} \, {}^{>}w_i(t) \, \alpha_i(0) +$$

$$+ \sum_{m=0}^{t} \pi_i^{t-m}(\alpha(0) \mid \theta, m) \cdot Z_{ij} h_{ij}(m) V_{j\theta}. \tag{5.56}$$

Using (5.54), (5.55) and (5.56), Eqn. (5.53) becomes

$$\pi_j^t(\alpha(0) \mid \theta, m) = \frac{\sum_{i=1}^{N} \left\{ \delta_{ij} \, {}^{>} w_i(t) \, \alpha_i(0) + \sum_{m=0}^{t} \pi_i^{t-m}(\alpha(0) \mid \theta, m) \, Z_{ij} h_{ij}(m) \, v_{j\theta} \right\}}{\psi(\theta, m \mid \alpha(0))}$$

$$(j = 1, 2, \ldots, N; \; t = 0, 1, \ldots; \; \theta = 1, 2, \ldots, R) \tag{5.57}$$

$$\pi_i^0(. \mid .) = \alpha_i(0); \; (i = 1, 2, \ldots, N),$$

where

$$\psi(\theta, m \mid \alpha(0)) = \Pr\{O(t) = \theta, \tau_i = m \mid \mathscr{E}(0)\} =$$

$$\sum_{i=1}^{N} \sum_{j=1}^{N} \left\{ \delta_{ij} \,{}^{>} w_i(t)\, \alpha_i(0) + \sum_{m=0}^{t} \pi_i^{t-m}(.|.) Z_{ij} h_{ij}(m)\, v_{j\theta} \right\}. \qquad (5.58)$$

In a matrix $(1 \times N)$ formulation, (5.57) can be written as

$$\boldsymbol{\Pi}^t(\alpha(0) \mid \theta, m) = \frac{[\alpha(0) \,{}^{>} W(t) + \sum_{m=0}^{t} \boldsymbol{\Pi}^{t-m}(.|.)(Z \boxdot H(m)) V_\theta]}{\{\theta, m \mid \alpha(0)\}} \qquad (5.59)$$

$$\theta = 1, 2, \ldots, R; \; t = 0, 1, 2, \ldots,$$

where

$$V_\theta = \operatorname{diag}[v_{j\theta}], \; (\theta = 1, 2, \ldots, R) \qquad (5.60)$$

$$\{\theta, m \mid \alpha(0)\} = [\alpha(0) \,{}^{>} W(t) + \sum_{m=0}^{t} \boldsymbol{\Pi}^{t-m}(.|.) \mid (Z \boxdot H(m)) V_\theta] \mathbf{1}, \qquad (5.61)$$

and $\mathbf{1} = [1, \ldots, 1]^T$ has $(N \times 1)$ components;

The point availability for the system as defined by the function given in (5.49), before each observable output, is

$$\mathscr{A}^\theta(t) = \sum_{j: j \in A_j \in A} \boldsymbol{\Pi}_j^t(\alpha(0) \mid \theta, m); \; \theta = 1, 2, \ldots, R. \qquad (5.62)$$

The interval availability [11, p. 240] is denoted by $\overline{\mathscr{A}^\theta(T)}$ for the interval $[0, T]$, and prior to each observable output, it is defined by

$$\overline{\mathscr{A}^\theta(T)} = \frac{1}{T} \sum_{t=0}^{T} \mathscr{A}^\theta(t), \; (\theta = 1, 2, \ldots, R). \qquad (5.63)$$

Suppose now that the unique source of information from the process is the sequence of outputs. The state of knowledge given in (5.59) is computed after each output has been received. When the process makes a transition, a new output is generated and the change in output may be noticed.

The sequence of states of knowledge, $\boldsymbol{\Pi}^1(\alpha(0) \mid \theta, m), \boldsymbol{\Pi}^2(\alpha(1) \mid \theta, m), \ldots$, is a semi-Markov process. The state of knowledge about the process, given the observable output θ and $\mathscr{E}(.)$, in matrix formulation is

$$\boldsymbol{\Pi}^t(\alpha(t-1) \mid \theta, m) = \alpha(t). \qquad (5.64)$$

Subsequently, we obtain

$$\Pr\{\sum_{j:j\in A_j\in A} \alpha_j(t) \in \hat{G}, \tau_i = m \mid \mathscr{E}(.)\} =$$

$$= \Pr\{\sum_{j:j\in A_j\in A} \alpha_j(t) \in \hat{G}, \tau_i = m \mid \alpha(t-1), \ldots, \alpha(0)\} =$$

$$= \Pr\{\sum_{j:j\in A_j\in A} \alpha_j(t) \in \hat{G} \mid \alpha(t-1)\} = \sum_{\theta:\Pi^t(\alpha(\cdot)\mid\theta,m)\in\hat{G}} \sum_{m=0}^{t} \Pr\{O(t)=\theta, \tau_i=m\mid\mathscr{E}(.)\} =$$

$$= \sum_{\theta:\Pi^t(\alpha(\cdot)\mid\theta,m)\in\hat{G}} \sum_{m=0}^{t} \psi(\theta, m \mid \alpha(.)), \tag{5.65}$$

where $\hat{G}$ represents the confidence availability interval given by the operational states. We can interpret (5.65) as a probability that by observing the output θ, the system described by an underlying semi-Markov process is in one of the operational states.

Examples: The following examples illustrate how to obtain reliability and availability prediction figures for completely or partially observable systems with the type of behaviour introduced in the previous sections.

Case 1: Consider a completely observable system with three fully operational states and two undergoing repair states. For computation purposes the probability transition matrix is

$$P = \begin{array}{c} \\ 1 \\ 2 \\ 3 \\ \\ 4 \\ 5 \end{array} \begin{bmatrix} 1 & 2 & 3 & 4 & 5 \\ 0.8431 & 0.0092 & 0.0310 & 0.0932 & 0.0235 \\ 0.1835 & 0.6192 & 0.0927 & 0.0946 & 0.0100 \\ 0.1545 & 0.1920 & 0.4920 & 0.0932 & 0.0683 \\ \hline 0.0 & 0.0 & 0.0 & 0.0 & 1.0 \\ 0.0 & 0.0 & 0.0 & 0.0 & 1.0 \end{bmatrix}.$$

The holding-time mass functions for the system under consideration are given in table 5.2 for $p_{ij} \geq 0$. The results of a computer simulation for the reliability function $\mathscr{R}(t)$ with three different initial state vectors $\alpha(0)$ are given in fig. 5.23.

Case 2: A more general system under partial observation is considered; the probability transition matrix for $r = 3$ and $s = 2$ is

$$Z = \begin{array}{c} 1 \\ 2 \\ 3 \\ 4 \\ 5 \end{array} \begin{bmatrix} 1 & 2 & 3 & 4 & 5 \\ 0.0 & 0.856 & 0.054 & 0.051 & 0.039 \\ 0.743 & 0.0 & 0.147 & 0.048 & 0.062 \\ 0.500 & 0.248 & 0.0 & 0.102 & 0.050 \\ 0.237 & 0.129 & 0.243 & 0.0 & 0.391 \\ 0.0 & 0.0 & 0.0 & 0.0 & 1.0 \end{bmatrix}$$

Table 5.2

Days m	$h_{11}(m)$ $h_{12}(m)$	$h_{13}(m)$ $h_{14}(m)$ $h_{15}(m)$	$h_{21}(m)$ $h_{22}(m)$	$h_{23}(m)$ $h_{24}(m)$	$h_{25}(m)$ $h_{31}(m)$	$h_{32}(m)$ $h_{33}(m)$	$h_{34}(m)$ $h_{35}(m)$
1	0.0342	0.0418	0.0031	0.1821	0.1514	0.1824	0.1613
2	0.0421	0.0523	0.0049	0.1120	0.0924	0.0925	0.0949
3	0.1525	0.1828	0.1124	0.0930	0.1134	0.0947	0.2144
4	0.1145	0.1516	0.1344	0.0750	0.0824	0.1220	0.1312
5	0.2612	0.0743	0.2184	0.0820	0.0944	0.0043	0.0074
6	0.0321	0.1204	0.0039	0.1520	0.1825	0.2144	0.1085
7	0.1140	0.0985	0.1185	0.1347	0.0320	0.1320	0
8	0	0.0003	0.0040	0.0025	0.1024	0.0987	0.0943
9	0.2164	0	0.0325	0	0.0003	0	0.1291
10	0.0032	0	0	0.0444	0	0.0072	0.0520
11	0.0200	0.0643	0	0.0932	0	0	0
12	0	0	0	0	0.0954	0	0
13	0	0	0.0002	0.0092	0.0040	0.0369	0
14	0.0098	0.0749	0	0	0	0	0.0069
15	0	0.1241	0	0.0005	0	0	0
16	0	0.0143	0	0	0.009	0	0
17	0	0	0	0	0	0.0100	0
18	0	0	0.0004	0	0.0421	0	0
19	0	0	0.1524	0	0	0	0
20	0	0.0004	0.0985	0	0	0	0
21	0	0	0	0	0	0	0
22	0	0	0.1085	0	0	0	0
23	0	0	0	0.0097	0	0.0040	0
24	0	0	0	0	0	0	0
25	0	0	0	0	0	0	0
26	0	0	0	0	0.0032	0	0
27	0	0	0	0	0	0	0
28	0	0	0	0	0	0	0
29	0	0	0	0	0	0	0
30	0	0	0.0079	0.0097	0.0032	0	0

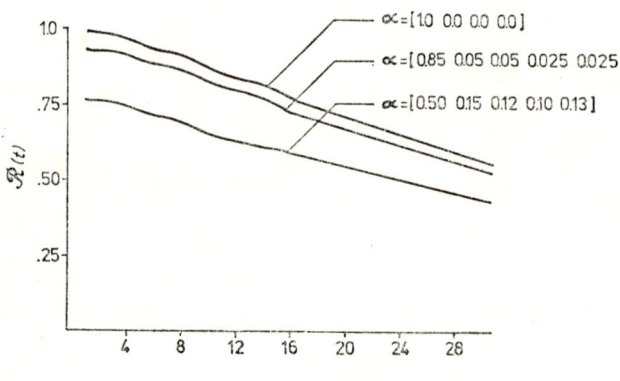

Fig. 5.23

MODELLING OF RELIABILITY, MAINTENANCE AND SAFETY

and the holding-time mass functions of the system for $z_{ij} > 0$ are given in table 5.3. The matrix of stochastic inspection is

$$V = [v_{j\theta}] = \begin{array}{c} \\ 1 \\ 2 \\ 3 \\ 4 \\ 5 \end{array} \begin{bmatrix} 1 & 2 \\ 0.7820 & 0.2180 \\ 0.9140 & 0.0860 \\ 0.8432 & 0.1568 \\ 0.4328 & 0.5672 \\ 0.6431 & 0.3569 \end{bmatrix}$$

and the initial state vector is $\alpha(0) = [0.74\ \ 0.12\ \ 0.03\ \ 0.07\ \ 0.04]$.

Computer simulation of the availability function $\mathcal{A}^\theta(t)$ for different observable outputs $\theta = 1,2$ yielded the results given in fig. 5.24 a, b. The values for interval availability are $\overline{\mathcal{A}^1(T)} = 0.6365$ and $\overline{\mathcal{A}^2(T)} = 0.6032$, respectively.

Table 5.3

Days m	$h_{12}(m)$ $h_{13}(m)$ $h_{14}(m)$	$h_{15}(m)$ $h_{21}(m)$	$h_{23}(m)$ $h_{24}(m)$ $h_{25}(m)$	$h_{31}(m)$ $h_{32}(m)$ $h_{34}(m)$	$h_{35}(m)$ $h_{41}(m)$	$h_{42}(m)$ $h_{43}(m)$	$h_{45}(m)$
1	0.0150	0.0120	0	0.0004	0.0031	0.1243	0
2	0.0180	0.0130	0	0.0071	0.0004	0.0947	0
3	0.2000	0.1600	0.0242	0	0.0005	0.1800	0.1530
4	0.3000	0.1800	0.0445	0.0095	0.0032	0.1244	0.2402
5	0.2000	0.2000	0.1894	0.1143	0.0174	0.1520	0.0825
6	0	0	0.0005	0.1324	0.1820	0.0030	0.0300
7	0	0	0.0010	0.0895	0	0	0.0170
8	0.1050	0.1310	0.2176	0.2032	0.1925	0.1004	0.1243
9	0.1140	0.1142	0.1462	0.1082	0.1220	0.0642	0.1320
10	0.0480	0.0510	0	0.0030	0.0045	0	0.1185
11	0	0.0790	0	0	0	0	0.0070
12	0	0.0598	0.2464	0.2121	0.118	0.1043	0
13	0	0	0	0	0	0.0825	0
14	0	0	0	0	0.1520	0	0
15	0	0	0	0	0.1242	0.0527	0.0955
16	0	0	0	0.0524	0	0	0
17	0	0	0	0.0433	0	0	0
18	0	0	0	0	0.0074	0	0
19	0	0	0	0	0	0	0
20	0	0	0	0	0.0310	0	0
21	0	0	0	0	0.0460	0	0
22	0	0	0	0	0.0020	0	0
23	0	0	0	0	0	0	0
24	0	0	0	0.0210	0	0	0
25	0	0	0	0	0	0	0
26	0	0	0.1000	0	0	0	0
27	0	0	0	0	0	0	0
28	0	0	0	0	0	0	0
29	0	0	0	0	0	0	0
30	0	0	0.0302	0	0	0	0.0055

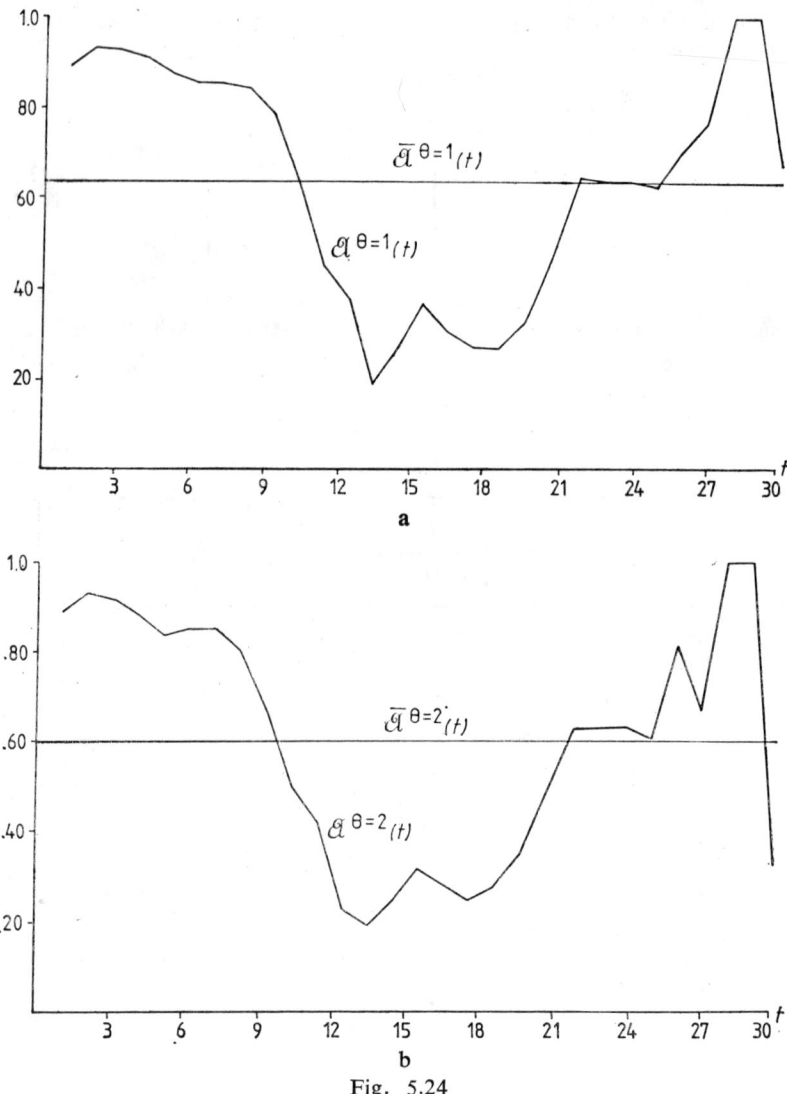

Fig. 5.24

5.6. Short-term reliability of a complex operating system

In the practice of systems engineering, the decision-maker is often interested in having a reliability criterion to be used for an adaptive control of complex systems (e.g. power systems). This use of the short-term reliability method associates a consistent criterion for actions desirable in implementation adaptive control (see Patton [179, 180, 181 and 182]). A breach of security for a complex power grid is defined to be unacceptably low voltage somewhere in the system, loss of system reliability, inadequacy of spinning generation capacity, etc. In order to evaluate the short-term reliability function of such a complex system, the transient steady-state probability analysis will be used.

The security function $\mathscr{S}(n)$ — or the short-term reliability function — is generally given by the expression

$$\mathscr{S}(n) = \sum_{i=0}^{L} p_i(n) Q_i(n), \quad (5.66)$$

where the general summation is over all possible L states of the system and $p_i(n)$ represents the probability that the complex system is in state i at time n. $Q_i(n)$ is the conditional probability that state i represents the breach of the complex system security at time n.

It must be mentioned that the system operating states are mutually independent and collectively exhaustive.

Probabilities $p_i(n)$ are calculated by using a set of $(L+1)$ equations of the following form.

$$p_i(n + \Delta n) = p_i(n) P_i(n) + \sum_{j=0}^{L} p_j(n) p_{ji}(n), \quad (i = 0, 1, 2, \ldots, L), \quad (5.67)$$

under the initial conditions that

$$p_0(0) = 1, \; p_i(0) = 0 \quad \text{for } i \neq 0. \quad (5.68)$$

In the above relation, $P_i(n)$ represents the probability that no transition exists out of state i in the time interval $(n, n + \Delta n)$; $p_{ji}(n)$ is the probability of transition from state j to state i in the time interval $(n, n + \Delta n)$.

If the "uptimes" and "downtimes" for individual components in a complex system (generators in the power grid) are exponentially distributed, then the system is given by the stationary Markov process

$$P_i(n) = 1 - \Delta n \sum_{\substack{j=0 \\ j \neq i}}^{L} \rho_{ij} \quad (5.69)$$

and

$$p_{ji}(n) = \rho_{ji} \Delta n, \quad (5.70)$$

for $\Delta n \to \varepsilon$ and ε is a small value of the time parameter n. The transition rate from state i to state j is denoted by ρ_{ij}.

In (5.67) we substituted expressions (5.69) and (5.70). By subtracting $p_i(n)$ from both sides of the resulting equation, dividing by Δn and taking the limit as $\Delta n \to 0$, we obtain the following set of differential equations.

$$\dot{p}_i(n) = -p_i(n) \sum_{\substack{j=0 \\ j \neq i}}^{L} \rho_{ij} + \sum_{\substack{j=0 \\ j \neq i}}^{L} p_j(n) \rho_{ji}, \quad (i = 0, 1, 2, \ldots, L). \quad (5.71)$$

Differential equations of the above form can be solved by numerical methods.

Considering now $(Q_i(n))$ as the conditional probability that state i becomes a breach of complex system security at time n, it will take on values in the range [0, 1]. It depends on the probability of various load levels.

In this case, the accuracy of the short-term reliability function is determined by the probabilities of various load levels at future times.

Let us consider a complex power system with g power generators running and two generators, a and b, on a standby basis [179]. The input parameters needed for calculating the short-term reliability function are λ_i, which represents forced outage rate of the i-th generator [outages/hour]; r_a, r_b are the startup rate of standby generators a and b, respectively [startups/hour], μ_i represents the forced outage repair rate of the i-th generator [repairs/hour].

The assumption is that no more than two generators can be on forced outage at the same time. The description of the operative states of the above complex system is given in table 5.4.

Table 5.4

State number	State code	State description
0	0	normal operation for the system
1	i	generator i on forced outage
2	ia	generator i on forced outage, standby generator a operating
3	ij	generators i and j on forced outage
4	ija	generators i and j on forced outage, standby generator a operating
5	ijb	generators i and j on forced outage, standby generator b operating
6	$ijab$	generators i and j on forced outage, standby generators a and b operating
7	$i(a)$	generator i and standby generator a on forced outage
8	$O(a)$	standby generator a on forced outage
9	$i(a)b$	generator i and standby generator a on forced outage, standby generator b operating

The associated set of state probability equations is

$$\dot{p}_0(n) = -p_0(n)\sum_{i=1}^{L}\lambda_i + \sum_{i=1}^{L}p_i(n)\,\mu_i + \sum_{i=1}^{L}p_{ia}(n)\,\mu_i + p_{0(a)}(n)\,\mu_a, \quad (5.72)$$

$$\dot{p}_i(n) = p_0(n)\,\lambda_i - p_i(n)\left[\mu_i + \sum_{\substack{j=0\\j\neq i}}^{L}\lambda_j + r_a\right] + \sum_{\substack{j=1\\j\neq i}}^{L}p_{ij}(n)\,\mu_j, \quad (i=1,2,\ldots,L) \quad (5.73)$$

$$\dot{p}_{ia}(n) = p_i(n)\,r_a - p_{ia}(n)\left[\mu_i + \lambda_a + \sum_{\substack{j=1\\j\neq i}}^{L}\lambda_j\right] + \sum_{\substack{j=1\\j\neq i}}^{L}[p_{ija}(n) + p_{ijab}(n)]\,\mu_j +$$

$$+ [p_{i(a)}(n) + p_{i(a)b}(n)]\,\mu_a, \quad (i=1,2,\ldots,L) \quad (5.74)$$

$$\dot{p}_{ij}(n) = p_i(n)\,\lambda_j + p_j(n)\,\lambda_i - p_{ij}(n)\,[\mu_i + \lambda_j + r_a + r_b], \quad (i=1,2,\ldots,L-1,$$
$$j=2,3,\ldots,L; \; j>i) \quad (5.75)$$

$$\dot{p}_{ija}(n) = p_{ia}(n)\,\lambda_j + p_{ja}(n)\,\lambda_i + p_{ij}(n)\,r_a - p_{ija}(n)\,[\mu_i + \mu_j + r_b]$$
$$(i=1,2,\ldots,L-1; \; j=2,3,\ldots,L; \; j>i) \quad (5.76)$$

$$\dot{p}_{ijb}(n) = p_{ij}(n)\,r_b - p_{ijb}(n)\,[\mu_i + \mu_j + r_a], \quad (i=1,2,\ldots,L-1;$$
$$j=2,3,\ldots,L; \; j>i) \quad (5.77)$$

$$\dot{p}_{ijab}(n) = p_{ija}(n)\,r_b + p_{ijb}(n)\,r_a - p_{ijab}(n)\,[\mu_i + \mu_j], \quad (i=1,2,\ldots,L-1;$$
$$j=2,3,\ldots,L=j>i), \quad (5.78)$$

$$\dot{p}_{i(a)}(n) = p_{ia}(n)\,\lambda_a - p_{i(a)}(n)[\mu_i + \mu_a + r_b] + p_{0(a)}(n)\,\lambda_i, (i=1,2,\ldots,L), \quad (5.79)$$

$$\dot{p}_{0(a)}(n) = \sum_{i=1}^{L}p_{i(a)}(n)\,\mu_i - p_{0(a)}(n)\left[\mu_a + \sum_{i=1}^{L}\lambda_i\right] + \sum_{i=1}^{L}p_{i(a)b}(n)\,\mu_i; \quad (5.80)$$

$$\dot{p}_{i(a)b}(n) = p_{i(a)}(n)\,r_b - p_{i(a)b}(n)[\mu_i + \mu_a], \quad (i=1,2,\ldots,L). \quad (5.81)$$

Differential eqn. (5.72) ÷ (5.81) can be solved numerically, provided that a set of initial conditions exists.

The above model will be applied to the case when the following assumptions were adopted [179]. The forced outage of a generating unit results in the loss of the total capacity of the unit; the times between forced outages and forced outage durations for a generating unit and the times to start standby generators are exponentially distributed; only two generating units at the most can suffer forced outages; two standby units only can be started during the time period studied.

The parameters $\mu_i, r_a, r_e, \lambda_a$, and λ_b are set to zero so as to obtain a model assuming no repair of generators on forced outage and no startup of standby generators during the period of investigation. Figure 5.25 a, b shows a 24-hour load curve forecast, under the assumption that the current time is midnight.

Computational results for the security (short-term reliability) function, using various models for calculating state probabilities are shown in fig. 5.25.

The model assumes that in all cases "normal" generation is constant over a time horizon of 24 hours and consists of the first twelve power generators

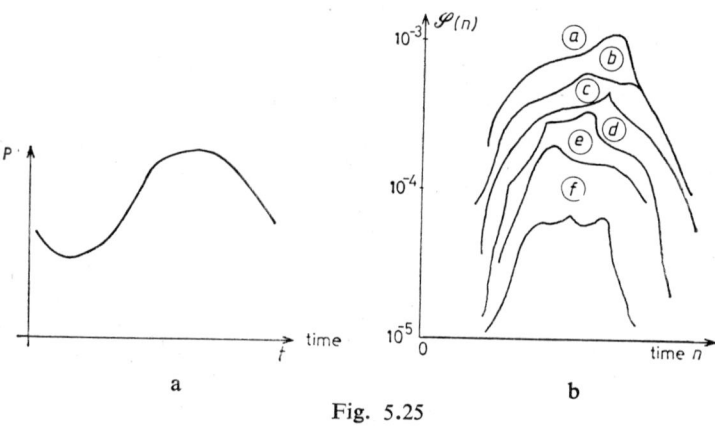

Fig. 5.25

with the characteristics given in table 5.5. The standby generators are generators 13 and 14. If generators 1—7 are presently running, the result for the short-term reliability function — the security forecast — is the solid line shown in fig. 5.26. The risk in the numerical value for function $\mathscr{S}(n)$ at 7 AM would indicate immediate connective action at the time. If an appropriate action would be to start units 8—10 at 7 AM, the plot $\mathscr{S}(n)$ resulting from this control and management action is given as a dashed line in the above figure.

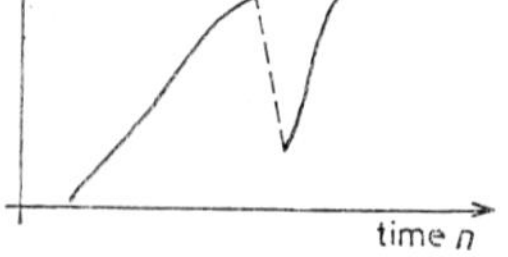

The above form for the security function (see eqn. (5.66)) is limited to consideration of steady-state breaches of security only. An alternative form of the security function which takes into account the steady state and transient breaches of security is given by the relation below (see Patton [180]).

$$\mathscr{S}(n) = \sum_{i=1}^{L} p_i(n - \Delta n) \left[1 - \sum_{\substack{j=1 \\ j \neq i}}^{L} R_{ij}(n - \Delta n), n \right] \cdot$$

$$Q_i(n) + \sum_{j=1}^{L} \sum_{\substack{i=1 \\ i \neq j}}^{L} p_j(n - \Delta n) R_{ji}(n - \Delta n, n) V_{ji}(n), \quad (5.82)$$

Fig. 5.26

where $p_i(n - \Delta n)$ represents the probability that the system is in state i at time $(n - \Delta n)$; $R_{ji}(.)$ is the probability of transition from state i to state j in the time interval $(n - \Delta n, n)$ provided that the system is in state i at time $(n - \Delta n)$; $Q_i(n)$ is the probability that state i constitutes a steady-state breach of security at time n; and $V_{ij}(n)$ is the probability that transition from state i to state j during the time interval $(n - \Delta n, n)$ results in a transient breach of security.

Table 5.5

Unit number	Capacity [MW]	Reliability parameters	
		λ_i [outages/hour]	μ [repairs/hour]
1	520	0.001 940	0.0250
2	440	0.000 456	0.0500
3	230	0.000 296	0.0556
4	230	0.000 296	0.0556
5	220	0.000 296	0.0556
6	162	0.000 285	0.0667
7	162	0.000 285	0.0667
8	111	0.000 137	0.0172
9	111	0.000 137	0.0172
10	111	0.000 137	0.0172
11	111	0.000 137	0.0172
12	67	0.000 217	0.0269
13	66	0.000 217	0.0269
14	66	0.000 217	0.0269
15	47	0.000 285	0.0156
16	47	0.000 285	0.0156
17	46	0.000 285	0.0156
18	27	0.000 285	0.0156

If, for practical reasons, we eliminate the implicit requirement that a state persists for Δn in order to constitute a steady-state breach of security, then for $V_{ji}(n) = 0$

$$\mathscr{S}(n) = \sum_{i=1}^{L} \left[p_i(n) \, Q_i(n) + \sum_{\substack{j=1 \\ j \neq i}}^{L} p_j(n - \Delta n) \, R_{ji}(n - \Delta n, n) \, V_{ji}(n) \right] \quad (5.83)$$

if $Q_i(n) = 1$.

Extended examples for calculating the short-term reliability of a power system, using the above equation, could be found in refs. [180], [181] and [182].

5.7. Mathematical foundations for safety analysis

5.7.1. Definition of a fault tree

The concept of *fault tree* is used to describe the logical interrelations between the TOP event and the BASIC events of the system. It can be represented as a finite directed graph without directed circuits. In a graph theoretical approach to the fault tree analysis, each vertex (node or basic event) may be in one of two operational states: true or false. Each gate AND, OR in the fault tree is described by a monotone boolean logic function. This function specifies the states of the gate (output event) in terms of its input predecessors. The basic events in the fault tree are considered independent of each other (independent variables).

Let us consider a fault tree with n basic events and let $z = (z_1, z_2, \ldots, z_n)$ be the vector of basic event outcomes, where

$$z_j = \begin{cases} 1 & \text{if basic event } j \text{ has occurred} \\ 0 & \text{otherwise} \end{cases}$$

and $\Psi(z) = \begin{cases} 1 & \text{if the TOP event occurs} \\ 0 & \text{otherwise.} \end{cases}$

$\Psi(.)$ is the boolean indicator function for the TOP event in a given fault tree.

The same as in the definition accepted for coherent structures, a basic event i is irrelevant to $\Psi(.)$ when, for all z, $\Psi(1_i, z) = \Psi(0_i, z)$, and $(1_i, z) = (z_1, z_2, \ldots, z_{i-1}, 1, z_{i+1}, \ldots, z_n)$, $(0_i, z) = (z_1, z_2, \ldots, z_{i-1}, 0, z_{i+1}, \ldots, z_n)$. Otherwise, an event i is relevant.

Assuming that a fault tree is a coherent structure (i.e. it does not have any irrelevant events), we can notice that

$$X \leq z \Rightarrow \Psi(X) \leq \Psi(z).$$

The boolean indicator function $\Psi(z)$ can be determined from either the min cut or min path sets.

DEFINITION 5.3. A *cut set* is a set of basic events whose occurrence causes the TOP event to occur.

DEFINITION 5.4. A *minimal cut set* is any cut set minimal with respect to being a cut.

If $\mathcal{K}$ is the family of min cut sets in a fault tree, then its cover is the set of relevant events. If $\mathcal{K} = (K_1, K_2, \ldots, K_k)$ is the set of min cuts of the system, then $\Psi(z) = \coprod_{l=1}^{k} \coprod_{i \in K_l} z_i$.

If $\mathscr{P} = (P_1, P_2, \ldots, P_p)$ denotes the family of min paths in the fault tree, then $\Psi(z) = \prod_{s=1}^{p} \coprod_{i \in P_s} z_i$, where $\coprod_{i=1}^{n} y_i = 1 - \prod_{i=1}^{n}(1 - z_i)$.

For a fault tree with no replications among min cut sets or min path sets, if the basic events are statistically independent, we can write, respectively,

$$\Pr\{\text{TOP EVENT}\} = \prod_{1 \leq l \leq k} \coprod_{i \in K_l} q_i, \quad \Pr\{\text{TOP EVENT}\} = \coprod_{1 \leq s \leq p} \prod_{i \in P_s} q_i.$$

The above relations are not valid in general. In cases when the basic events are statistically independent, the bounds for the reliability indicator of a fault tree are as is shown in table 5.6.

Table 5.6

No.	Evaluation for $\Pr\{\text{TOP EVENT}\}$	Observations
1	$\coprod_{1 \leq s \leq p} \prod_{i \in P_s} q_i \leq \Pr\{\text{TOP EVENT}\} \leq \prod_{1 \leq l \leq k} \coprod_{i \in K_l} q_i$	Basic events are statistically independent
2	$\Pr\{\text{TOP EVENT}\} = \Pr\left\{\bigcup_{l=1}^{k} E_l\right\}$	E_l is the event that all basic events in min cut set K_l occur. All basic events are statistically independent.
3	$\Pr\{\text{TOP EVENT}\} = \sum_{r=1}^{k} (-1)^{r-1} S_r$	$S_r = \sum_{1 \leq i_1 < i_2 < \cdots < i_r < k} \Pr\{E_{i_1} \cap E_{i_2} \cap \cdots \cap E_{i_r}\}$ Basic events are statistically independent
4	$\Pr\{\text{TOP EVENT}\} \leq S_1 = \sum_{s=1}^{k} \prod_{i \in K_s} q_i$ $\Pr\{\text{TOP EVENT}\} \geq S_1 - S_2$ $\Pr\{\text{TOP EVENT}\} \leq S_1 - S_2 + S_3$	— ,, —

If events in the fault tree are associated (positively dependent), the bounds on the probability of the TOP EVENT are given by

$$\max_{1 \leq l \leq k} \prod_{i \in K_l} q_i \leq \Pr\{\text{TOP EVENT}\} \leq \min_{1 \leq s \leq p} \prod_{i \in P_s} q_i. \tag{5.84}$$

The symbols used in the technique of fault tree construction are given in figs. 5.27 a and b.

A more specific fault tree diagram is represented in fig. 5.27c. It should be mentioned that this resultant fault tree lends itself to quantification and, thus, a quantitative description of the frequency of occurrence of the undesired event is possible as well as judgments about which contributing events are quantitatively and/or qualitatively important.

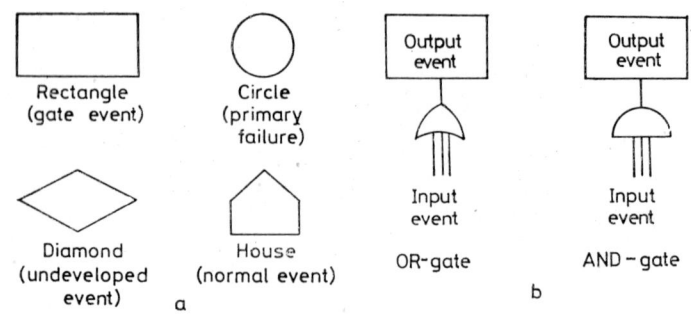

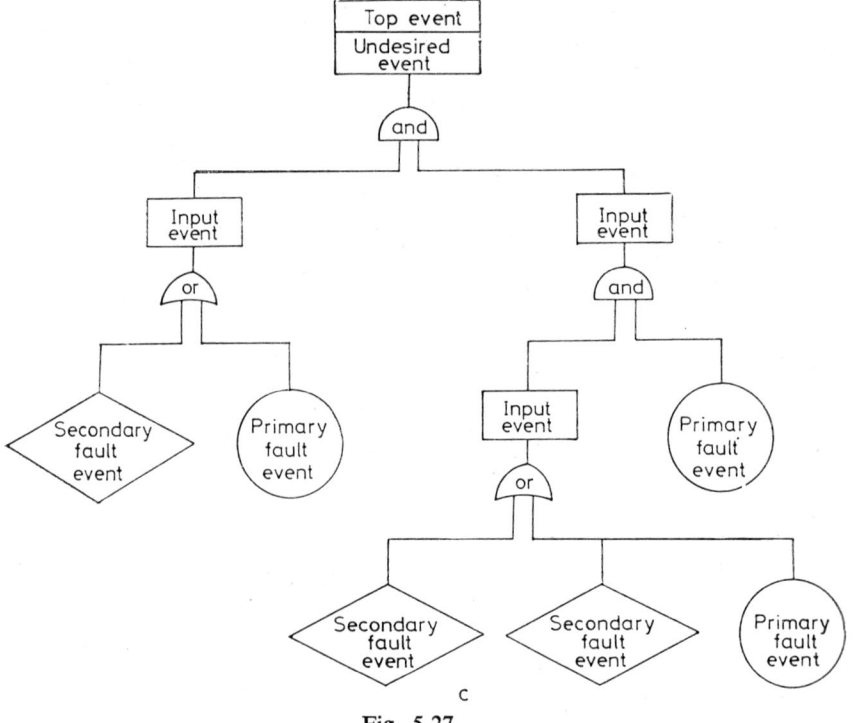

Fig. 5.27

The quantitative and qualitative analysis of complex systems *via* a fault tree allows us to identify
  1) How the system fails,
  2) The functional relationship of system failure,

3) The level of resilience covered by the used design concept;
4) Alternative concepts of basic design for different complex systems;
5) An overview picture concerning specific aspects of system operation and failure;
6) Comprehensive Cause-Effect descriptions;
7) Numerical quantification of the complex system failures;
8) Measures against the unknown.

In his study [266] of fault tree methodology for complex situations, Young emphasized that 1) The fault tree analysis methodology is interactive in its nature; it should begin by trying to provide a basic understanding which can be expanded as the process continues; 2) The fault tree process and analysis answer questions about failure importance, minimal cut sets, etc.; 3) The fault tree analysis is useful for the systems engineering analyst, when used to increase understanding and evaluation of the system; it also has an engineering value when fault tree diagrams are obtained.

5.7.2. Algorithms for evaluating min cut sets in a fault tree analysis

Few algorithms are presented in the literature (see [257], [258], [259]) for computing min cut sets in a fault tree. As was mentioned by Chatterjee [234] "it is neither possible nor desirable to obtain by inspection all the min cut set of fault trees with hundreds of gates and basic events". We shall present now two such algorithms — one "upward" another "downward" — which were developed by Chatterjee. In the literature the "downward" algorithm is referred to as MOCUS (see Fussell [219]).

The DOWNWARD-algorithm starts from the top gate and expands minimally. Consequently, "it goes down the tree and stops when all gates are expanded to basic events" [234]. Before using this algorithm, we must index the gates from the bottom such that gates which have only basic events as inputs have the lowest indices.

If $\mathcal{M} = \{M_1, M_2, \ldots, M_n\}$ is a family of sets and $\widetilde{\mathcal{M}} = \{\widetilde{M}_1, \widetilde{M}_2, \ldots, \widetilde{M}_n\}$ is the family of minimal sets of $\mathcal{M}$ for Π (the top gate index), then the DOWNWARD-algorithm is executed in five steps.

Step I: Set $\widetilde{\mathcal{M}} = \{(\Pi)\}$;

Step II: At any general step, we have a set $\widetilde{\mathcal{M}} = \{\widetilde{M}_i; i = 1, 2, \ldots, n\}$. A gate I is associated with a set $\widetilde{M}_i$ (the last set in the collection) and with a logical function which expresses its output in terms of inputs.* If $(I_1, I_2, \ldots, I_n)$ are inputs to the logical function, then a family of sets $\mathcal{N}$ out of $\widetilde{M}_i$ is constructed such that $\mathcal{N} = \{(\widetilde{M}_i - I) \cup \mathcal{B}_l\}$, where $\mathcal{B}_l$ is one of the min cut sets of the logic function

---

* The logical function is completely specified by the min cut sets.

of I on $(I_1, I_2, \ldots, I_k)$. It is clear that the cardinality of $\mathcal{N}$ is the same with that of the min cut sets of the logical function of I.

Step III*: Set $\mathcal{M} = \{\widetilde{\mathcal{M}} - M_i\} \cup \mathcal{N}$;
Step IV: Reduce the set $\mathcal{M}$ to a minimal family $\widetilde{\mathcal{M}}$;
Step V: GO TO Step II if $\mathcal{M}$ is a cover of $\widetilde{\mathcal{M}}$. Otherwise STOP.

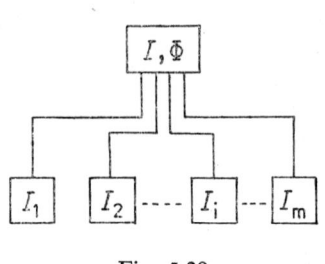

Fig. 5.28

The UPWARD-algorithm starts at the bottom of the tree and develops upward such that "at every step... we find out the min cut sets of the intermediate events defined by a gate" [234].

From fig. 5.28, we can see that at a general step for a gate I, the immediate inputs are $I_1, I_2, \ldots, I_m$.

We shall accept the following notation.
a) $\widetilde{\mathcal{M}}_{I_j} = (1_{\widetilde{M}_j}, 2_{\widetilde{M}_j}, \ldots, p_{j\widetilde{M}_j})$ denote the family of min cut sets of the event I_j; p_j denotes the number of min cut sets of event I_j; b) $\mathbb{B} = (\mathcal{B}_1, \mathcal{B}_2, \ldots, \mathcal{B}_q)$ denote the family of min cut sets which is defined on $(I_1, I_2, \ldots, I_m)$; and c) $\mathcal{M}_I = \{\bigcup_{j \in \mathcal{B}} \widetilde{M}_j \mid \mathcal{B} \in \mathbb{B}; \widetilde{M}_j \in \widetilde{\mathcal{M}}_j\}$ denote the family of binary indicated event sets of I.

In an alternative notation, we have a tractable expression such that
$$\mathcal{M}_I = \{1_{M_I}, 2_{M_I}, \ldots, l_{M_I}, \ldots\}, \quad l_{M_I} = \bigcup_{i \in \mathcal{B}_r} j_i \widetilde{M}_I,$$
where $l = \sum_{k=1}^{r=1} (\prod_{j \in \mathcal{B}_k} p_j) + \prod_{i \in \mathcal{B}_r} j_i$, $\vee j_i$, $(j_i = 1, 2, \ldots, p_i; r = 1, 2, \ldots, q)$.

A binary indicated cut set (BICS) is a set of basic events in an artery set. It is the minimal set of components which causes the top event to occur through a specific artery in the fault tree graph. It is to be mentioned that BICS are not necessarily minimal cut sets, because of possible replication of gates and basic events.

In his work [234], Chatterjee gives an answer to the problem "given a family of subsets, how do you find the family of minimal sets?", and evaluated methods to identify whether a BICS is a min cut set (MCS) or not.

Procedure 1: Order the sets in the increasing order of cardinalities. For BICS of size $-r$ we have to find out whether any of the MCS of size i ($1 \leqslant i \leqslant \leqslant r - 1$) is included in it or not.** If $(a_1, a_2, \ldots, a_k)$ is a min cut set and we want to check $(b_1, b_2, \ldots, b_n)$ BICS, then assume that for $k < n$, $a_1 < a_2 < \ldots < a_k$, $b_1 < b_2 < \ldots < b_n$.

Let us define now BICS *vs*. MCS by the following algorithm (see Chatterjee [234]).

---

* This step indicates that the set M_i must be replaced by the family of sets which takes the analysis one step down the tree.

** First, check element by element if smaller set elements are contained in it or not. Arrange these sets in increasing order.

Step I: $a_1 < b_1$. If YES, then BICS under inspection is an MCS. STOP If not then GO TO Step II.

Step II: Find i such that $a_1 \geq b_i$ and i is the smallest such number. If $a_1 = b_p$, $p < n$, then GO TO Step III. Otherwise STOP. If $p = n$, then BICS is not an MCS for $k=1$.

Step III: For any general step we have to identify if $a_r \in (b_{m+1}, \ldots, b_n)$, where $r \leq k$. Repeat Steps I and II. At Step III, set $r = r + 1$ and $m = p$ as found in Step II.

Procedure 2: A BICS fails to become an MCS, if and only if a proper subset of it is already an MCS or a cut set.

A rough examination of BICS with respect to various bounds on min cut sets leads to the following results: 1) the number of BICS and maximum size of BICS are upper bounds on the number of min cut sets and maximum size of min cut sets; 2) the number of BICS of size $-r$ is the upper bound to the number of min cut sets of size $-r$; 3) a lower bound LB for the number of min cut sets of size $-r$ is given by $LB = \max\left\{0, \left\{\text{number of BICS of size } r - \sum_{i=1}^{r-1} \text{number of min cut sets of size } i\right\}\right\}$; 4) the actual storage requirement for a computer analysis of the DOWNWARD and UPWARD algorithms may have a sharper bound given by $\sum_{r=1}^{n} r \times$ (number of BICS of size $-r$).

5.7.3. Modular representation of fault trees

Owing to the complexity of large-scale trees, which involve a large number of gates and basic events, the systems engineer can use a modular decomposition concept. Birnbaum and Esary explored for the first time the concept of module in the framework of a general theory of coherent structures. Such a decomposition is possible even for a fault tree because coherent structures and fault trees are mathematically equivalent.

DEFINITION 5.5. A *module* of a fault tree is a set of basic events M, together with an indicator function χ_M such that $\Psi(z) = \Gamma[\chi_M(z)^M, z^{M'}]$, where Γ is non-decreasing and z^M implies that the coordinates of z are restricted to M.

LEMMA 5.1. *Decompose a fault tree into statistically independent modules. If* a) *we can find a modular decomposition* $\{(M_1, \chi_1), \ldots, (M_r, \chi_r)\}$ *such that* $\chi_1(z), \ldots, \chi_r(z)$ *are statistically independent, and* b) z_i *for* $i \in M_s$, $s = 1, 2, \ldots, r$, *may be associated (positively dependent), then*

$$\Pr\{\Psi(z) = 1\} = g[\Pr\{\chi_1(z) = 1\}, \ldots, \Pr\{\chi_r(z) = 1\}] \leq g_\Gamma^{\cdot}[u_{\chi_1}(q), \ldots, u_{\chi_r}(q)],$$

where $u_{\chi_s}(q)$ *is the* min-max *upper bound (see eqn. (5.84)) for module* M_s *and* g_Γ. *is the expected value of* $\Gamma[\chi_1, \chi_2, \ldots, \chi_r]$.

The finest modular representation for a fault tree is given by the following algorithm.

Step 1: Obtain the easy modules; combine them and obtain the family of min cut sets.

Step 2: Obtain the min cut disjoint modules, which are disjoint from the rest of the min cut sets.

Step 3: For each such module as found in Step 2, obtain the min path disjoint module.

Step 4: If a decomposition is executed in Step 3, then GO TO Step 2. Otherwise, GO TO Step 5. Do this for each module, considered as a system, as obtained in Step 3.

Step 5: Obtain the lowest level k out of n gates. Combine them and treat them as one event.

Step 6: Given the family of min cut sets for the fault tree, obtain the k out of n modules that appear at the bottom of the finest modular representation.

Step 7: If the graph representation for the fault tree is totally disjointed then the system is prime. STOP. Otherwise, obtain the connected subgraphs of the graph.

Step 8: Do this for each connected subgraph of the graph. Obtain the essential graph and a set R for a connected subgraph of the graph.

Step 9: If R is the empty set, then the system is prime. STOP.

Step 10: Obtain the asymmetric primes at the lowest level of the system under consideration. If found, combine them and go to Step 5. Otherwise, go to Step 8 for other connected subgraphs of the graph.

5.7.4. Fault tree analysis using techniques of boolean difference

Owing to its specific structure, a fault tree can be treated using the concept of boolean difference [218]. The boolean difference $\mathbb{E} = \mathbb{E}(x_1, x_2, \ldots, x_i, \ldots x_n)$ with respect to literal x_i is defined by

$$\frac{d\mathbb{E}}{dx_i} = \mathbb{E}(x_1, \ldots, x_{i-1}, 0, x_{i+1}, \ldots, x_n) \oplus$$

$$\oplus \mathbb{E}(x_1, \ldots, x_{i-1}, 1, x_{i+1}, \ldots, x_n) = \mathbb{E}\big|_{x_i=0} \oplus \mathbb{E}\big|_{x_i=1},$$

where $\oplus$ denotes the exclusive OR operation.

In the concept of boolean difference a fault tree is a coherent structure such that

$$\frac{d\mathbb{E}}{dx_i} = \begin{cases} 0 & \text{if } \mathbb{E} \text{ is unconditionally independent of } x_i; \\ 1 & \text{if } \mathbb{E} \text{ is unconditionally dependent on } x_i; \\ \text{an expression in which } x_i \text{ is missing otherwise,} \end{cases}$$

and the value of x_i has no effect on the value of $\mathbb{E}$ when $d\mathbb{E}/dx_i = 0$ and always affects the value of $\mathbb{E}$ when $d\mathbb{E}/dx_i = 1$.

For the general case when $d\mathbb{E}/dx_i$ is an expression, each solution to the constraint $d\mathbb{E}/dx_i = 1$ yields a set of values of the remaining literals such that the value of $\mathbb{E}$ is determined solely by the value of x_i. For a given fault tree and its associated failure expression, the boolean difference "provides an analytical technique for determining the dependence of an entire network on the status of a specified unit" [218]. The probability that x_i fails and induces system failure is given by

$$\Pr\{x_i \cdot d\mathbb{E}/dx_i\} = \Pr\{x_i\} \cdot \Pr\{d\mathbb{E}/dx_i\},$$

and the probability that a network is critically dependent on component x_i is given by the probability that $d\mathbb{E}/dx_i = 1$.

A higher order boolean difference investigates multiple component and subsystem failures and is defined as

$$\frac{d\mathbb{E}}{d(x_i, x_j)} = \mathbb{E}\,|_{x_i=x_j=0} \oplus \mathbb{E}\,|_{x_i=x_j=1}.$$

The boolean difference indicator provides the analytical framework for calculating the dependence of an entire network on the status of a specified component or subsystem. The probability that x_i fails and induces system failure is given by

$$\Pr\{x_i \cdot d\mathbb{E}/dx_i\} = \Pr\{x_i\}\,\Pr\{d\mathbb{E}/dx_i\}.$$

For a simple network as that given in fig. 5.29, the corresponding fault tree is given in fig. 5.30.

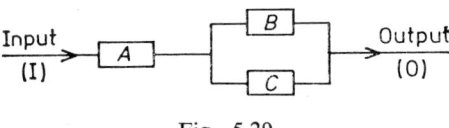

Fig. 5.29

If event $\mathbb{E} = 1$ denotes network failure, then for the fault tree given above we can write $\mathbb{E} = A + BC$.

The boolean difference for the above expression is

$$\frac{d\mathbb{E}}{dA} = \mathbb{E}\,|_{A=1} \oplus \mathbb{E}\,|_{A=0} = (1 + BC) \oplus (0 + BC) = 1 + BC$$

$$d\mathbb{E}/dA = 1\,(\overline{BC}) + \overline{1}\,(BC) = (\overline{BC}) + 0 = \overline{B} + \overline{C}.$$

The solutions to the equation

$$\frac{d\mathbb{E}}{dA} = \overline{B} + \overline{C} = 1$$

are $\overline{B} = 1$, $\overline{C} = 1$ or $\overline{BC} = 1$.

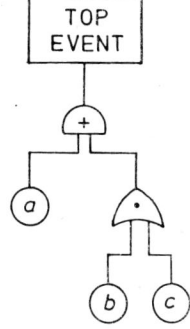

Fig. 5.30

The meaning of these solutions is that the network status is critically dependent on the status of A when either $B\,(B=0$ or $B=1)$ or $C(C=0, \overline{C}=1)$

has not failed or neither of them has ($\bar{B} = \bar{C} = 1$). When elements B and C have failed, the status of A cannot influence network operation.

Similarly, $\dfrac{d\mathbb{E}}{dB} = \bar{A}C$; $\dfrac{d\mathbb{E}}{dC} = \bar{A}B$.

If components B and C are viewed as a subsystem, then $\dfrac{d\mathbb{E}}{d(B, C)} = (A + 0.0) \oplus (A + 1.1) = A \oplus 1 = \bar{A}$.

The probability that A is critical to the network is $\Pr\{d\mathbb{E}/dA\} = \Pr\{\bar{B} + \bar{C}\} = 1 - bc$, and the probability of A failure and simultaneous network failure is given by $\Pr\{A\} \Pr\{d\mathbb{E}/dA\} = a(1 - bc)$.

For component B, we can write $\Pr\{d\mathbb{E}/dB\} = \Pr\{\bar{A}\,C\} = (1 - a)\,c$, $\Pr\{B\} \Pr\{d\mathbb{E}/dB\} = b(1 - a)\,c$, and, for component C, $\Pr\{d\mathbb{E}/dC\} = (1 - a)\,b$, $\Pr\{C\} \Pr\{d\mathbb{E}/dC\} = c(1 - a)\,b$.

5.7.5. Fault tree analysis and measures of event importance

In engineering practice, specialists are often faced with statements like "component a is more important than component b". The same statement is valid for min cut sets, subsystems, etc. We shall deal now with the concepts of marginal, structural, competitive and diagnostic importance.

The measure of *marginal importance* deals with the change in the system reliability due to some specified change in component reliability performance. In several respects the marginal importance is a measure of the system resilience; it is "a measure of the weakening or strengthening of the system due to a change in component reliability".

If $h(\cdot)$ is the reliability function, then the marginal importance of component i is given by

$$I_h^m(i) = \frac{\partial h(p)}{\partial p_i} = h(1_i, p) - h(0_i, p)$$

with the following properties:

a) $0 \leqslant I_h^m(i) \leqslant 1$; b) $I_h^m(i)$ does not depend on p_i; c) $I_h^m(i)$ is non-decreasing in p_j when $j \notin \{\bigcup_{P \in P_i} P\}$, where P_i is a family of min path sets containing i; d) $I_h^m(i)$ is non-increasing in p_j when $j \in \{\bigcap_{P \in P(i)} P\}$.

The problem of evaluating $I_h^m(i)$ consists in evaluating $h(p)$ by using bounds on $h(1_i, p)$ and $h(0_i, p)$ to bound $I_h^m(i) = h(1_i, p) - h(0_i, p)$.

The following computational results are obtained.
a) For statistically independent components,

$$I_h^m(i) \leq \sum_{P_r \in P(i)} (\prod_{R \in b[P-P_r]} \prod_{j \in R} p_j) \cdot \prod_{\substack{j \in P_r \\ j \neq i}} (1 - p_j),$$

where $b[\cdot]$ denotes the blocking clutter *.

b) For independent events, the marginal importance for the event i is given by

$$I_h^m(i) \leq \sum_{P_r \in P_i} [\min_{P_s \in P - P_r} \{\prod_{j \in P_s} p_j\}] \cdot \prod_{\substack{j \in P_r \\ j \neq i}} (1 - p_j).$$

c) For systems with associated components (basic events), the marginal importance is given by

$$I_h^m(i) \leq \sum_{P_r \in P(i)} [\min_{P_s \in P - P_r} \{\prod_{i \in P_s} p_i\}] \cdot [\prod_{\substack{j \in P \\ j \neq i}} (1 - p_i)].$$

The *structural importance* of a component is due to its location in the structure and is defined as

$$I_\psi^m(i) = \frac{n(i)}{2^{n-1}} = \frac{1}{2^{n-1}} \sum [\Psi(1_i, z) - \Psi(0_i, z)],$$

where $n(i)$ represents the number of critical cut sets of i and is given by

$$n(i) = \sum_z [\Psi(1_i, z) - (0_i, z)].$$

The *competitive importance* for a component i, $I_h^c(i)$, is given by the probability that i causes system failure when the system eventually fails.
In a mathematical formulation,

$$I_h^c(i) = \int_0^\infty [h(1_i, \overline{F(t)}) - h(0_i, \overline{F(t)})] \cdot dF_i(t)$$

with the properties that $0 \leq I_h^c(i) \leq 1$ and $\sum I_h^c(i) = 1$ and $F_i(t)$ represents the failure distribution.

The structural importance is given by the relation

$$I_\psi^c(i) = \sum_r \frac{1}{n} \frac{n_r^*(i)}{\binom{n-1}{r-1}},$$

---

* A family R on a finite set N is called a clutter if $R \neq \emptyset$, $R \neq \{\emptyset\}$ and no element of R is properly contained in any other element of R.

where $n_r^*(i) = n_{n-r+1}(i)$ denotes the number of critical path sets in the fault tree graph.

The structural competitive importance for a min cut set K is given by

$$I_\psi^c(K) = \frac{1}{n} \sum_{r=1}^n \frac{|K| n_r |K|}{\binom{n-1}{r-1}}.$$

Chatterjee [234] proved that

$$I_\psi^c(i) \geq \sum_{K \in K(i)} \frac{I_\psi^c(K)}{|K|}.$$

The *diagnostic importance* for a component i, $I_h^d(i)$, is defined as the probability that component i is failed with one of its min cuts, given that the system is failed such that

$$I_h^d(i) = \frac{\Pr\left\{\begin{array}{l}\text{component } i \text{ has failed with at least}\\ \text{one of its min cuts}\end{array}\right\}}{\Pr\{\text{system failed}\}}$$

and $0 \leq I_h^d(i) \leq 1$, $I_h^d(i) \leq \sum_{K \in K(i)} I_h^{dc}(K)$, where

$$I_h^{dc}(K) = \frac{\Pr\{\text{min cut set } K \text{ occurred}\}}{\Pr\{\text{system failed}\}}.$$

The bounds on $I_h^d(i)$ are as follows.

a) For a system with statistically independent events,

$$\frac{\prod_{P \in b[K(i)]} \prod_{i \in P} p_i}{\prod_{K \in K} \prod_{i \in K} p_i} \leq I_h^d(i) \leq \frac{\prod_{K \in K(i)} \prod_{i \in K} p_i}{\prod_{P \in P} \prod_{i \in P} p_i}.$$

b) For associated basic events,

$$\frac{\{\max_{K \in K(i)} (\prod p_i)\}}{\{\min_{P \in P} (\prod_{i \in P} p_i)\}} \leq I_h^d(i) \leq \frac{\{\min_{P \in b[K(i)]} (\prod_{i \in P} p_i)\}}{\{\max_{K \in K} (\prod_{i \in K} p_i)\}}.$$

The results concerning the concept of importance in fault tree analysis are reviewed in table 5.7.

Table 5.7

Concept of importance	Symbol	Range	Computational relation	Computability
Marginal importance	$I_h^m(i)$	$0 \leq I_h^m(i) \leq 1$	$\dfrac{n(i)}{2^{n-1}}$	Possible to obtain bounds
Competitive importance	$I_h^c(i)$	$0 \leq I_h^c(i) \leq 1$ $\sum_{i=1}^n I_h^c(i) \leq 1$	$\sum_{r=1}^n \dfrac{n_r(i)}{n\binom{n-1}{r-1}}$	Difficult but can be simulated
Diagnostic importance	$I_h^d(i)$	$0 \leq I_h^d(i) \leq 1$	$C(i)/C$; $C(i)$ is the number of cut sets containing at least one min cut set of i and C is the total number of cut sets for the system.	Easy to obtain bounds

5.7.6. A linguistic representation of a fault tree

Olman and Worrell [236] presented a linguistic representation of a fault tree that is similar to that used in the definition of the ALGOL computer language. Worrell [237] presented a similar approach where the characters are those of a standard FORTRAN language.

DEFINITION 5.6. Each event of a fault tree must be given a name.

DEFINITION 5.7. A declarator consists of a unique combination of alphabetic characters followed by the sign ($). Declarators and their explanation are given in table 5.8.

DEFINITION 5.8. In a linguistic representation, a fault tree consists of a fault tree header followed by a number of event definitions.

DEFINITION 5.9. A similar tree is a logically connected sequence of events, each of which is defined in the input language. It is used as a model from which one or more similar sequences of events are generated.

Throughout the remainder of this chapter the complement will be denoted by ($\neg$), the union by ($\cup$), and the intersection by ($\cap$).

In an ALGOL approach used to describe a fault tree, the following basic symbols are used (see also table 5.8):

$\langle$basic symbol$\rangle$ :: = $\langle$letter$\rangle$| $\langle$digit$\rangle$| $\langle$special symbol$\rangle$| $\langle$delimiter$\rangle$
$\langle$letter$\rangle$ :: = A|B|C|D|E|F| ... |X|Y|Z|
$\langle$digit$\rangle$ :: = 0|1|2|3 ... |8|9|

⟨special symbol⟩::= — ⟨delimiter⟩::= ⟨logical operator⟩| ⟨separator⟩| ⟨bracket⟩| ⟨declarator⟩
⟨logical operator⟩:: ∪ | ∩ | ¬
⟨separator⟩::= , | ·|$| COMMENT $|= |
⟨bracket⟩::(|)
⟨declarator⟩::⟨gate declarator⟩| ⟨primary event declarator⟩ | ⟨relationship declarator⟩| ⟨fault tree declarator⟩
⟨gate declarator⟩::= ⟨standard gate declarator⟩| ⟨special gate declarator⟩
⟨standard gate declarator⟩:: AG $ |OG$| EOG$| PAG$| IG$
⟨special gate declarator⟩:: SG$
⟨primary event declarator⟩:: BE$| EE$| DE$| UE$| CE$
⟨relationship declarator⟩:: IN$| OUT$| SI$| SO$
⟨fault tree declarator⟩::= FAULT TREE $

Table 5.8

Declarator	Explanation
FAULT TREE$	Fault tree header
A G $	AND gate definition
O G $	OR gate definition
E O G $	EXCLUSIVE OR gate definition
P A G $	Priority AND gate definition
I G $	INHIBIT gate definition
B E $	Basic event definition
C E $	Conditioning event definition
S G $	Special gate definition
E E $	External event definition
I N $	Input declaration
O U T $	Output declaration
S I $	Similar input declaration
S O $	Similar output declaration

Letters are used for forming names, digits have no individual meaning and a special symbols $ is used to form names in the same manner as letters and digits. Delimiters have a fixed meaning which sometimes is obvious or else will be given in the definition of the language. We shall define a "comment" for including text among the symbols of a fault tree. A given sequence of basic symbols may be replaced by a period without affecting the fault tree.

The sequence .COMMENT$ ⟨any sequence not containing $⟩ $ is equivalent to a period.

We can use names for the identification of fault trees and the events that define the fault tree such that
⟨name symbol⟩::= ⟨letter⟩| ⟨digit⟩| ⟨special symbol⟩
⟨name⟩::= ⟨name symbol⟩| ⟨name⟩ ⟨name symbol⟩.

As is already known, a fault tree is made up of events which indicate the presence of various failures which may occur in the system being modeled. Information about related events in the tree is contained in lists.
⟨event⟩::= ⟨primary event⟩| ⟨intermediate event⟩
⟨list⟩::= ⟨intermediate event list⟩| ⟨event list⟩
⟨event name⟩::= ⟨name⟩
⟨primary event⟩::= ⟨event name⟩
⟨intermediate event⟩::= ⟨event name⟩.

A primary event represents either the existence of a given condition (CE$, EE$) or the failure of a component of the analysed system which cannot be (BE$) or has not been (DE$, UE$) represented in the model.

An intermediate event could be considered to be the output of a gate and represents the failure of a larger part of the system. It can also be defined as a logical combination of primary events.

Intermediate event lists are used for output declaration. They ensure that primary events in a fault tree are not given as output events from an event.
⟨intermediate event list⟩::= ⟨intermediate event⟩| ⟨intermediate event list⟩, ⟨intermediate event⟩
⟨event list⟩::= ⟨event⟩| ⟨event list⟩, ⟨event⟩.

Logic expressions explicitly specify the logic to be used in combining the events which are given as input:
⟨logic primary⟩::= ⟨event⟩| (⟨logic expression⟩)
⟨logic secondary⟩::= ⟨logic primary⟩| ¬ ⟨logic primary⟩
⟨logic term⟩::= ⟨logic secondary⟩| ⟨logic term⟩| ∩ ⟨logic secondary⟩
⟨logic expression⟩::= ⟨logic term⟩| ⟨logic expression⟩ ∪ ⟨logic term⟩.

We can use declarations to assign names to the events in a fault tree and also to define the logic required to obtain the output intermediate event from the input events.

⟨relationship declaration⟩::= ⟨input declaration⟩|
⟨output declaration⟩| ⟨similar input declaration⟩|
⟨similar output declaration⟩
⟨declaration⟩::= ⟨primary event declaration⟩|
⟨intermediate event declaration⟩| ⟨relationship declaration⟩.

Declarations can also be used to specify how an event is related to other events in the fault tree.

In order to assign a name and a type to the defined primary event, we can use a primary event declaration.

⟨primary event declaration⟩: : = ⟨primary event declarator⟩ ⟨primary event⟩.

An intermediate event declaration assigns a name to an intermediate event. It also specifies implicitly (through the type) or explicitly (with a logic expression) the logical combination of input events which define the output event.

⟨standard intermediate event declaration⟩: : = ⟨standard gate declarator⟩ ⟨intermediate event⟩

⟨special intermediate event declaration⟩: : = ⟨special gate declarator⟩ ⟨intermediate event⟩ = ⟨logic expression⟩

⟨intermediate event declaration⟩ :: = ⟨standard intermediate event declaration⟩ ⟨special intermediate event declaration⟩

The intermediate event is defined by the AND gate, OR gate or the EXCLUSIVE OR gate.

The relationship declarations will specify other events of the fault tree that are logically related to the event being defined.

⟨input declaration⟩: : = IN$ ⟨event list⟩

⟨output declaration⟩: : = OUT$ ⟨intermediate event list⟩

⟨prefix⟩: : = ⟨name⟩

⟨similar input declaration⟩: : = SI$ ⟨prefix⟩| ⟨event list⟩

⟨similar output declaration⟩: : = SO$ ⟨prefix⟩| ⟨intermediate event list⟩

⟨input declaration list⟩: : = ⟨input declaration⟩| ⟨input declaration list⟩ · ⟨input declaration⟩

⟨output declaration list⟩: : = ⟨output declaration⟩| ⟨output declaration list⟩ · ⟨output declaration⟩

⟨similar input declaration list⟩: : = ⟨similar input declaration⟩| ⟨similar input declaration list⟩ · ⟨similar input declaration⟩

⟨similar output declaration list⟩: : = ⟨similar output declaration⟩| ⟨similar output declaration list⟩ · ⟨similar output declaration⟩

⟨relationship list⟩: : = ⟨input declaration list⟩| ⟨input declaration list⟩ · ⟨output declaration list⟩

⟨similar relationship list⟩: : = ⟨input declaration list⟩ · ⟨similar output declaration list⟩| ⟨input declaration list⟩ · ⟨similar input declaration list⟩ · ⟨output declaration list⟩| ⟨similar input declaration list⟩ · ⟨output declaration list⟩.

A similar tree is defined as a logically connected sequence of events. Each event is given in the input language. A generated sequence of events can be connected to the fault tree only through the output from the top or bottom of the intermediate event.

An event definition is expressed as

⟨event definition⟩: : = ⟨primary event definition⟩|⟨intermediate event definition⟩

⟨primary event definition⟩ : : = ⟨primary event declaration⟩ · ⟨output declaration list⟩.

The primary event of any fault tree is completely defined by a primary event definition.

Intermediate events of a fault tree are completely defined by an intermediate event definition.

⟨standard intermediate event definition⟩: : =
⟨standard intermediate event declaration⟩ · ⟨relationship list⟩|
⟨standard intermediate event declaration⟩ ·
⟨similar relationship list⟩

⟨special intermediate event definition⟩ : : =
⟨special intermediate event declaration⟩ · ⟨relationship list⟩| ⟨special intermediate event declaration⟩ ·
⟨input declaration list⟩ ·
⟨similar output declaration list⟩

⟨intermediate event definition⟩: : = ⟨standard intermediate event definition⟩| ⟨special intermediate event definition⟩

As was already emphasized, "a fault tree is a mathematical model of a real or imaginary system which is suitable for probabilistic analysis. A fault tree is composed of events which represent the way the various primary events of the system are combined".

⟨fault tree name⟩: : = ⟨name⟩
⟨fault tree header⟩: : = FAULT TREE$ ⟨fault tree name⟩ ⟨fault tree body⟩: : = ⟨event definition⟩| ⟨fault tree body⟩ · ⟨event definition⟩

The fault tree header provides the name by which the fault tree is to be identified while the fault tree body provides the definitions of all the events in the fault tree.

Finally, the linguistic representation of a fault tree is expressed by

⟨fault tree⟩: : = ⟨fault tree header⟩ · ⟨fault tree body⟩.

To achieve the symbolic manipulation of set equations, we can use the Set Equation Transformation System (SETS) developed by Worrell at Sandia Laboratories [237]. According to Worrell, "The Set Equation Transformation System allows the generation of set equations directly, or by logical combination of other set equations through a process of substitution. It also provides for the reduction of set equations by the application of set identities, and the permanent retention of set equations for use at a later time. The operations allowed in an equation are the set operations of intersection, union and complement. Thus, the system provides a comprehensive capability for generating and manipulating set equations symbolically. Moreover, since the processing that can be accomplished using SETS is valid for any Boolean algebra, the system is useful for processing the logic equations derived from fault trees."

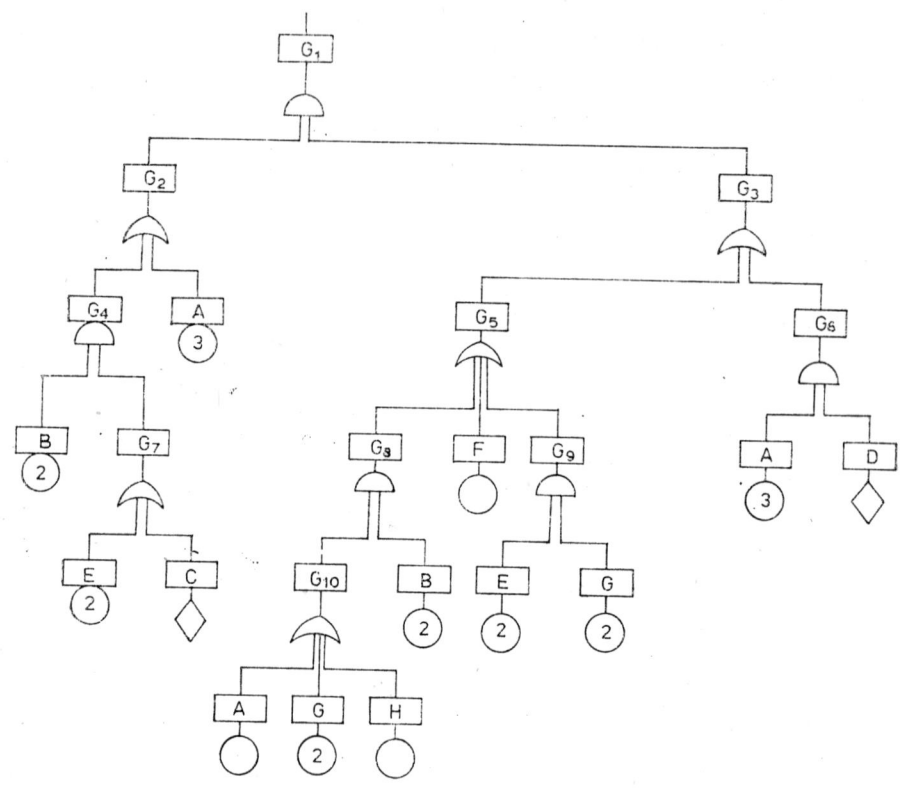

Fig. 5.31

In what follows, we shall explore a linguistic representation of a fault tree. Let the fault tree in fig. 5.31 be given. In order to find the fundamental ways in which the intermediate event G1 can occur, the first step is to build up a linguistic representation of it. The block set equations for the above fault tree are

$G1 = G2 \wedge G3$ $G6 = A \wedge D$
$G2 = G4 \vee A$ $G10 = A \vee G \vee H$
$G3 = G5 \vee G6$ $G7 = E \vee C$

G 4 = B ∧ G 7   G 8 = G 10 ∧ B
G 5 = G 8 ∨ F ∨ G 9   G 9 = E ∧ C.

The linguistic representation is given below.

FAULT TREE
AG$ G 1.   IN$ G 2, G 3.
OG$ G 2.   IN$ G 4, A. OUT$ G 1.
OG$ G 3.   IN$ G 5, G 6. OUT$ G 1.
BE$ A.   OUT $ G 2, G 6, G 10.
AG$ G 4.   IN$ B, G 7. OUT$ G 2.
OG$ G 5.   IN$ G 8, F, G 9. OUT$ G 3.
AG$ G 6.   IN$ A, D. OUT$ G 3.
BE$ B.   OUT$ G 4, G 8.
OG$ G 7.   IN$ E, C. OUT$ G 4.
UE$ D.   OUT$ G 6.
AG$ G 8.   IN$ G 10, B. OUT$ G 5.
BE$ F.   OUT$ GS.
AG$ G 9.   IN$ G, E. OUT$ G 5.
BE$ E.   OUT$ G 7, G 9.
UE$ C.   OUT$ G 7.
OG$ G 10.   IN$ A, G, H. OUT$ G 8.
BE$ G.   OUT$ G 9, G 10.
BE$ H.   OUT$ G 10.

The table of literal occurrence is

literal	number of occurrences
A	4
B	7
C	3
D	1
E	4
F	3
G	3
H	2

We can see that there are eight different literals in the set equation for G 1— RED (Reduce Equation procedure), given in a factored form.

G 1 — RED = B ∧ ((H ∨ G ∨ F) ∧ (E ∨ C) ∨ A) ∨ A ∧ (E ∧ G ∨ E ∨ D)

In an expanded form G 1 — RED can be written as in the table below.

Term number	Number of literals	G_1 – RED
1	2	$A \wedge B$ ∨
2	2	$A \wedge F$ ∨
3	2	$A \wedge D$ ∨
4	3	$B \wedge E \wedge H$ ∨
5	3	$B \wedge E \wedge G$ ∨
6	3	$B \wedge E \wedge F$ ∨
7	3	$B \wedge C \wedge H$ ∨
8	3	$B \wedge C \wedge G$ ∨
9	3	$B \wedge C \wedge F$ ∨
10	3	$A \wedge E \wedge G$

This form is similar to that of cut sets in the given fault tree.

5.7.7. Automated construction of fault trees

The construction of a fault tree is a time consuming activity. Additionally, a given engineering system has not a unique representation in a fault tree form. Fussell [173] gives a method for synthesizing, together with proper editing, a fault tree for system-independent component failure information (component failure transfer functions).

In a methodology for automated construction of fault trees a system and its associated boundary conditions must be defined. This methodology is valid principally for electrical systems. The electrical system is represented into the computer as an associated schematic diagram. The diagram indicates wiring layout, mechanical coupling between components and the correlation between system components and the appropriate catalogued library information.

DEFINITION 5.10. A system boundary condition defines the situation for which the fault tree is to be constructed. The TOP event is the most important condition.

Fussell attempted to construct an automatic fault tree as is shown below.

a) the initial configuration of the system is described by additional system boundary conditions;

b) the configuration must represent the system in the unfailed state;

c) the system boundary conditions depend on the TOP event;

d) the initial conditions are system boundary conditions that define the operating condition of the system;

e) all components that have more than one operating state generate an initial condition;

f) the existing system boundary conditions include any fault event which is declared to exist;

g) in certain cases, partial development of the TOP event is also required as a system boundary condition.

DEFINITION 5.11. A component failure transfer function describes one mode of failure for a component and is considered as a mini-fault tree; the set of these functions determines the degree of resolution of the final fault tree.

A failure transfer function consists of seven parts:

1) *output event* — the mode of failure being considered;

2) *output logic gate* — the logic with which the failure transfer function is coupled into the fault tree with other appropriate failure transfer functions having the same output event;

3) *internal events* — fault events requiring further logical development within the failure transfer function;

4) *internal logic gates* — the logical development of the internal events as required by the output and input events;

5) *input events* — they can be either primary events or undeveloped fault events;

6) *discriminator* — it designates which failure transfer functions may coexist in the final fault tree;

7) *coordinator* — indicates which failure transfer function in a given set is to be used in the fault tree.

A *component coalition scheme* is a procedure "for determining which components share an alliance with respect to a current flow. The component coalitions are determinable from the system schematic diagram alone and are independent of the particular components involved, with the exception of the power supplies" [173].

First-order fault events are developed manually to the level of higher order fault events and are used only as TOP events.

Second-order fault events are fault events that state a fault condition of the system that extends beyond any single component.

Third-order fault events are fault events that cause a component to "behave failed" because subsystems, not simply another individual components, are causing the component to behave failed.

Fourth-order fault events are fault events that result in component C behaving failed because other components have direct input to component C.

In describing how the computer constructs fault trees, Fussell emphasized that "The automated fault tree construction technique is similar to conventional fault tree construction techniques in that it starts with the TOP event and the development then proceeds through intermediate gates to the primary failure of the components. The construction is complete when the terminal events of every branch are primary events."

A few computer programs are known to have been written in FORTRAN IV for constructing fault trees ([251], [252]) (for the case of electrical systems see [173]). A flowchart for such a computer code (DRAFT) is given in fig. 5.32.

5.7.8. Vesely model for fault tree evaluation

In the Vesely model for fault tree evaluation, known as "Kinetic Tree Theory", it is assumed that the primary events in the fault tree are independent [16]. One primary failure may happen at any number of places in the fault tree; those primary failures which are unique are assumed to be independent. An assumption in the above model is that the mode failures (critical paths) of the fault tree are known and can be calculated using the previously described methods.

If we consider only a primary failure of the fault tree, then we can define the following basic data: $\lambda(t)\,dt$ represents the probability of the failure occurring

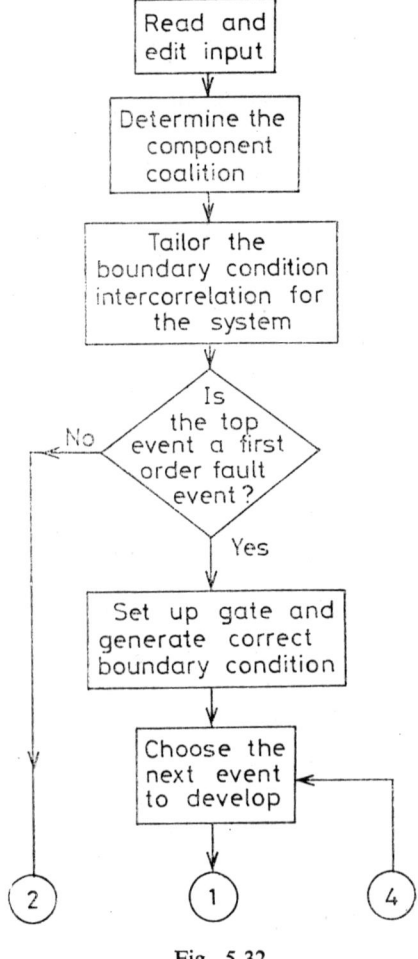

Fig. 5.32

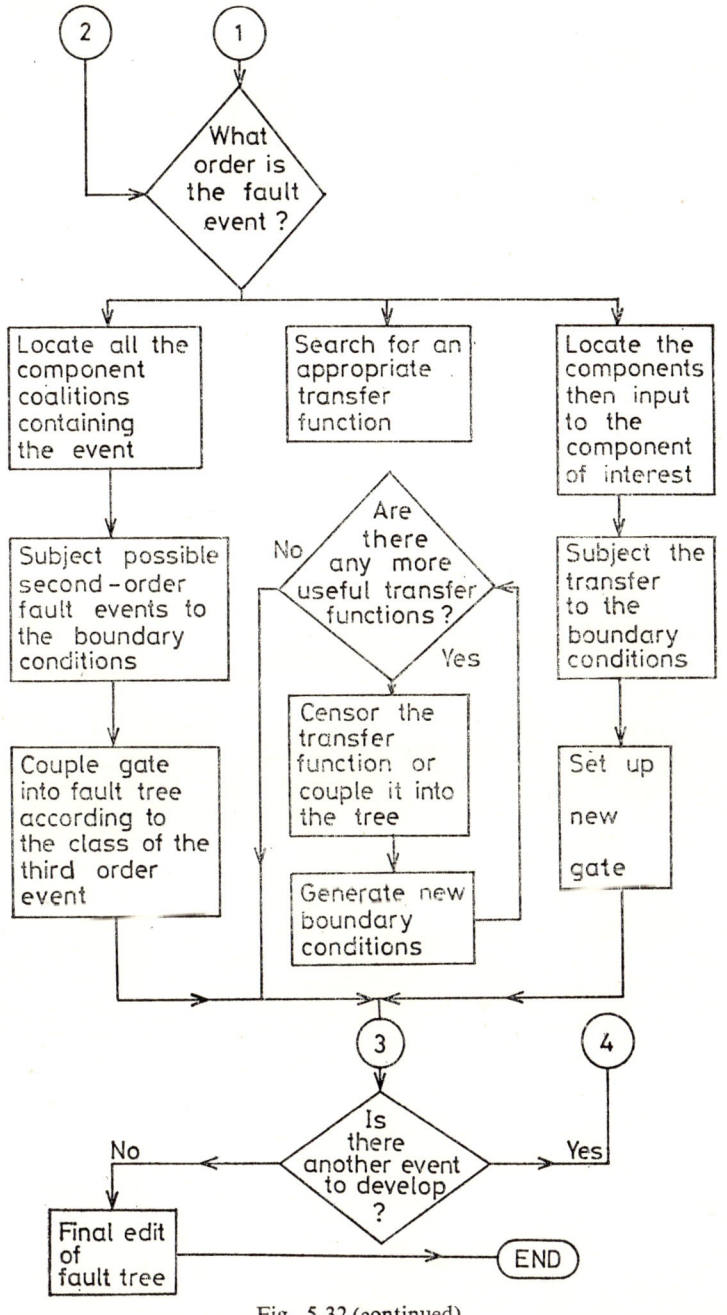

Fig. 5.32 (continued)

in the period $(t; t + dt)$, given that the failure does not exist at time t; $\mu(t)dt$ is the probability of the failure being repaired (maintained) in the period $(t; t + dt)$, given that the failure exists at time t.

The quantities $\lambda(t)$ and $\mu(t)$ represent the failure rate and repair rate for the primary failure, respectively, and they are assumed to be known for every primary failure of the fault tree. We shall now define the following quantities:

a) the probability of primary failure first occurring in the period $(t; t + dt)$, given that it is not existing at time t', is $a(t', t) = \exp\left\{-\int_{t'}^{t} \lambda(t'') \, dt''\right\} \lambda(t) \, dt$, $t' \leq t$;

b) the probability that the primary failure does not occur from time t' to time t is $f(t', t) = \exp\left\{-\int_{t'}^{t} \lambda(t'') \, dt''\right\}$, $t' \leq t$;

c) the probability that the primary failure is repaired at time t to $(t + dt)$, given that it exists at time t', is $b(t', t) \, dt = \exp\left\{-\int_{t'}^{t} \mu(t'') \, dt''\right\} \mu(t) \, dt$, $t' \leq t$.

The following primary failure characteristics are also essential for a fault tree evaluation. $w(t)$ represents the expected number of times the primary failure occurs at time t per unit time; $q(t)$ is the probability of the primary failure existing at time t.

The expected number of times for the primary failure to occur in the period (t', t) is given by $w(t', t) = \int_{t'}^{t} w(t'') \, dt''$.

For a considered fault tree if the initial condition that at $t = 0$ the primary failure does not exist, then the quantity $w(t)$ is given by the relation $w(t) = a(0, t) +$
$+ \int_{0}^{t} w(t'') \, dt'' \int dt' \cdot b(t'', t') \cdot a(t', t)$.

If the failure happened to be non-repairable (e.g. $b(t', t) = 0$), then for the above equation we can write $w(t) = a(0, t)$.

The value of $w(t)$ can also be calculated using the $q(t)$-characteristic such that $w(t) = [1 - q(t)] \lambda(t)$. In this case, $w(t) = 1 - w(t)/\lambda(t)$.

Analysing the above situation, we can see that quantities $w(t)$ and $q(t)$ as well as $\lambda(t)$, $\mu(t)$, $a(\cdot)$, $b(\cdot)$ and $f(\cdot)$ are complete functions which characterize the probabilistic behaviour of the failure for all time.

A mode failure which has been previously defined as a min cut set represents a "compounded" type of failure which consists of n primary failures (primary events).

The probability that the mode failure exists at time t, $Q(t)$, is given by $Q(t) = \prod_{i=1}^{n} q_i(t)$, where $q_i(t)$ is the existence probability for the i-th primary failure of the min cut set (mode failure).

The probability that the mode failure does not exist at time t is $P(t) = 1 - Q(t)$, and it can be determined from the primary failure information.

It is clear that if the mode failure exists at time t, then the top failure (TOP event) also exists at time t. Using this model we can identify those critical mode failures by which the TOP event is most likely to occur.

If $\Lambda(t)$ represents the mode failure rate, then it is defined as the probability of the mode failure occurring in period $(t; t + \mathrm{d}t)$ given that the mode failure does not exist at time t. $\Lambda(t)\,\mathrm{d}t = \Pr\{d_{t+\mathrm{d}t}|\,u_t\}$, where $d_{t+\mathrm{d}t}$ is the event of the mode failure existing at time $(t + \mathrm{d}t)$, and u_t is the event of the mode failure not existing at time t.

Using results from the basic theory of probability, the above equation becomes

$$\Lambda(t)\,\mathrm{d}t = \frac{\Pr\{d_{t+\mathrm{d}t}, u_t\}}{\Pr\{u_t\}},$$

where $\Pr\{d_{t+\mathrm{d}t}, u_t\} = \sum_{i=1}^{n} w_i(t)\,\mathrm{d}t \prod_{\substack{l=1 \\ l \neq i}}^{n} q_l(t)$ and $\Pr\{u_t\} = 1 - Q(t)$.

The computational relation for $\Lambda(t)$ becomes

$$\Lambda(t) = \frac{\sum_{i=1}^{n} w_i(t) \prod_{\substack{l=1 \\ l \neq i}}^{n} q_l(t)}{1 - Q(t)}.$$

Using the above value for the mode failure rate, the first occurrence distribution for the mode failure $A(t', t)$ can be written as $A(t', t) = \exp\left\{-\int_{t'}^{t} \Lambda(t'')\,\mathrm{d}t''\right\}$ $\Lambda(t)$, $t' \leqslant t$.

The non-occurrence probability measure for the mode failure $F(t', t)$ is given by $F(t', t) = \exp\left\{-\int_{t'}^{t} \Lambda(t'')\,\mathrm{d}t''\right\}$, $t' \leqslant t$, and represents the probability of the mode failure not-occurring in the time interval (t', t).

If $W(t)$ represents the expected number of times the mode failure occurs at time t per unit time, then

$$W(t) = [1 - Q(t)]\,\Lambda(t),$$

or, in terms of the primary failure information,

$$W(t) = \sum_{i=1}^{n} w_i(t) \prod_{\substack{l=1 \\ l \neq i}}^{n} q_l(t).$$

The quantities $W(t)$, $\Lambda(t)$ and $Q(t)$ show the effects of maintenance activities and environmental impact on any particular mode failure (min cut set).

Using the above results, the systems engineer can determine the characteristics of the top failure in a fault tree.

Consider that the mode failures are already calculated and there are N such mode failures. Let $Q_0(t)$ be the probability that the top failure (TOP event) exists at time t with the complement $(1 - Q_0(t))$, i.e. the probability that the top failure is not existing at time t. If d_i represents the event that the i-th mode failure exists at time t, then $\Pr\{d_i\} = Q_i(t)$.

TOP event exists if and only if one or more mode failures exist so that

$$Q_0(t) = \Pr\left\{\bigcup_{i=1}^{N} d_i\right\},$$

where $\bigcup$ denotes the union of events.

The quantity $Q_0(t)$ can be calculated with the relation

$$Q_0(t) = \sum_{i=1}^{N} \Pr\{d_i\} - \sum_{i=2}^{N}\sum_{j=1}^{i-1} \Pr\{d_i d_j\} + \ldots + (-1)^{N-1} \Pr\{d_1, d_2, d_3 \ldots d_N\}.$$

If we consider the simultaneous existence of m min cut sets given by the general event $\{d_1 d_2 \ldots d_N\}$, as the primary failures are assumed independent, then $\Pr\{d_1 d_2 d_3 \ldots d_N\} = \prod^{+1,\ldots,m} q(t)$, where $\prod^{1,\ldots,m}$ represents the product of unique primary failure quantities where the primary failure occurs in at least one of the mode failures $1,\ldots,m$.

Using the above considerations, quantity $Q_0(t)$ becomes

$$Q_0(t) = \sum_{i=1}^{N} Q_i(t) - \sum_{i=2}^{N}\sum_{j=1}^{i-1} \prod^{+i,j} q(t) + \ldots + (-1)^{N-1} \prod^{+1,\ldots,N} q(t).$$

For large-scale fault trees, assuming a large number of primary events, the quantity $Q(t)$ reaches the "steady state", in other words, constant values in a very short computer time.

As Vesely mentioned, "for these situations, $Q_0(t)$ need only be calculated as a function of time until it assumes its respective steady state value, or this steady state value can only be calculated using these steady state values for the $q(t)$ of the primary failures" [16].

We can write a sequence of bounds on $Q_0(t)$ such that

$$Q_0(t) \leq \sum_{i=1}^{N} Q_i(t)$$

$$Q_0(t) \geq \sum_{i=1}^{N} Q_i(t) - \sum_{i=2}^{N}\sum_{j=1}^{i-1} \prod^{+i,j} q(t).$$

$$\vdots$$

Using this approximation technique, $Q_0(t) \leq 1 - \prod_{i=1}^{N}(1 - Q_i(t))$, which gives an upper limit, i.e. a safe and conservative estimate for $Q_0(t)$.

The top failure intensity $W_0(t)$ represents the expected number of times the top failure occurs at time t per unit time so that

$$W_0(t)\,dt = \Pr\left\{A \bigcup_{i=1}^{N} \theta_i\right\},$$

where $\bigcup_{i=1}^{N} \theta_i$ is the event of one or more of the θ_i occurrences (the event of the i-th mode failure occurring in period $(t;\, t+dt)$).

The event $A = u_1 u_2 \ldots u_N$ is an event of all the mode failures not existing at time t (u_i represents the event of the i-th mode failure not existing at time t).

$$W_0(t)\,dt = \Pr\left\{\bigcup_{i=1}^{N} \theta_i\right\} - \Pr\left\{B \bigcup_{i=1}^{N} \theta_i\right\},$$

where $\Pr\left\{\bigcup_{i=1}^{N} \theta_i\right\} = \sum_{i=1}^{N} \Pr\{\theta_i\} - \sum_{i=2}^{N}\sum_{j=1}^{i-1} \Pr\{\theta_i\theta_j\} + \ldots + (-1)^{N-1}\Pr\{\theta_1\theta_2\ldots\theta_N\}$

and $\Pr\{\theta_1\theta_2\ldots\theta_N\} = W(t;\,1,\ldots,m)\,dt \prod^{+1,\ldots,m} q(t)$.

The above product is the product of the existence probabilities of primary failures other than the k common primary failures. The quantity $W(t;\,1,\ldots m)$ represents the failure intensity for a mode failure which has as its primary failures the primary failures which are common members of all the mode failures $1, 2, \ldots, m$.

The following quantities are required for the calculation of $W_0(t)\,dt$:

$$\Pr\left\{\bigcup_{i=1}^{N} \theta_i\right\} = \sum_{i=1}^{N} W_i(t)\,dt - \sum_{i=2}^{N}\sum_{j=1}^{i-1} W(t;\,i,j)\,dt \times \prod^{+i,j} q(t) +$$

$$+ \sum_{i=3}^{N}\sum_{j=2}^{i-1}\sum_{k=1}^{j-1} W(t;\,i,j,k)\,dt \times \prod^{+i,j,k} q(t) \ldots + (-1)^{N-1}W(t;\,1,\ldots,N)\,dt \times \prod^{+1,\ldots,N} q(t)$$

and $\Pr\left\{B \bigcup_{i=1}^{N} \theta_i\right\} = \sum_{i=1}^{N} \Pr\{\theta_i B\} - \sum_{i=2}^{N}\sum_{j=1}^{i-1}\Pr\{\theta_i\theta_j B\} + \ldots + (-1)^{N-1}\Pr\{\theta_1\theta_2\ldots\theta_N B\},$

where $B = \bigcup_{i=1}^{N} d_i$.

In the above relation, $\Pr\{\theta_1\theta_2\ldots\theta_m X\}$ is the probability of m min cut sets simultaneously occurring in period $(t;\, t+dt)$ with one or more of the other mode failures already existing at time t (event X).

Let us define $\Pr\{\theta_1\theta_2\ldots\theta_m X\} = W_x(t;\,1,2,\ldots,m)\,dt$, where $W_x(\cdot)$ represents the rate of occurrence of the m mode failures, simultaneously occurring at t with one or more of the other mode failures already existing at time t.

With this notation we can write

$$\Pr\left\{B\bigcup_{i=1}^{N}\theta_i\right\} = \sum_{i=1}^{N} W_B(t;i)\,dt \cdot \sum_{i=2}^{N}\sum_{j=1}^{i-1} W_B(t;i,j)\,dt + \ldots +$$
$$+ (-1)^{N-1} W_B(t;1,\ldots,N)\,dt,$$

and because event B denotes a union of events, the above equation includes

$$W_B(t;1,\ldots,m)\,dt = \sum_{i=1}^{N} \Pr\{\theta_1, \theta_2, \ldots, \theta_m d_i\} - \sum_{i=2}^{N}\sum_{j=1}^{i-1} \Pr\{\theta_1 \ldots \theta_m d_i d_j\} +$$
$$+ \ldots + (-1)^{N-1} \Pr\{\theta_1 \ldots \theta_m d_1 d_2 \ldots d_N\}.$$

The general term $\Pr\{\theta_1 \ldots \theta_m d_1 d_2 \ldots d_N\}$ in the above expression can be calculated by the relation

$$\Pr\{\theta_1 \ldots \theta_m d_1 \ldots d_n\} = W(t;1,\ldots,m-1,\ldots,n)\,dt \prod_{1,\ldots,m}^{1,\ldots,n} q(t).$$

In the above relation, the term $W(t;1,\ldots,m-1,\ldots,n)$ is defined as the failure intensity for a mode failure whose primary failures are the primary failures common to all m mode failures $(1,\ldots,m)$ deleted from any of the mode failures $1,\ldots,n$.

The product symbol $\prod_{1,\ldots,m}^{1,\ldots,n}$ is defined as the product of unique primary failure quantities, where the primary failure is a member of any of the mode failures $1,\ldots,n$ or is a member of the mode failures $1,\ldots,m$, but it is not a common member of these m mode failures.

The probability of the m mode failures simultaneously occurring in period $(t; t+dt)$ is finally given by

$$W_B(t;1,\ldots,m)\,dt = \sum_{l=1}^{N} W(t;1,\ldots,m-1)\,dt \cdot \prod_{1,\ldots,m} q(t) -$$
$$- \sum_{l=2}^{N}\sum_{k=1}^{l-1} W(t;1,\ldots,m-l,k)\,dt \cdot \prod_{1,\ldots,m}^{l,k} q(t) + \ldots,$$

and $W_B(t; a_1, a_2, \ldots, a_n) = \sum_{l=1}^{N} W(t; a_1, \ldots, a_n - l) \cdot \prod_{a_1,\ldots,a_n}^{l} q(t) -$
$$- \sum_{l=2}^{N}\sum_{k=1}^{l-1} W(t; a_1, \ldots, a_n - l, k) \cdot \prod_{a_1,\ldots,a_n}^{l,k} q(t) + \ldots.$$

Finally, the top failure intensity $W_0(t)$ can be calculated by

$$W_0(t) = \alpha(t) - \beta(t),$$

where

$$\alpha(t) = \sum_{i=1}^{N} W_i(t) - \sum_{i=2}^{N} \sum_{j=1}^{i-1} W(t; i,j) \cdot \prod^{+i,j} q(t) + \sum_{i=3}^{N} \sum_{j=2}^{i-1} \sum_{k=1}^{j-1} W(t; i,j,k) \cdot \prod^{+i,j,k} q(t) - \ldots,$$

$$\beta(t) = \sum_{i=1}^{N} W_B(t; i) - \sum_{i=2}^{N} \sum_{j=1}^{i-1} W_B(t; i,j) + \ldots,$$

and

$$W_B(t; i_1 \ldots i_n) = \sum_{l=1}^{N} W(t; i_1, \ldots, i_n - l) \prod_{i_1, \ldots, i_n}^{l} q(t) -$$

$$- \sum_{l=2}^{N} \sum_{k=1}^{l-1} W(t; i_1 \ldots i_n - l, k) \prod_{i_1, \ldots, i_n}^{l,k} q(t) + \ldots.$$

For large-scale fault trees with a great number of min cut sets, we can calculate bounds for $W_0(t)$. Various combinations of upper and lower bounds lead to numerical measures as follows.

$$W_0(t)_{\max} = \sum_{i=1}^{N} W_i(t)$$

$$W_0(t)_{\min} = \alpha(t)_{\min} - \beta(t)_{\max},$$

where

$$\alpha(t)_{\min} = \sum_{i=1}^{N} W_i(t) - \sum_{i=2}^{N} \sum_{j=1}^{i-1} W(t; i,j) \prod^{i,j} q(t)$$

$$\beta(t)_{\max} = \sum_{l=1}^{N} W(t; i-l) \prod_{i}^{l} q(t).$$

The top failure rate $\Lambda_0(t)$ is defined as the quantity $\Lambda_0(t) = \dfrac{W_0(t)}{1 - Q_0(t)}$, where $W_0(t)$ and $Q_0(t)$ are already calculated and $\Lambda_0(t)$ is therefore known. A simple upper bound for $\Lambda_0(t)$ is given by relation $\Lambda_0(t) \leq \sum_{i=1}^{N} W_i(t) \Big/ \prod_{i=1}^{N} (1 - Q_i(t))$. The characteristics $\Lambda_0(\cdot)$, $W_0(\cdot)$ and $Q_0(\cdot)$ characterize a fault tree completely.

A general computer flow chart for a program which calculates a fault tree is given in fig. 5.33.

5.7.9. Error analysis and fault trees

In many applications, statistical data used in fault tree evaluation have a high degree of uncertainty. Failure rates can be considered to be random variables which take values in a given field according to the components of each system.

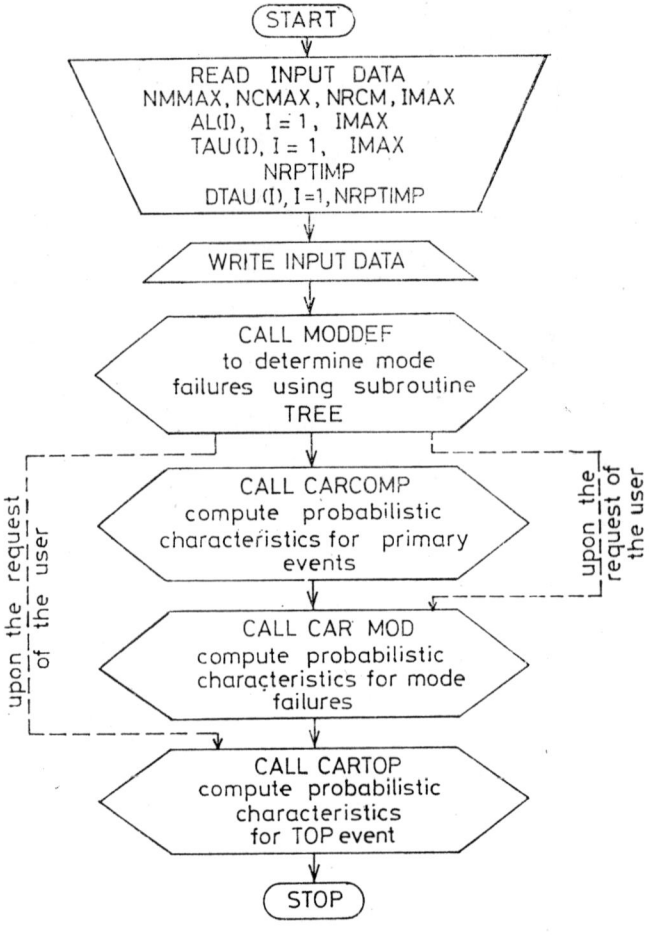

Fig. 5.33

The process whereby failure distribution probabilities are determined for a given system depending on the individual distributions of its components is called *error propagation analysis*. A Monte Carlo simulation model may be used for this purpose. The probability distributions chosen for each primary event are combined so as to obtain distributions corresponding to the whole system.

Chapter 6

On risk-sensitive Markovian decision models for complex systems maintenance

Markovian models for maintenance modelling are well known in the literature Broadly speaking, all maintenance models attempt to optimize (either maximize or minimize) the expected value of a decision process (maintenance strategies) when the system experiences a Markovian behaviour. The degree of model sophistication varies according to some specific decomposition rules over the structure topology. Real world maintenance situations can be modelled when the deterioration/ replacement process has a finite or infinite horizon or when the Markovian "core process" (Markov or semi-Markov) is completely or partially observable. It is not infrequent in the open literature that the decision-maker (or the maintenance engineer) is implicitly considered to be an expected value (risk-neutral) person. There are few cases, if any, in which Markovian models for optimal maintenance strategies involve a risk-sensitive (risk-averse or risk-preference) decision-maker. However, Kamien and Schwartz [56] and McCall [57] consider maintenance strategies under risk for the steady state case. In his survey of the use of mathematical models in plant maintenance decision-making, Turban [17] observed that "the future of maintenance management cannot rely on judgement alone, it is too costly". He also noticed that "the properties of most existing models need improvement". In this respect, an increased potential of maintenance models may be obtained by regarding the attitude of the decision-maker toward risk, in taking different maintenance decisions, as part of the model. A utility function and a risk coefficient will be used in the model discussed in this chapter.

We shall introduce first the class of Risk-Sensitive Markovian Decision Models (RSMDM) in order to analyse optimal maintenance strategies on coherent systems.

6.1. Preliminaries on the risk-preference phenomenon

Regarding risky propositions [65], the attitude of the decision-maker toward risk can be incorporated into a mathematical form, as a utility function which is sometimes called the risk-preference function (see Boyd and Matheson [58]). A preference ordering of these uncertain propositions can form a set of "lotteries" (Fishburn [60] and Howard [61]).

A *"lottery"* is a technical term which denotes an uncertain event whose outcomes are described by prizes or appropriate relative values that may be directly comparable or not (see also Charlwood [62]). If v is the outcome of a lottery, then it is possible to build up a utility function $u(v)$ (see ref. [61]).

A *certain equivalent* ($\tilde{v}$) for any given lottery is a fixed amount such that the decision-maker is indifferent about either receiving that amount or keeping the lottery, such that

$$u(v) = u(\tilde{v}). \tag{6.1}$$

In each state of MDP, we can define a lottery, where the possible results are given by the maintenance alternatives; prizes associated with each lottery represent the expected cost of implementing repair/replacement actions.

The *risk premium* v_p is given by

$$v_p = \bar{v} - \tilde{v}, \tag{6.2}$$

where $\bar{v}$ is the expectation of v, and is defined [61] as the amount of expectation that the individual (i.e. the maintenance engineer) is willing to forgo in order to avoid risk. If $v_p = 0$, then the decision-maker is risk-indifferent; for $v_p \neq 0$, the decision-maker is risk-sensitive ($v_p > 0$ is risk-averse and $v_p < 0$ is risk-preference).

A local measure of risk-preference has been introduced by Pratt [63] as the *risk-aversion coefficient* γ given by the equation

$$\gamma = -\frac{u''(v)}{u'(v)} = -\frac{d}{dv}\log u'(v), \tag{6.3}$$

where v is a random variable which represents a real-value outcome of a lottery from the MDP. A constant risk-aversion attitude is well described by the exponential utility function [61]

$$u(v) = \frac{1 - e^{-\gamma v}}{1 - e^{-\gamma}}, \tag{6.4}$$

which has the "delta property" (see Howard [14]) "that if all prizes in a lottery are increased by any amount Δ, the certain equivalent of the lottery will also increase by Δ". $(\tilde{v} + \Delta) = \Delta + \tilde{v}$.

In (6.4), $\gamma < 0$ implies risk-preference, $\gamma > 0$ risk-aversion and $\gamma = 0$ risk-indifference (expected value decision-maker).

The problem of choosing among alternatives for optimal maintenance strategies as an economic decision under uncertainty and risk-preference is represented by the utility functional

$$U(f_i) = \min_{k \in K_i} \left\{ \sum_{j \in \mathscr{S}} f_{ij}^k u_j(.) \right\}, \tag{6.5}$$

where f_{ij}^k represents a probability connection between two lotteries under a decision alternative k, and $u_j(.)$ is the utility of the j-th lottery. f_i^k is the probability function governing the i-th lottery when a maintenance decision k has been chosen.

6.2. A risk-sensitive Markov decision model for maintenance strategies in technical systems

Let us assume that there exists a complex technical system (defined as in Chapter 3) which experiences a Markov behaviour with a finite number of states corresponding to the performance index of the total structure function $\Phi_\omega^*(X)$; the system is completely observable. After a transition has been observed in the performance index of the system we can identify the new state of the system. In each state $(0, 1, \ldots, L)$ of the system, corresponding to the deterioration level and as a result of an inspection, the decision-maker is able to choose a maintenance decision alternative (i.e. repair or replacement) which moves the system to a more desirable state. When the system achieves zero performance level, corresponding to state L, a replacement maintenance action will make the system "as new". For the above case, the probability set p_{ij}^k, i.e. the transition rate probabilities observe the Borel measurability [10], and hence the process is Markovian in its behaviour.

Deterioration dynamics. The states of the system are in a lexicographical ordering corresponding to the real value of the total structure function $\Phi_\omega^*(.)$. The state 0 represents the process when the total structure function has maximum value of performance N, and state L is a "trapping state", i.e. an absorbing one ($p_{LL} = 1$), when the total structure function has the value zero (i.e. the system does not operate). The sequence of totally ordering states forms a Markov chain with a finite number of states. The deterioration dynamics for the system is represented in fig. 6.1.

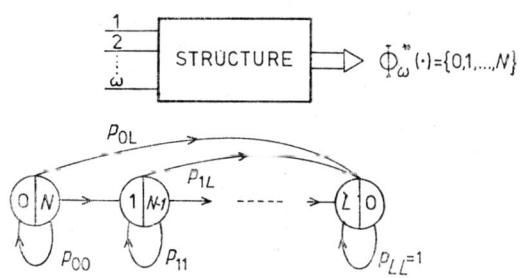

Fig. 6.1

The behaviour of the system subject to degradation is given by $\varphi_{ij}(t)$, i.e. the probability that the process is in state j at time t, given that at time zero it was in state i. Formally, we can write

$$\varphi_{ij}(t) = \Pr\{S(t) = j \mid S(0) = i, \mathscr{E}(t)\}. \tag{6.6}$$

Because the state-space of the above deterioration process has a total ordering, using known relations from Markov processes [8], we subsequently have

$$\varphi_{ij}(t+1) = \sum_{k=0}^{L} \varphi_{ik}(t) p_{kj}, \tag{6.7}$$

where p_{kj} is the transition rate probability for the process ($p_{kj} = \Pr\{S(t+1) = j \mid S(t) = k, \mathscr{E}(t)\}$) and

$$\varphi_{ij}(t+1) > 0, \quad \begin{pmatrix} i = 0, 1, \ldots, L-1 \\ j = i+1, \ldots, L \end{pmatrix}$$

$$\varphi_{ii}(t+1) \geq 0, \quad (i = 0, 1, \ldots, L-1)$$

$$\varphi_{LL}(t+1) = 1 \tag{6.8}$$

$$\varphi_{ij}(t+1) = 0 \quad \text{otherwise.}$$

As the process is under a Markov deterioration, it can be completely defined by the transition probabilities p_{ij}, ($i, j = 0, 1, \ldots, L$). The matrix relation between the transition probabilities P, and multistep transition probabilities $\Phi(t)$ is given by the well-known relation

$$\Phi(t) = P^t. \tag{6.9}$$

Decision alternatives. Consider that the system under a Markov deterioration is in a repair environment. We include in our model risk sensitivity of the decision-maker in the case of maintenance decision alternatives and consider that the inspection time as well as repair or replacement time are negligible, compared with the lifetime of the system under operation. The set of decision alternatives in each state are given by do nothing (0), do inspection with repair for the whole system (1), do inspection with replacement of the system (2). We must recall that in this model the inspection process is deterministic, i.e. after an observation has been made we can classify the performance state of the system (the observation space has the same cardinality as the system state space).

The decision "do nothing" keeps the system in a deterioration state. By a repair activity we "translate" the process from state i to state j (lexicographically greater than i), increasing the level of performance in the total structure function. By replacing the whole system we shall bring it to state 0 and the system will become "as new". We suppose that the new system has the same "technological age" and the same Markov process characteristics as the renewed one.

The controlled process. The deterioration process of a coherent system with maintenance decision alternatives is an MDP; however in the next paragraph we shall include risk-sensitivity of the decision-maker subject to maintenance alternatives. The transition probabilities of the controlled process can be written as is shown below.

— no repair/replacement actions take

$$p_{ij}^0 = \begin{cases} p_{ij} & \begin{pmatrix} i = 0, 1, \ldots, L-1 \\ j = i+1, \ldots, L \end{pmatrix} \\ 0 & \text{otherwise} \end{cases}$$

$$p_{ii}^0 \geq 0 \quad (i = 0, 1, 2, \ldots, L);$$

— replacement of the whole system:

$$p_{ij}^1 = \begin{cases} 1 & (i = 1, 2, \ldots, L: j = 0) \\ 0 & \text{otherwise}; \end{cases}$$

— repair of several components of the system:

$$p_{ij}^2 = \begin{cases} \alpha_i > 0 & \begin{pmatrix} i = 1, 2, \ldots, L \\ j = i - 1, \ldots, 0 \end{pmatrix} \\ 0 & \text{otherwise}, \end{cases}$$

where α_i is the probability that the process moves from state i to state j $(j \prec i)$ with the additional relation that

$$\sum_{j=i-1}^{0} p_{ij}^2 = 1, \quad (i = 1, 2, \ldots, L).$$

The decision process—cost and utility structure. In order to construct a mathematical model for the maintenance management of the system, we have to assess costs associated with a maintenance action or penalty due to the fact that the system performs to a less desirable value for $\Phi_\omega^*(X)$.

The *inspection cost* has a value for each state of the system. Different costs will be associated with an inspection procedure when the system is in one of the states $0, 1, \ldots, L$. Thus,

$a_{ij} > 0$ — inspection cost for the system in state i, given that the system "chooses" for the next transition state j, $(i = 1, 2, \ldots, L-1; j = 1 + 1, \ldots, L)$,

$a_{ij} = 0$ — otherwise.

The *repair cost* is given by

$n_{ij} > 0$ — the repair cost for the system in state i, if the repair action "translates" the system to state j (increasing the performance level of the total structure function)

$$\Phi_\omega^*(.), \quad (i = 1, 2, \ldots, I, \; j = i - 1, \ldots, 0),$$

$n_{ij} = 0$ — otherwise.

The *replacement cost* refers to the whole system and depends on the performance state at the time the replacement decision is taken:

$m_{i0} > 0$ — the replacement cost of the system in state i; by a replacing action, the system starts the new cycle from state zero; $i = 1, 2, \ldots, L$,

$m_{i0} = 0$ — otherwise.

A *penalty cost* is associated with the system when we leave it in a degraded state. Let us define $\bar{s}_{ij}(t)$ as the expected number of times state j is entered through time t, given that the system started in state i at time zero.

$$\bar{s}_{ij}(t) = \sum_{m=0}^{t} \varphi_{ij}(m), \quad (t = 0, 1, 2, \ldots).$$

The expected penalty cost figures for the system are given by

$$d_{jk} = f_j + (1 - \delta_{ij}) \sum_{k=j+1}^{L} p_{jk}\Delta_{jk}, \quad \begin{pmatrix} j = 1, 2, \ldots, L-1 \\ k = j+1, \ldots, L \end{pmatrix}, \quad (6.10)$$

$$d_i(t) = \sum_{j=i+1}^{L} \sum_{k=j+1}^{L} \bar{s}_{ij}(t)\left[f_j + (1 - \delta_{ij}) \sum_{k=j+1}^{L} p_{jk}\Delta_{jk}\right], \quad (i = 0, 1, 2, \ldots, L), \quad (6.11)$$

where f_j is the penalty if the system is in state j, p_{jk} is the transition rate probability for the deterioration process from state j to state k, Δ_{jk} is the unit step penalty cost per system transition and performance level of the total structure function, from state j to any lexicographically greater state k. The quantity d_{jk} represents the penalty cost of state j before the process moves to any other state k; $d_i(t)$ depends on the expected number of times state i is entered through time t, when the system is running for t time periods and gives sufficient information for the penalty in state i after t time periods. Penalty and inspection costs [23] associated with any state $i > 0$ will be included in a single relation

$$t_{ij} = a_{ij} + d_{ij}, \quad (i = 1, 2, \ldots, L-1; \; j = i+1, \ldots, L). \quad (6.12)$$

The overall cost for a maintenance action is given by

$$c_{ij} = \begin{cases} c_{ij}^0 = t_{ij} & \text{for } \begin{pmatrix} i = 0, 1, \ldots, L-1 \\ j = i+1, \ldots, L \end{pmatrix} \\ c_{ij}^1 = t_{ij} + m_{i0} & \text{for } \begin{pmatrix} i = 0, 1, \ldots, L \\ j = 0 \end{pmatrix} \\ c_{ij}^2 = t_{ij} + n_{ij} & \text{for } \begin{pmatrix} i = 1, 2, \ldots, L \\ j = i-1, \ldots, 0 \end{pmatrix} \end{cases}. \quad (6.13)$$

The utility cost functional for the process is given by

$$u(\tilde{v}_i(t+1)) = \sum_{j=0}^{L} p_{ij} u(c_{ij} + \tilde{v}_j(t)), \quad (6.14)$$

$$(i = 0, 1, \ldots, L; \; t = 0, 1, 2, \ldots).$$

Using notation similar to that given in ref. [35], we can write

$$u_i(t+1) = \sum_{j=0}^{L} p_{ij} u(c_{ij} + \tilde{v}_j(t)), \quad (6.15)$$

$$(i = 0, 1, \ldots, L; \; t = 0, 1, 2, \ldots).$$

Since the decision-maker is considered to be constantly risk-averse (exponential utility function), from the "delta property" given in the previous paragraph, we have

$$u_i(t+1) = \sum_{j=0}^{L} p_{ij} \exp(-\gamma c_{ij}) u_j(t), \quad (6.16)$$

$$(i = 0, 1, \ldots, L; \; t = 0, 1, 2, \ldots).$$

If $q_{ij} = p_{ij} \exp(-\gamma c_{ij})$ is considered to be the *"disutility measure"* for RSMDP, then we can write the alternative relation

$$u_i(t+1) = \sum_{j=0}^{L} q_{ij} u_j(t), \quad (i = 0, 1, \ldots, L;\ t = 0, 1, 2, \ldots), \qquad (6.17)$$

or, in matrix formulation, we have

$$u(t+1) = Q u(t), \quad (t = 0, 1, 2, \ldots). \qquad (6.18)$$

DEFINITION 6.1. An $(N \times N)$ matrix $Q = [q_{ij}]$ is reducible if there is a permutation

$$Q = \left[\begin{array}{c|c} B & 0 \\ \hline C & D \end{array}\right]$$

and B and D are square matrices; otherwise, Q is called irreducible. The next theorem is given in ref. [64].

THEOREM 6.1. (Frobenius). *An irreducible non-negative matrix* $Q = [q_{ij}]$ *always has a positive characteristic value* λ, *i.e. a simple root of the characteristic equation. The moduli of all the other characteristic values do not exceed* r. *To the "maximum" characteristic value* r, *there corresponds a characteristic vector with positive coordinates.*

Howard and Matheson [35] proved that for a risk-sensitive Markov process,

$$r u = Q u, \qquad (6.19)$$

where u is an eigenvector for Q corresponding to the largest eigenvalue r.

We have to choose a maintenance decision in each state of the process in accordance with different risk-aversion coefficients γ. For large t ($t \to \infty$), there exists an optimal gain for the process

$$\tilde{g} = -\frac{1}{\gamma} \ln r, \qquad (6.20)$$

which depends only on γ and r.

To find an optimal solution for the above problem, we shall give here a policy iteration solution, where the policy evaluation phase is modified from [35]. We avoid solving a set of nonlinear equations (6.16) by performing a utility vector normalization. After a normalization procedure is executed, the new utility vector will have the following components:

$$\hat{u}_i = \frac{-(\operatorname{sgn}\gamma) u_i}{u_L}, \quad (i = 0, 1, \ldots, L). \qquad (6.21)$$

This normalization procedure ensures the final convergence on the policy iteration algorithm [35]. Both possibilities of the policy evaluation phase were investigated (solving a set of nonlinear equations by the Newton-Raphson method, or normalization over the eigenvector u corresponding to the largest eigenvector of Q) and it was found that the second method was extremely advantageous (see ref. [128]). For both methods, see the flowcharts on pages 184, 185 and 186 respectively. We give below an algorithm for solving optimal maintenance strategies in case of RSMDP.

Step 0: Set the initial parameters of the process; also set $\gamma = -1$.
Step 1: Choose an initial policy for maintenance strategy.
Step 2: Calculate disutility matrix coefficients Q.
Step 3: Calculate eigenvalues and eigenvectors for Q.
Step 4: Choose the largest eigenvalue r and corresponding eigenvector u
Step 5: Normalize the utility vector for $\hat{u}$ (relation (6.21)).
Step 6: Minimize the utility cost functional $U_i^k = \sum_{j=0}^{L} q_{ij}^k \hat{u}_j$, for each state i and each decision k.
Step 7: If the condition of the policy iteration algorithm is satisfied *, then go to Step 8. If it is not satisfied, then go to Step 1.

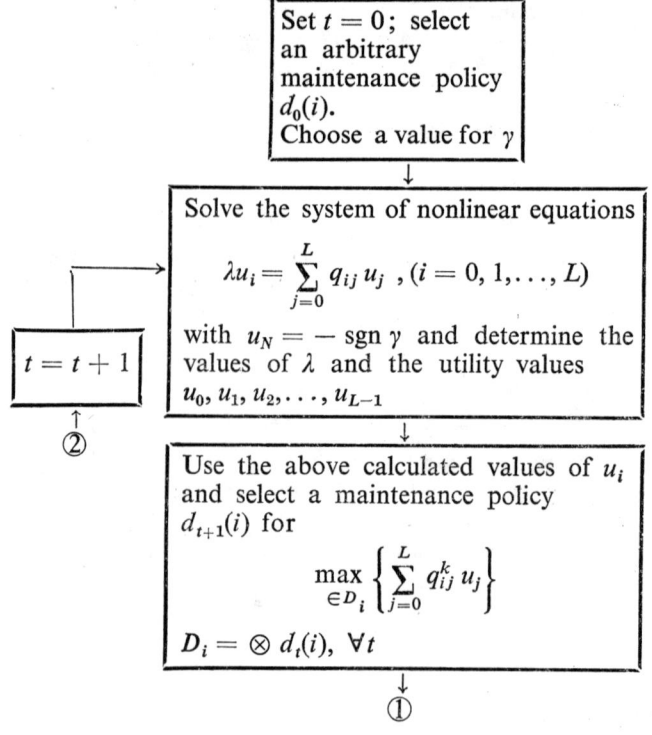

* If on two successive iterations the same policy is obtained, then that policy is optimal.

RISK-SENSITIVE MARKOVIAN DECISION MODELS

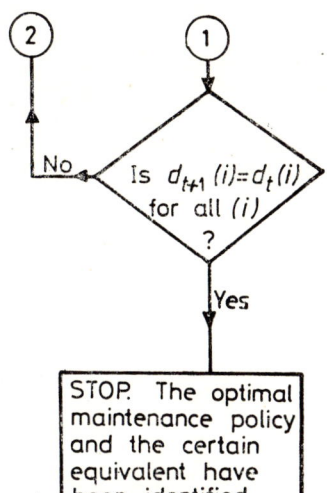

```
        ②           ①
         ↑           ↓
          ╲ No   Is d_{t+1}(i)=d_t(i)
           ←──────  for all (i)
                      ?
                    │ Yes
                    ↓
            ┌──────────────────┐
            │ STOP. The optimal│
            │ maintenance policy│
            │ and the certain  │
            │ equivalent have  │
            │ been identified  │
            └──────────────────┘
```

Set $t = 0$, select an arbitrary maintenance policy $d_0(i)$.
Choose a value for γ

↓

Solve the system of nonlinear equations

$$e^{-\gamma(\tilde{g}+\tilde{v}_i)} = \sum_{j=0}^{L} p_{ij}\, e^{-\gamma(c_{ij}+\tilde{v}_j)}$$

and find the values of $\tilde{g}$ and $\tilde{v}_0, \tilde{v}_1, \ldots, \tilde{v}_{L-1}$ with the condition that $\tilde{v}_L = 0$

$t = t+1$

↑
②

↓

Use the previously calculated values for $\tilde{v}_i$ and select the maintenance policy $d_{t+1}(1)$ for

$$\max_{k \in D_i} \left\{ -\frac{1}{\gamma} \ln\left[\sum_{j=0}^{L} p_{ij}^k\, e^{-\gamma(c_{ij}^k + v_j)} \right] \right\}$$

$D_i = \otimes d_t(i), \forall t$

↓
①

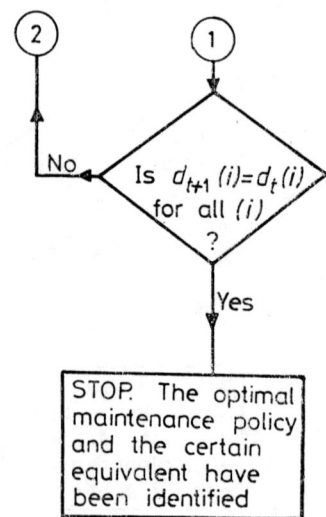

Step 8: Write the optimal and suboptimal solution for the risk-aversion coefficient γ.

Step 9: Increase the value of γ. If $\gamma = 1$, then go then to Step 10; If $\gamma < 1$, $\gamma = \gamma + \varepsilon$, then go to Step 0; If $\gamma = 0$, use a usual procedure for MDP. ε is an increment for the risk-aversion coefficient.

Step 10: Stop.

For Step 3, we can use an ICL (NAG) library routine (SUBROUTINE FO2 AGF) available for ICL 1905 computer software library or any kind of subroutine available which fulfills the requirements of the given application.

Structure of the control limit replacement rule. The problem of control limit replacement rule is well known in the literature about systems engineering (see Kolesar [18]). A control limit replacement rule is of the form "replace the system if and only if $S(t) > i^*$", where i^* is called the control limit and represents some integer between 0 an L. For large and complex systems, the rule for an optimal replacement is very important. In our model, we shall correlate the deterioration performance of the system for the set of pertinent observations through which we can classify the performance levels and the maintenance cost structure of the system. However, our approach differs from that given in the conservative literature [12], [18]; instead of minimizing the expected cost, our optimization procedure is based on the utility cost functional when the decision-maker is constant risk-averse (exponential utility function) and the local risk-aversion measure is given by γ. We shall introduce the discount factor β and mimic the approach to the problem of time-risk preference given by Jaquette [66] for RSMDP. The problem of time-risk preference in MDP has also been discussed by Porteus [67]. At present, the problem of time-risk preference is far from being solved. For a discussion of the time-risk problem, see Pollard [68].

The only maintenance decision alternatives to be considered here are either "do nothing" or "inspection with replacement" ($k = 0, 1$). In deriving the desired results for a risk-sensitive control limit replacement rule, we use theoretical results similar to those proposed in ref. [66].

First, it is necessary to make several assumptions and formulate general theorems for RSMDP with discounting.

As was mentioned before, the process is stationary and it is running over an infinite time horizon; we have complete information on all states and times. The class of replacement rules for a time-risk-preference MDP with β fixed is denoted by R_s, and contains 2^L elements for each fixed pair (γ, β). A strategy σ is defined by a sequence $\{f_t, t = 0, 1, 2, \ldots\}$ or by

$$\sigma = (f_0, f_1, \ldots, f_t, \ldots), \tag{6.22}$$

where f_t is a decision vector for each state at time t (see [10]); i.e. $f_t(i)$, which is the i-th element of f_t, is a maintenance action in state i at time t.

DEFINITION 6.2. [66]. An optimal utility policy in an MDP is one whose total discounted cost

$$R(t) = \sum_{t=0}^{\infty} \beta^t \, r(\sigma(t)) \circ z_\sigma(t) \tag{6.23}$$

yields minimum expected cost utility, using an exponential utility function (constant risk-aversion decision-maker). In the above definition, $r(\sigma(t))$ represents the maintenance cost vector for the system under the strategy $\sigma(.)$ at time t. $z_\sigma(t)$ is a zero vector except for a 1 in component i if $z_\sigma(t) = i$, where $z_\sigma(t)$ denotes the state of the process at time t when strategy σ is used to choose maintenance actions on the technical system. The sign "0" indicates componentwise multiplication of two vectors of the same dimension. In the remainder of this section, we shall consider only the case of risk-preference decision-maker ($\gamma > 0$); the case ($\gamma < 0$) could be approached in a similar way. We shall discuss now a policy utility.

DEFINITION. 6.3 A policy σ^* is called β-utility optimal with constant risk-aversion γ, if $U^\beta_{\sigma^*}(\gamma) < U^\beta_\sigma(\gamma)$ for all σ and

$$U^\beta_\sigma(\gamma) = - E[\exp(-\gamma R(\sigma))]. \tag{6.24}$$

To operate with time-risk preference MDP, Jaquette [66] introduced a new optimality criterion, the *moment optimal criterion*. A close analysis of the new criterion for MDP indicates that a moment optimality is equivalent to infinite horizon risk-sensitive optimality with discounting if the aversion (preference) to risk is sufficiently small [66, p. 9]. We shall now give some conditions for the process which leads to an optimal control limit replacement rule for the case of a RSMDP.

Condition 1: There exists a real number γ_0 ($\gamma_0 > 0$) and an f such that f^∞ is β-utility optimal with risk aversion γ for all γ ($0 \leqslant \gamma \leqslant \gamma_0$).

Condition 2: For a fixed value of γ, number $N(\gamma)$ is defined as the smallest integer t such that $\beta^t \gamma < \gamma_0$.

Condition 3: $\sum_{j=k}^{L} p_{ij}$ is non-decreasing in i for each $k = 0, 1, \ldots, L$.

Condition 4: $\{r^{(0)}_i\}_{i=0}^{L}$ is a non-decreasing bounded sequence; the replacement action in state L is mandatory. Condition 1 states that the decision process is risk-sensitive and a β-utility policy shall be used for the optimization process. Condition 2 states inherent conditions for moment optimality criterion. Condition 3 states that the probability that the system makes a transition into any set of states $\{k, k+1, \ldots, L\}$ connected to the "trapping state" is non-decreasing in i as the system moves toward the terminal state L. Condition 4 states that if we do not replace the system, then the expected cost per occupancy in state i is non-decreasing in i. From Conditions 1 and 2, and using the basic results in [66] for a vector space $\mathscr{L}$ of negatives of Leplace transforms of bounded random variables, under the transformation dummy argument γ (this implies that $U_\sigma(\gamma)$ are in $\mathscr{L}$),

$$U_\sigma(\gamma) = -E\left[\exp\left(-\gamma \sum_{t=0}^{\infty} \beta^t r(\sigma(t)) \circ z_\sigma(t)\right)\right]. \tag{6.25}$$

Hence we have the following

LEMMA 6.1. *Under Conditions 1 and 2, the maintenance policy σ is no better than the policy $\sigma^{N(\gamma)} f^\infty$ for any risk-aversion coefficient less or equal to γ.*

For a complete proof of this lemma, the reader is referred to [66, p. 12–13].

An optimal β-utility policy $\sigma^*(\gamma)$ can be constructed to be piecewise constant; this can be proved by examining $\sigma^*(\gamma)$ as the risk-aversion coefficient γ increases.

A time-risk preference MDP as described in this section does not state that a stationary policy is utility-optimal. However, Howard and Matheson [35] proved that, for the RSMDP, optimality holds for $\beta = 1$. For $0 < \beta < 1$, if a stationary policy is β-utility optimal (in the sense of Definition 6.3), the policy must be f^∞ (i.e. $\sigma = (f, f, \ldots)$), independent of the aversion to risk factor γ.

From the results given here, we can construct β-utility optimal policies. We shall recall the following theorem, whose proof was given in ref. [66, p. 15].

THEOREM 6.2. *Under Conditions 1 and 2, if Lemma 6.1 holds and if policy $\sigma^*(\gamma)$ with a fixed β is chosen as an ultimately stationary piecewise constant function of γ, then a computation algorithm for β-utility optimal policy exists.*

The algorithm is described in the following steps:

Step 0: Set up γ and β as fixed.
Step 1: Choose γ_0 and calculate the utility functional $U_{f^\infty}(\beta^{N(\gamma)} \gamma)$.
Step 2: Calculate the utility functional

$$u = L_{f_{N(\gamma)-1}} U_{T^{N(\gamma)}\sigma^*}(\beta^{N(\gamma)-1} \gamma) = $$
$$= \min_K \{\exp\{-\gamma r(k)\} \circ P(k) \cdot U_{f^\infty}(\beta^{N(\gamma)} \gamma)\},$$

for $\beta^{N(\gamma)} \gamma < \gamma_0$, where u represents the utility state vector.

Step 3: Continue the above procedure backwards until all f_t have been constructed.
Step 4: Increase γ and β for pertinent limits and go to Step 1.
If $\gamma = \gamma_{\max}$, $\beta = \beta_{\max}$, then go to Step 5.
Step 5: Stop.

We are now ready to prove the following

THEOREM 6.3. *Under Conditions 1–4, there exists a control limit replacement rule $\tilde{R}_s$ that is optimal over all $\tilde{R}_s \in R_s$.*

Proof. Conditions 1–4 and the fact that u_i is non-decreasing (see Theorem 6.2) ensure the existence of an optimal state-dependent control replacement rule that calls for replacement for $i \geqslant i^*$, and do nothing otherwise for any fixed pair (γ, β).

6.3. A risk-sensitive semi-Markov decision model for maintenance strategies in technical systems. Finite-horizon decision-making

In this paragraph we shall present a semi-Markov model. The optimization process has some complications caused by the randomly transformed holding-time in any given state of the process. We shall accept the general outlines for the process as stated earlier, i.e. the process is considered to be completely observable such that upon inspection, we can identify exactly the new state of the system. The transition rate probabilities p_{ij} are defined as before for a stationary Markov process. We introduce now the holding-time mass function probabilities which are given by

$$h_{ij}(m) = \Pr\{S(t+1) = j \mid S(t) = i, \tau_i = m, \mathcal{E}(t)\}, \quad (i,j = 0, 1, \ldots, L). \quad (6.26)$$

We deal with the case of the semi-Markov process when the deterioration process is allowed to select its time of transition first and then select a destination conditional on the transition time. A transition can be observed, but the state of the process is known after performing an observation (see fig. 6.2). First, we consider the case of a finite horizon process which involves maintenance decisions for an optimal solution. According to [8], the core process is described by

$$\bar{c}_{ij}(m) = p_{ij} h_{ij}(m), \quad (i,j = 0, 1, \ldots, L: m = 0, 1, 2, \ldots). \quad (6.27)$$

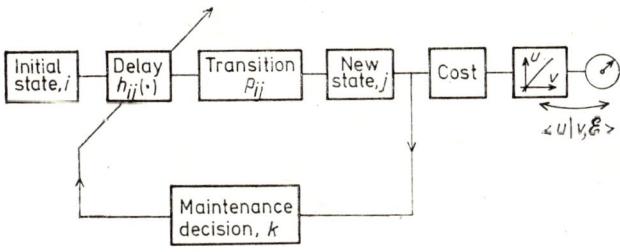

Fig. 6.2

The structure of the maintenance cost of the process is y_{ij}^k, $(k = 0, 1, 2)$, which expresses the labour cost of a maintenance action k, and c_{ij}^k (see the RSMDP model) is a fixed cost for a maintenance action k. Hence, we can write the following utility cost functional

$$u(\tilde{v}_i(t)) = \sum_{j=0}^{L} p_{ij} \sum_{m=0}^{t} h_{ij}(m) u(my_{ij} + c_{ij} + \tilde{v}_j(t-m)), \quad (6.28)$$

$$(i = 0, 1, 2, \ldots, L; \, t = 0, 1, 2, \ldots).$$

Using an exponential utility function and its "delta property", the above relation becomes

$$u(\tilde{v}_i(t)) = \sum_{j=0}^{L} p_{ij} \sum_{m=0}^{t} h_{ij}(m) \cdot \exp\left[-\gamma(my_{ij} + c_{ij})\right] u(\tilde{v}_j(t-m)), \quad (6.29)$$

$$(i = 0, 1, \ldots, L; \; t = 0, 1, 2, \ldots).$$

For convenience, we adopt the notation used in the previous section, and then

$$u_j(t) = \sum_{j=0}^{L} p_{ij} \sum_{m=0}^{t} h_{ij}(m) \exp(-\gamma(my_{ij} + c_{ij})) u_j(t-m), \quad (6.30)$$

$$(i = 0, 1, \ldots, L; \; t = 0, 1, 2, \ldots).$$

In a matrix formulation, the above relation can be written as

$$U(t) = \sum_{m=0}^{t} ((P \,\square\, H(m)) \,\square\, E(m)) \, U(t-m), \quad (6.31)$$

where $E(m) = [e_{ij}(m)]$ and $e_{ij}(m) = \exp\left[-\gamma(my_{ij} + c_{ij})\right]$.

For a matrix notation of the "core process" $C(m) = P \,\square\, H(m)$, the matrix form of the utility cost functional is

$$U(t) = \sum_{m=0}^{t} (C(m) \,\square\, E(m)) \, U(t-1). \quad (6.32)$$

We can define a "disutility matrix" subject to the risk-sensitive SMDP,

$$Q^*(m) = C(m) \,\square\, E(m), \quad (6.33)$$

and therefore (6.32) becomes

$$U(t) = \sum_{m=0}^{t} Q^*(m) \, U(t), \quad (t = 0, 1, \ldots). \quad (6.34)$$

The maintenance engineer has to choose among a finite set of maintenance alternatives at each stage of the process when the system experiences a semi-Markov behaviour. From the above relation and using a dynamic programming technique for the process optimization, we can write (6.30) in a matrix form.

$$U(t) = \min_{k} \left\{ \sum_{m=0}^{t} ((P^k \,\square\, H^k(m)) \,\square\, E^k(m)) \, U(t-m) \right. \quad (6.35)$$

$$U(0) = [u_j(0)]; \quad u_j = -(\mathrm{sgn}\,\gamma), \quad (j = 0, 1, \ldots, L). \quad (6.36)$$

We have to choose an optimal maintenance procedure which minimizes the utility cost functional in each state of the semi-Markov process.

From the above relations and by a specialization to a Markov process ($h_{ij}(m) = \delta(m-1)$), where $\delta(m-1)$ is a unit step function, we obtain the results given in ref. [35]. If $y_{ij} = 0$, then it is obvious that the above results lead to RSMDP.

Infinite horizon case. In the above case we have considered that the system has a finite horizon time to run and the semi-Markov process (the core process) is time varying. However, for many real systems the time to failure is small in comparison with the life time of the system. We may be interested to obtain an optimal maintenance strategy over an infinite time horizon ($t \to \infty$), for the case when appropriate maintenance actions in each state of the system must be taken. In order to obtain a solution for the case when $t \to \infty$ for SMDP when the decision maker is risk-sensitive (exponential function), we have to use the spectral properties of irreducible non-negative matrices from Section (6.2) (also see [64]) and the fact that a semi-Markov process has a special structure as defined by Pyke [39] (see also Chapter 5). Howard and Matheson [36] proposed a "policy iteration" solution to this problem. A similar algorithm will be given here.

The utility cost functional can be written as

$$u_i(t) = \sum_{j=0}^{L} \sum_{m=0}^{t} q_{ij}^*(m) u_j(t-m), \quad (i=0,1,\ldots,L; \ t=0,1,\ldots). \tag{6.37}$$

The limiting behaviour is analysed by using the largest eigenvalue for the disutility matrix $Q^*(m)$ and we can write

$$\lim_{t \to \infty} \frac{1}{r^t} u_i(t) = \lim_{t \to \infty} \sum_{j=0}^{L} \sum_{m=0}^{t} \frac{1}{r^{t-m}} \cdot \frac{1}{r^m} q_{ij}^*(m) u_j(t-m) \tag{6.38}$$

$$(i=0,1,\ldots,L; \ t=0,1,2,\ldots).$$

Using a geometric transformation of the above relation, for the case when $t \to \infty$, then

$$q_{ij}^{*g}\left(\frac{1}{r}\right) = \sum_{m=0}^{t} \frac{1}{r^m} q_{ij}^*(m) \tag{6.39}$$

and hence (6.38) becomes

$$\lim_{t \to \infty} \frac{1}{r^t} u_i(t) = \lim_{t \to \infty} \sum_{j=0}^{L} q_{ij}^{*g}\left(\frac{1}{r}\right) \frac{1}{r^{t-m}}. \tag{6.40}$$

Using the properties of irreducible matrices (given in ref. [64], and emphasized in [35]), the final relation is

$$u_i = \sum_{j=0}^{L} q_{ij}^{*g}\left(\frac{1}{r}\right) u_j, \quad (i=0,1,\ldots,L). \tag{6.41}$$

This is a steady-state solution to the problem in the utility form. In (6.41) r represents the largest eigenvalues for $Q^*\left(\dfrac{1}{r}\right)$. A policy improvement rule is given by the relation

$$U_i^k = \sum_{j=0}^{L} q_{ij}^{*gk}\left(\frac{1}{r}\right) u_j, \quad (i=0,1,\ldots,L). \tag{6.42}$$

In the following algorithm we formulate the solution to optimal maintenance strategies for risk-sensitive SMDP.

Step 0: Set up the initial parameters of the process. Set $\gamma = -1$.
Step 1: Choose an initial policy for maintenance strategy.
Step 2: Calculate the geometric transform for the disutility matrix.
Step 3: Solve the set of equations

$$u_i = \sum_{j=0}^{L} q_{ij}^{*g}\left(\frac{1}{r}\right) u_j, \quad (i=0, 1, \ldots, L).$$

Step 4: Minimize the utility cost functional

$$U_i^k = \sum_{j=0}^{L} q_{ij}^{*gk}\left(\frac{1}{r}\right) u_j, \quad (i=0, 1, \ldots, L)$$

for each state i and each decision k, where u_j are from Step 3.

Step 5: If the condition for "policy iteration" algorithm is satisfied, go to Step 6. (Note: If on two successive iterations the same policy is obtained, then this is optimal). If not satisfied, then go to Step 1.

Step 6: Write optimal and suboptimal solutions for the risk-aversion coefficient γ.

Step 7: Increase the value of γ. If $\gamma = 1$, then go to Step 8. If $\gamma < 1$, $\gamma = \gamma + \varepsilon$, then go to Step 0. If $\gamma = 0$, then use a normal procedure for SMDP (N.B. ε is an increment for the risk-aversion coefficient).

Step 8: STOP.

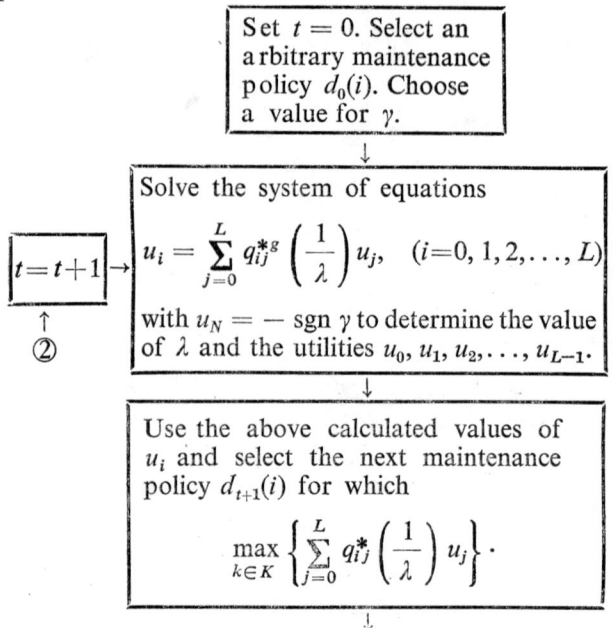

RISK-SENSITIVE MARKOVIAN DECISION MODELS

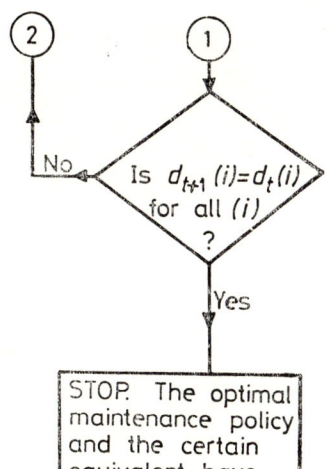

Is $d_{t+1}(i)=d_t(i)$ for all (i) ?

— No → ②
— Yes ↓

STOP. The optimal maintenance policy and the certain equivalent have been identified

---

Set $t = 0$. Select an arbitrary maintenance policy $d_0(i)$. Choose a value for γ.

↓

Solve the system of equations

$$e^{-\gamma \tilde{v}_i} = \sum_{j=0}^{L} q_{ij}^{*g}(e^{\gamma \tilde{g}}) e^{-\gamma \tilde{v}_j}, \quad (i=0, 1, \ldots, L)$$

and find the values for $\tilde{g}$ and $\tilde{v}_0, \tilde{v}_1, \ldots, \tilde{v}_{L-1}$ with the condition that $\tilde{v}_L = 0$.

$t = t+1$ ↑ ②

Use the above calculated values of $\tilde{v}_i$ and select a maintenance policy $d_{t+1}(i)$ for

$$\tilde{V}_i^k = \max_{k \in k_i} \left\{ -\frac{1}{\gamma} \ln \sum_{j=0}^{L} q_{ij}^{*gk}(e^{\gamma \tilde{g}}) e^{-\gamma \tilde{v}_j} \right\}.$$

Repeat the procedure for each state i to find an optimal policy.

↓ ①

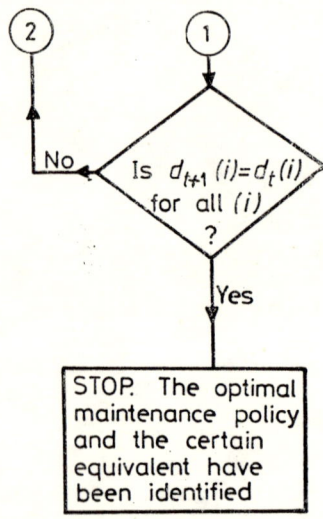

6.4. A special case: partially observable risk-sensitive Markov decision models for maintenance strategies

Observable systems were exhausted in the analysis of previous cases. However, as was emphasised in Chapter 3, there are systems where it is not possible to observe the operational states of the components, but these states can be inferred only by performing a set of independent observations through a "noise channel" (a probabilistic observation of states). A pictorial representation of such a Markov process is given in fig. 6.3. Because of the "noise channel", this model will consider rather

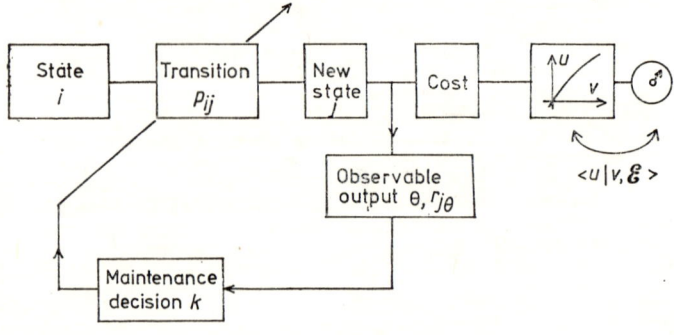

Fig. 6.3

differently the dynamics of $\Pi(t)$, i.e. the state vector at time t, and our main objective here is to build up a finite horizon maintenance model when the decision-maker is risk-sensitive.

State space. The states of the core process are lexicographically ordered, in compliance with the definitions given for lexicographic coherent structures (see Chapter 3). The number of states corresponds to the performance levels for the total structure function $\Phi_\omega^*(X)$. The system is subject to deterioration and the probability transition rate is

$$p_{ij} > 0, \quad (i = 0, 1, \ldots, L; \quad j = i+1, \ldots, L),$$

$$p_{LL} = 1$$

$$p_{ij} = 0 \quad \text{otherwise};$$

$$\sum_{j=0}^{L} p_{ij} = 1, \quad (i = 0, 1, \ldots, L).$$

The state probability vector is considered to be known, as

$$\pi(t) = [\pi_i(t)], \quad (i = 0, 1, \ldots, L; \quad t = 0, 1, 2, \ldots),$$

$$\pi_i(t) \geq 0,$$

$$\Pi = \text{set}\left[\pi_i; \pi_i \geq 0; \sum_{i=0}^{L} \pi_i = 1\right].$$

Observation space and system dynamics. We assume that after an inspection has been performed, the observable outputs are also lexicographically ordered and that the outputs are probabilistically correlated to the internal state of the process such that

$$r_{j\theta} > 0, \quad (j = 0, 1, \ldots, L; \quad \theta = j, j+1, \ldots, M),$$

$$r_{j\theta} = 0 \quad \text{otherwise};$$

$$\sum_{\theta=1}^{M} r_{j\theta} = 1, \quad (j = 0, 1, \ldots, M).$$

Following refs. [13] and [31], and relying on Bayes' rule for the dynamic inferrence over the partially observable Markov process, where the inspection process has been explicitly introduced, the new state of knowledge is given by

$$\pi'_j = T_j(\pi \mid \theta, k) = \frac{\sum_i \pi_i p_{ij}^k r_{j\theta}^k}{\sum_{i,j} \pi_i p_{ij}^k r_{j\theta}^k} \tag{6.43}$$

if at the starting time the initial state vector was given by π and k represents a pertinent maintenance alternative in state i before the process visits a new state j.

Maintenance cost and utility structure. The model given in this section allows us to employ explicit maintenance (repair or replacement) and inspection (by means of probabilistic observation) costs. If c_{ij}^k represents the cost of a repair replace-decision in state i and $a^k_{j\theta}$ represents the inspection cost which correlates the internal state j of the process and the observable output θ, then we can write the following utility cost functional:

$$u(\tilde{v}_\pi(t)) = \sum_{i=0}^{L} \pi_i(t) \sum_{j=0}^{L} p_{ij} \sum_{\theta=1}^{M} r_{j\theta} \cdot u(c_{ij} + a_{j\theta} + \tilde{v}_{T(\pi|\theta)}(t-1)). \qquad (6.44)$$

Using an alternative notation and considering that $w_{ij\theta} = c_{ij} + a_{j\theta}$, then for an exponential utility function we can write

$$u_\pi(t) = \sum_{i=1}^{L} \pi_i(t) \sum_{j=0}^{L} p_{ij} \sum_{\theta=1}^{M} r_{j\theta} \cdot \exp(-\gamma w_{ij\theta}) u_{T(\pi|\theta)}(t-1), \quad (t = 0, 1, \ldots). \quad (6.45)$$

In a matrix formulation, (6.45) becomes

$$U_\pi(t) = \sum_{\theta=1}^{M} \pi^t P R_\theta E_\theta U_{T(\pi|\theta)}(t-1), \qquad (6.46)$$

where $R_\theta = \text{diag}[r_{j\theta}]$, $E_\theta = [e_{ij\theta}] = [e_\theta]$ and

$$e_{ij\theta} = e_\theta = \exp(-\gamma w_{ij\theta}). \qquad (6.47)$$

We can show that the utility functional above remains linear in π when the lexicographic order over the states in the core process and the observations performed after inspection were assumed. If k is an appropriate maintenance decision, then we can write

$$u_\pi(t) = \min_k \left\{ \sum_{i=0}^{L} \pi_i \sum_{j=0}^{L} p^k_{ij} \sum_{\theta=1}^{M} r^k_{j\theta} e^k_\theta u_{T(\pi|\theta, k)}(t-1) \right\} \qquad (6.48)$$

where $u_\pi(0) = u^0$ is the value of terminating the process in each internal state. However,

$$u_\pi(0) = \min_k \left\{ \sum_{i=0}^{L} \pi_i \sum_{j=0}^{L} p^k_{ij} \sum_{\theta=1}^{M} r^k_{j\theta} \cdot u(w_{ij\theta}) \right\}. \qquad (6.49)$$

But as $u(w_{ij\theta}) = \exp(-\gamma w_{ij\theta}) = ct$, the equation

$$u_\pi(0) = \sum_{i=0}^{L} \pi_i \alpha^0_i, \qquad (6.50)$$

holds, where

$$\alpha^0_i = \sum_{\theta=1}^{M} \sum_{j=0}^{L} p_{ij} r_{j\theta} e_\theta. \qquad (6.51)$$

We give now a general

LEMMA 6.2. *For a risk-sensitive decision process, if x is the value of an outcome and $u(x)$ is the utility associated with it, and if the utility function $u(\cdot)$ is an exponential one, then $u(x) \to x$ with some appropriate scale change if the risk-aversion coefficient tends to zero $(\gamma \to 0)$.*

Proof. Let $u(.)$ be given by the relation (see also fig. 6.4)

$$U(x) = K[1 - e^{-\gamma x}], \tag{6.52}$$

where here K is a constant for an appropriate scale change. Let us set K such that $u(1) = 1$ and then we have $K = [1 - e^{-\gamma}]^{-1}$. We must show that $u(x) \to x$ as $\gamma \to 0$.

By expanding (6.52) in a Taylor series,

$$U(x) = [1 - e^{-\gamma}]^{-1} [1 - e^{-\gamma x}] = \frac{1 - \left(1 - \gamma x + \frac{(\gamma x)^2}{2!} + \cdots\right)}{1 - \left(1 - \gamma + \frac{\gamma^2}{2!} + \cdots\right)} \tag{6.53}$$

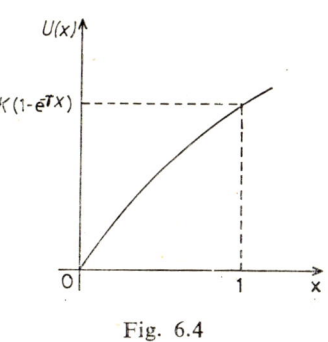

Fig. 6.4

and if $\gamma \cong 0$, then

$$U(x) = \frac{\gamma x}{\gamma} = x. \tag{6.54}$$

The proof is complete.

From the above lemma, we can see that a risk-sensitive partially observable Markov process for $\gamma = 0$ is the case given in ref. [31], with the relaxation over the lexicographic order for Markov states and observable outputs after an inspection of a system has been performed.

In order to obtain an optimal solution of maintenance strategies for partially observable Markov processes when the maintenance engineer is a constant risk-preference person, we shall follow [31] for the "one-pass algorithm". For the δ-th component in α-vector, the utility cost functional is

$$u_\pi(t) = \min_\delta \left[\sum_{i=0}^{L} \alpha_i^\delta(t) \pi_i \right]. \tag{6.55}$$

After the process has made a transition, the utility functional becomes

$$u_{T(\pi|\theta,k)}(t-1) = \frac{\min_\delta \left\{ \sum_{j=0}^{L} \alpha_j^\delta(t) \sum_{i=0}^{L} \pi_i p_{ij}^k r_{j\theta}^k \right\}}{\sum_i \sum_j \pi_i p_{ij}^k r_{j\theta}^k}. \tag{6.56}$$

From (6.56), we can see that the utility functional is linear in π and therefore

$$u_{T(\pi|\theta)}(t-1) = \pi \alpha(t-1). \tag{6.57}$$

Let us define a vector $l_t(\pi, k, \theta)$ that is equal to the corresponding α-vector index for the region containing the transformed information vector $T(\pi|\theta, k)$. Therefore (6.56) can be written as

$$u_{T(\pi|\theta,k)}(t-1) = \frac{\sum_j \alpha_j^l t(\pi, k, \theta) \sum_i \pi_i p_{ij}^k r_{j\theta}^k}{\sum_i \sum_j \pi_i p_{ij}^k r_{j\theta}^k}. \quad (6.58)$$

Using (6.58), then (6.48) becomes

$$u_\pi(t) = \min_k \left[\sum_{i=0}^{L} \pi_i \sum_{j=0}^{L} p_i^k \sum_{\theta=1}^{M} r_{j\theta}^k e_\theta^k \alpha_j^l t(\pi, k, \theta)(t-1) \right], \quad (t = 0, 1, 2, \ldots). \quad (6.59)$$

The above relation proves again that $u_\pi(.)$ is linear in π. For each maintenance decision alternative k and each output θ from the inspection process, the function $l_t(\pi, k, \theta)$ is a finitely valued function of π.

Computing $u(t)$. In order to compute the α-vectors and the corresponding mappings of these vectors onto the set of maintenance control alternatives, we must construct the following algorithm.

Assume: We have calculated $\alpha^\delta(t-1)$ and consequently

$$u_{T(\pi|\theta,k)}(t-1).$$

Find: An algorithm to calculate the vectors $\alpha^\delta(t)$ from the above information.

We need to pick up an information vector, say π^0, and then from (6.59) calculate the optimum control alternative and the corresponding α-vector when t control intervals remain in the process. Let us call $k^*, \alpha^*(t)$ the optimal maintenance alternative and α-vector respectively, after performing the above statement. The results from the set theory, as those given in ref. [13], enforce the relation

$$T(\pi|\theta, k^*) \alpha^l (t-1) \leq T(\pi|\theta, k) \alpha^\delta (t-1) \quad (6.60)$$

for all δ, and consequently

$$\sum_{i,j} \pi_i p_{ij}^{k^*} r_{j\theta}^{k^*} (\alpha_j^l (t-1) - \alpha_j^\delta (t-1)) \geq 0, \forall \delta. \quad (6.61)$$

For each possible output, there exists a set of these inequalities $\pi \alpha^*(t) \leq \pi \alpha_k(t)$, where $\alpha_k(t)$ denotes the α-vector for the maintenance alternative k of the initial state vectors π^0 and $\alpha^*(t)$, such that control alternative k^* must be optimal.

$$\sum_{i=0}^{L} \pi_i(\alpha_i^*(t) - \alpha_{k,i}(t)) \leq 0, \quad \forall k, \quad (t = 0, 1, \ldots), \quad (6.62)$$

and $\pi_i \geq 0$, $(i = 0, 1, \ldots, L)$, $\sum_{i=0}^{L} \pi_i = 1$.

The above relations specify the region in the space of information vector over which the α-vector $\alpha^*(t)$ defines the optimal utility function $u_\pi(t)$; in addition, alternative k^* is optimal in this region.

The algorithm is given in a diagrammatic form in fig. 6.5.

one-pass algorithm

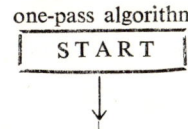

STEP 1

Pick an initial state vector and calculate the optimal control alternative and corresponding α-vector k^* and $\alpha^*(t)$ using (6.59).

STEP 2

Construct a complete list of inequalities that define the region, using

(1) $\quad U_\pi(t) = \min\limits_{k} \left[\sum\limits_{i=0}^{L} \pi_i \sum\limits_{j=0}^{L} p_{ij}^k \sum\limits_{\theta=1}^{M} r_{j\theta}^k e_\theta^k \alpha_j^{l_t(\pi,\,\theta,\,k)}(t-1) \right]$

(2) $\quad \sum\limits_{i,j} \pi_i p_{ij}^{k^*} r_{j\theta}^{k^*} e_\theta^{k^*} [\alpha_j^l(t-1) - \alpha_j^\delta(t-1)] \geq 0, \quad \forall\, \delta$

(3) $\quad \sum\limits_{i} \pi_i [\alpha_i^*(t) - \alpha_{k,i}(t)] \leq 0, \quad \forall\, k$

(4) $\quad \sum\limits_{i} \pi_i = 1$

STEP 3

Use the following LP to calculate the maximum set of inequalities that define the region for $\alpha^*(t)$

$$\max_{\pi}\, \pi b^k$$

s.t.

$\pi b^m \leq 0, \quad (m = 1, 2, 3, \ldots)$

$\pi_i \geq 0$

$\sum\limits_{i} \pi_i = 1$

From each boundary of the region, construct a new α-vector using (1, Step 2) and store the following in a list for each vector, its corresponding optimal control alternative and one information vector for which it is the α-vector.

STEP 4

Store the indices of the α-vectors that are neighbours to the region under consideration to limit the number of inequalities of type (2, Step 2) during the $(t+1)$ horizon calculation. If there are only α-vectors on the list whose region has not been calculated, pick a new α-vector and return to Step 2. Otherwise, the complete specification of the optimal control policy has been calculated.

Fig. 6.5

6.5. Risk-sensitive MDP with probabilistic observation of states; an infinite-horizon case

The model given in this section allows the study of optimal maintenance strategies of systems with Markov behaviour, where the states of the "core process" are partially observable. By relaxing the constraint given in Section 6.4, no restriction is apparent in the definition of probability rate transition and inspection matrices. It seems that the analysis using a decision tree approach may have certain advantages when branching and bounding in the decision tree are possible.

A definition of the above procedure was given by Lawler and Wood [90]:

"Among the most general approaches to the solution of constrained optimisation problems is that of 'branching-and-bounding' (or, in the words, of Bertier and Roy, 'séparation et évaluation progressive'). Like dynamic programming, branching and bounding is an intelligently structured search of the space of all feasible solutions. Most commonly, the space of all feasible solutions is repeatedly partitioned into smaller and smaller subsets and a lower bound (in the case of minimisation) is calculated for the cost of the solutions within each subset. After each partitioning, those subsets with a bound that exceed the cost of a known feasible solution are excluded from all further partitionings. The partitioning continues until a feasible solution is found such that its cost is no greater than the bound for any subset".

Lemmas 6.3 and 6.4 provide the computational procedure for the bounds of partially observable RSMDP.

Let us consider that $u_i = \lim_{t \to \infty} u_i(t)$, where $u_i(t)$ are successive solutions of

$$u_i(t) = \max_k \left\{ \sum_{j=1}^{L} p_{ij}^k c_{ij}^k u_j(t-1) \right\}, \quad (i = 1, 2, \ldots, L; \ t = 0, 1, \ldots). \ u_j(0) = 1, (j = 1, 2, \ldots, L),$$

and k is a maintenance decision alternative. We can also find u_i, $\forall i$, using results from ref. [35] that for large t ($t \to \infty$), the utility of any state will be multiplied by r (the largest eigenvalue for Q, the disutility matrix), at each successive stage.

LEMMA 6.3. *An upper bound for an RSMDP with probabilistic observation of states exists and is given by*

$$f(\pi) \leq \sum_i \pi_i u_i.$$

Proof. For the utility functional $f(\pi, \cdot)$, we can write by induction

$$f(\pi, 1) = \max_k \left\{ \sum_i \sum_j \pi_i p_{ij}^k e_{ij}^k \right\} \leq \sum_i \pi_i \max_k \left\{ \sum_j p_{ij}^k e_{ij}^k \right\} \leq \sum_i u_i(1).$$

Suppose that we have $f(\pi, t) \leq \sum_i \pi_i u_i(t)$.

Then, we can write the following utility functional.

$$f(\pi, t+1) = \max_k \{ \sum_i \pi_i \sum_j p_{ij}^k e_{ij}^k u(\pi_i p_{ij}^k r_{j0} f(\pi', t)) \} \leq$$

$$\leq \sum_i \pi_i \max_k \{ \sum_j p_{ij}^k e_{ij}^k u(\pi_i p_{ij}^k r_{j0} \tilde{v}_j(t)) \} \leq$$

$$\leq \sum_i \pi_i \max_k \{ \sum_j p_{ij}^k e_{ij}^k u(\tilde{v}_j(t)) \} \leq \sum_i \pi_i u_i(t+1).$$

or $f(\pi, t+1) \leq \sum_i \pi_i u_i(t+1)$.
Then,

$$f(\pi) = \lim_{t \to \infty} f(\pi, t) \leq \lim_{t \to \infty} \sum_i \pi_i u_i(t) \leq \sum_i \pi_i \lim u_i(t) \leq \sum_i \pi_i u_i.$$

LEMMA 6.4. *A lower bound for the utility functional of the above process exists and is given by*

$$f(\pi) > \max_k \pi [I - r^k P^k]^{-1} u^k,$$

where $u_i^k = \sum_j p_{ij}^k u(r_{ij}^k) = \sum_j p_{ij}^k \exp(-\gamma r_{ij}^k) = \sum_j q_{ij}^k$, *and r^k is the largest eigenvalue for the disutility matrix Q^k over a completely observable RSMDP.*

Proof. A possible policy is to make the same decision every period in the future, when $t \to \infty$. Suppose that we call the decision k. According to the results given by Howard and Matheson [35], for a large t, we need to multiply the utility vector by r (the largest eigenvalue for Q), when the process moves from one step to the other. At any future transition, r^k must be multiplied by the transition probability matrix P^k for any feasible decision k.

Then $\quad f(\pi) \geq \pi u^k + \pi r^k P^k u^k + \pi r^k r^k P^k P^k u^k + \ldots \geq \pi [I - r^k P^k]^{-1} u^k \geq$

$$\geq \max_k \pi [I - r^k P^k]^{-1} u^k.$$

6.5.1. A Branch and Bound algorithm

The utility of the upperbound on the return is obtained by using a modified decision set, assuming that the states are observed with certainty ($R_\theta = I$). The utility of the lowerbound on the return is given by the utility of the total expected return of any reasonable maintenance policy.

If the node has no continuation, then we have $u(\text{UPB}(II_k 0_\theta)) = \sum_j T_j(\pi/k, \theta) u(\tilde{v}_j)$,

where $\tilde{v}_j$ is the certain equivalent of the lottery in the case of completely observable state j. Similarly, $u(\text{LOB}(II_k 0_\theta)) = \max_m [T(\pi/m, \theta) (I - r^m P^m)^{-1} u^m]$. If the node has a branch originating from it, the above computational routine must be performed on all the terminal nodes. In this case, we have

$$u(\text{UPB}(I)) = \max_k \{ \sum_\theta \{\theta|\pi\} u[w_{II_k 0_\theta} + \text{UPB}(II_k 0_\theta)] \} =$$

$$\max_k \{ \sum_\theta \{\theta|\pi\} \exp(-\gamma w_{II_k 0_\theta}) \cdot u(\text{UPB}(II_k 0_\theta)) \},$$

and

$$u(\text{LOB}(I)) = \max_k \left\{ \sum_\theta \{\theta|\pi\} \, u[w_{II_k O_\theta} + \text{LOB}(II_k O_\theta)]\right\} =$$

$$= \max_k \left\{ \sum_\theta \{\theta|\pi\} \exp(-\gamma w_{II_k O_\theta}) \cdot u(\text{LOB}(II_k O_\theta))\right\},$$

where $\{\theta|\pi\} = \sum_{i,j} \pi_i p_{ij} r_{j\theta}$, and $e_{ij\theta} = \exp(-\gamma w_{II_k O_\theta}) = e_\theta$.

Finally, we have the following utility functionals.

$$u(\text{UPB}(I)) = \max_k \left\{ \sum_\theta \{\theta|\pi\} \, u(\text{UPB}(II_k O_\theta)) \, e_\theta \right\}$$

$$u(\text{LOB}(I)) = \max_k \left\{ \sum_\theta \{\theta|\pi\} \, u(\text{LOB}(II_k O_\theta)) \, e_\theta \right\}.$$

The criterion of decision branching for further tree extension has considerable effect on the efficiency of the Branch and Bound algorithm (convergence and computation time). We can use a Fibonnacci search method to find potential nodes in the decision tree to be extended.

As a bounding evaluation function for the optimal decision in any state we hall consider that "The utility of the lower bound for the decision $k \in K_i$, $\forall i$, is reater than the utility of the upper bound for any other decision l for that branch ($l \neq k$) (lexicographical ordering)".

6.5.2. A Fibonnacci search method for Branch and Bound algorithm of a Markovian decision process with logical conditions

As already shown, in the Branch and Bound algorithm for the MDP given above we need a decision criterion for further branching in view of finding the optimal solution. (Two criteria have been given and emphasized before.) In this paragraph we shall show how a Fibonnacci search method (see Beveridge et al. [102, pp. 193]) can be used for the selection of new branching trees (Step 4 in the Branch and Bound algorithm given below).

Let us consider that we can identify a number of H terminal nodes and define the Fibonnacci number as

$$F_N = 1 + H. \qquad (6.63)$$

Suppose that in the range of interest, we must compare H cases in order to determine the optimum node to be expanded, where H satisfies the above equation. The ordering for search of those nodes can be done in the increasing value of the cardinality of nodes. Once the system has been ordered, it is only necessary to label the various possible cases from 1 to H, consecutively. Then, the range of interest becomes $(H + 1)$ in length. Any given case will be identified by an integer h, where $1 \leq h \leq H + 1$. For N experiments, each experiment corresponds to an integral value of H, and hence to a particular case in the decision tree. The position of the first experiment, (the first searching procedure) is given by (see [102, pp. 194])

$$l_1 = \frac{F_{N-2}}{F_N}(H+1), \tag{6.64}$$

or, using (6.63), by $l_1 = F_{N-2}$. The experiments to be performed correspond to $h = l_1$ and $h = 1 = H - l_1$. For the general case, the node to be searched is given by the integral distance in the interval $(1 + H)$: $l_j = F_N - (j+1)\dfrac{1+H}{F_N}$ or, using results from above, by $l_j = F_N - (j+1)$ and l_j is always an integer.

Artificial locations are used for the case when $1 + H \neq F_N$, which will be added at one end of the sequence to bring the total to a Fibonacci number, such that $1 + H + F = F_N$, where now $F_N > 1 + H > F_{N-1}$, and F is the number of pseudo-experiments which have to be added. The search algorithm should then be applied as in the previous description.

For the MDP with logical conditions, for each node ($II_k\mathcal{M}d$), we can use $y(.) = \mathrm{UPB}(.) - \mathrm{LOB}(.)$ as a criterion for comparison, where $\mathrm{UPB}(.)$ $\mathrm{LDB}(.)$ have been calculated in earlier stages of the Branch and Bound algorithm (Step 3).

Let us consider a fictitious example for an MDP with logical conditions with $L = 3$ number of states, $M = 2$ independent observations and $\mathcal{M} = 4$ cause-effect models. The number of decision alternatives in each state is $k = 2$. We can easily find that $H = 8$, and because $1 + h \neq F_N$ (see [102, pp. 182]), then $F = 4$ and $F_N = 13$ for $N = 6$. Following (6.64), $l_1 = 5$, and, accordingly, $h = 5$ and $h = 4$. But as for this example the experiment $h = 4$ incidentally corresponds to the pseudo-experiment $(F = 4)$, then the node $(1, 2, 1)$ corresponding to $h = 5$ should be further extended in the B & B algorithm.

6.5.3. Algorithm description

Step 1: Find all labelled nodes in the tree.
Step 2: Perform the above routine on the labelled nodes.
Step 3: If at level *1* a dominating decision k is found, then the procedure terminates; otherwise, repeat the procedure with the new tree. Better upper and lower bounds on the utility functional will increase the computational efficiency. If $u(L_i^A) \geqslant u(L_i^B)$, then $L_i^A \geqslant L_i^B$ or $A \succ B$ (policy A is preferred to B), and policy A dominates B, then policy B need not be considered for any further analysis. Now, we need to find policies which are not dominated.

The algorithm is sequential in character. As the process makes the transition part of the previously done computation can be used for the next stage. We shall give now an example in order to illustrate the steps involved in applying the model given above.

Example. Let us consider a system under maintenance control alternatives. For the sake of simplicity, we shall consider that $L = 2$ and $M = 3$. The decision-maker is a risk-sensitive person and follows an exponential utility function.

The data for the problem are given below.

Alternative 1 "do nothing"	Alternative 2 "do repair"
$P^1 = \begin{bmatrix} 0.4 & 0.6 \\ 0.7 & 0.3 \end{bmatrix}$	$P^2 = \begin{bmatrix} 0.5 & 0.5 \\ 0.6 & 0.4 \end{bmatrix}$
$R^1 = \begin{bmatrix} 0.3 & 0.3 & 0.4 \\ 0.5 & 0.2 & 0.3 \end{bmatrix}$	$R^2 = \begin{bmatrix} 0.4 & 0.2 & 0.4 \\ 0.3 & 0.5 & 0.2 \end{bmatrix}$
$v^1 = \begin{bmatrix} 3 \\ 2 \end{bmatrix}$	$v^2 = \begin{bmatrix} 2 \\ 1 \end{bmatrix}$

$$\pi = [0.6 \quad 0.4]$$

v^1 and v^2 represent the loss from the system when a "do repair" or a "do nothing" decision must be taken.

First iteration. The probability state vector after a transition has happend is given in tables 6.1 and 6.2. For a total risk-aversion person ($\gamma = 1$), we have the following data:

$$c_{ij}^1 = \begin{bmatrix} 1 & 2 \\ 3 & 5 \end{bmatrix}, \qquad c_{ij}^2 = \begin{bmatrix} 3 & 4 \\ 2 & 3 \end{bmatrix}$$

$$u(v^1) = \begin{bmatrix} 0.04978 \\ 0.13533 \end{bmatrix}, \qquad u(v^2) = \begin{bmatrix} 0.13533 \\ 0.36787 \end{bmatrix}$$

$$\underline{k = 1} \qquad \underline{k = 2}$$
$$w_{II_k O_\theta} = [3 \quad 2 \quad 2], \qquad [4 \quad 2 \quad 1]$$

Table 6.1 · · · · · · · · · · · · · · · · · · Table 6.2

Alternative 1

$T(\pi/\theta, 1)$	State	
	1	2
$\theta = 1$	0.49	0.51
$\theta = 2$	0.86	0.14
$\theta = 3$	0.82	0.18

Alternative 2

$T(\pi/\theta, 2)$	State	
	1	2
$\theta = 1$	0.80	0.20
$\theta = 2$	0.27	0.73
$\theta = 3$	0.90	0.10

Let us suppose that policy A, which we are going to follow, gives the utility vector

$$L^A = \begin{bmatrix} 4 \\ 2 \end{bmatrix} \Rightarrow u(L^A) = \begin{bmatrix} 0.01831 \\ 0.13533 \end{bmatrix}.$$

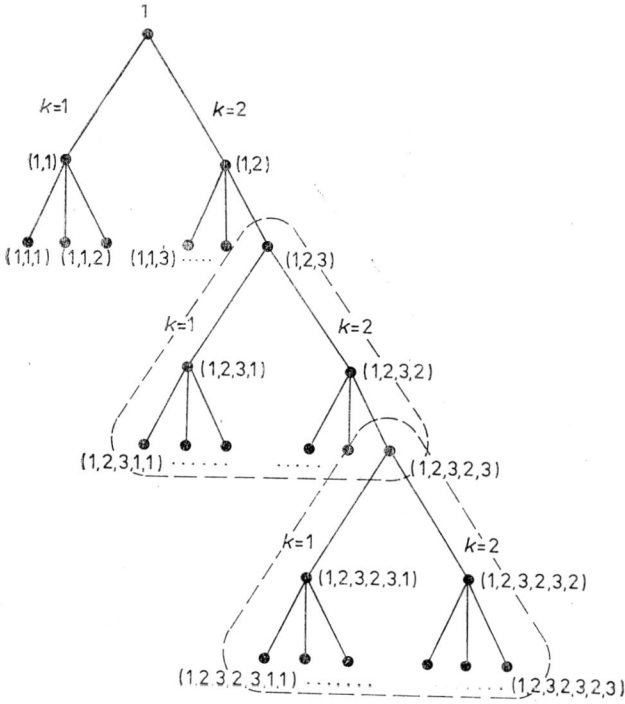

Fig. 6.6

The upper and lower bounds of the nodes in the decision tree are given in table 6.3. We need to proceed to a second iteration such that a new tree may be built (see fig. 6.6). We choose (1, 2, 3) as a new branching node. At the second iteration, the probability state vector for alternatives one and two is given in tables 6.4 and 6.5. The bound over the nodes are given in table 6.6. At the second iteration, the process does not reach the optimal criterion. Then, we need to proceed to the third iteration. The probability state vector at the third iteration is given in tables 6.8 and 6.9. The bounds over the nodes are given in table 6.7. At the fourth iteration, the bounds for the nodes in the new tree are given in table 6.10. At the fourth iteration, LOB (1, 2) > UPB(1,1) and then decision 2 ("do repair") is preferred to decision 1 when the system is in state 1. For the case of a total risk-averse decision-maker, he should always follow a repair maintenance action when the system is in state 1.

Table 6.3
First iteration

Node	Upper bound	Lower bound
1, 1, 1	0.0882775	0.070969
1, 1, 2	0.0711675	0.047565
1, 1, 3	0.073734	0.0510756
1, 2, 1	0.2027666	0.0522458
1, 2, 2	0.2702032	0.0861816
1, 2, 3	0.1841634	0.0428842
1, 1	$2.8219 \cdot 10^{-3}$	$1.9968 \cdot 10^{-3}$
1, 2	0.01604	$4.1916 \cdot 10^{-3}$
1	0.01604	$1.9968 \cdot 10^{-3}$

Table 6.4
Alternative 1

$T(\pi/\theta, 1)$	State	
	1	2
$\theta = 1$	0.43	0.57
$\theta = 2$	0.92	0.08
$\theta = 3$	0.89	0.11

Table 6.5
Alternative 2

$T(\pi/\theta, 2)$	State	
	1	2
$\theta = 1$	0.87	0.13
$\theta = 2$	0.16	0.84
$\theta = 3$	0.96	0.04

Table 6.6
Second iteration

Node	Upper bound	Lower bound
1, 1	$2.99 \cdot 10^{-3}$	$2.125 \cdot 10^{-3}$
1, 2	0.0182770355	$4.7696 \cdot 10^{-3}$

Table 6.7
Third iteration

Node	Upper bound	Lower bound
1, 1	$3.1404 \cdot 10^{-3}$	$2.21533 \cdot 10^{-3}$
1, 2	0.197465943	$5.1597 \cdot 10^{-3}$

Table 6.8
Alternative 1

$T(\pi/\theta, 1)$	State	
	1	2
$\theta = 1$	0.38	0.62
$\theta = 2$	0.96	0.04
$\theta = 3$	0.93	0.07

Table 6.9
Alternative 2

$T(\pi/\theta, 2)$	State	
	1	2
$\theta = 1$	0.92	0.08
$\theta = 2$	0.87	0.13
$\theta = 3$	0.98	0.02

Table 6.10
Fourth iteration

Node	Upper bound	Lower bound
1, 1, 1	0.10282696	0.0908624
1, 1, 2	0.0532089	0.01811892
1, 1, 3	0.0557754	0.02655014
1, 2, 1	0.1539388	0.0276716
1, 2, 2	0.16556602	0.02788587
1, 2, 3	0.13998616	0.0206504
1, 1	$2.68105 \cdot 10^{-3}$	$1.4499 \cdot 10^{-3}$
1, 2	0.105869	$2.2734 \cdot 10^{-3}$
1	0.105869	$1.4499 \cdot 10^{-3}$

Practical aspects and extensions of risk-sensitive Markovian models for complex systems maintenance. The maintenance models given in this chapter could be serviceable in practical applications when large quantities of money are dependent on decisions which involve risk. Examples could be concerned with replacement decisions of computers in a big firm or other expensive equipment in technical systems. We can couple the above models with a sensitivity analysis for small deviations in the transition matrix of the core process or changing in maintenance cost strategies. In the following we shall perform a sensitivity analysis of the limiting probability vector for the case of small deviations in the transition probability matrix of the deterioration process. We shall use the theoretical results of Takahashi [28] in order to calculate the deviations on the limiting state vector when the process runs for a long-time horizon.

Let us consider two deterioration processes described by regular Markov chains, $\mathscr{D}$ and $\mathscr{D}'$, with the finite state space $0, 1, \ldots, L$. The probability transition matrices are given by $P = [p_{ij}]$ and $P' = [p'_{ij}]$, such that $(1 + \varepsilon)^{-1} p_{ij} \leqslant p'_{ij} \leqslant (1+\varepsilon) p_{ij}$, $(i, j = 0, 1, \ldots, L;\ i \neq j)$, where ε is a positive constant.

THEOREM 6.4 [28]. *If the above relation holds, then the limit state probability π' corresponding to $\mathscr{D}'$ is given by*

$$\frac{\pi_k}{\pi_k + (1 + \varepsilon)^{2L}(1 - \pi_k)} \leqslant \pi'_k \leqslant \frac{\pi_k}{\pi_k + (1 + \varepsilon)^{-2L}(1 - \pi_k)}$$

for every k, $k = 0, 1, 2, \ldots, L$.

The use of exponential utility functions in measuring risk sensitivity has several computational advantages for the policy iteration algorithm, because of the "delta" property. This function also expresses the constant risk-aversion of the decision-maker. The systems analyst can offer to the model user more information about the changing in the optimal maintenance strategies, either for different risk-aversion coefficients or for different input data in the transition matrix.

An example for a machine replacement. We shall consider an example adapted from Kao [27], namely, a piece of equipment whose performance at any time t, ($t = 0, 1, 2, \ldots$), can be characterized by one of the following states: initial, good, marginal and terminal. These states ($\Phi^*_\omega(X) = 0, 1, 2, 3$) are denoted by 0, 1, 2 and 3, respectively. The underlying state of the equipment is always known with certainty. The deterioration and repair processes of the equipment can be approximated by a Markov process with the transition probabilities

$$P^0 = \begin{bmatrix} 0 & 0.80 & 0.15 & 0.05 \\ 0 & 0 & 0.97 & 0.03 \\ 0 & 0 & 0 & 1.0 \\ 0 & 0 & 0 & 1.0 \end{bmatrix} \quad P^1 = \begin{bmatrix} 0.4 & 0.3 & 0.2 & 0.1 \\ 0.85 & 0.15 & 0 & 0 \\ 0.75 & 0.20 & 0.05 & 0 \\ 0 & 0.50 & 0.40 & 0.10 \end{bmatrix}.$$

The cost of replacement and repair is given in table 6.11. The decision-maker is characterized by an exponential utility function. The computation results are summarized in table 6.12. A pictorial representation of results is given by curve 1 in fig. 6.7.

Table 6.11
Repair and Replacement Information for the Example

ITEM	The underlying state i			
	0	1	2	3
Occupancy cost for the system	1	2	3	7
Repair cost (Alternative 1)	2	5	10	25
Fixed replacement cost	5	10	20	30
Repair cost (Alternative 2)	3	6	9	20

Table 6.12
Results for RSMDP

Risk aversion coefficient	Gain for the maintenance process	Optimal maintenance policy				Number of iterations
−0.1	5.00058	2	1	2	1	3
−0.8	4.98539	2	1	2	1	3
−0.6	4.99999	2	1	2	1	3
−0.4	5.00000	2	1	1	1	2
−0.2	5.00000	2	1	1	1	2
0.2	2.50854	0	1	1	1	3
0.4	3.35222	0	1	1	1	3
0.6	3.17949	1	1	1	1	2
0.8	2.98928	1	1	1	1	2
+1.0	2.84361	1	1	1	1	2

(N.B. — 0 — "do nothing", 1 — "repair the system", 2 — "replace the system".

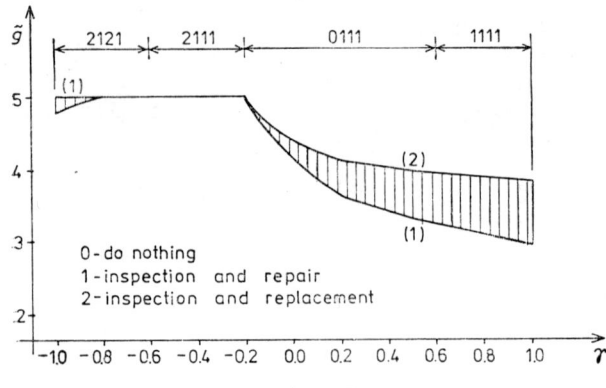

Fig. 6.7

If we choose a repair activity of a different kind (which finally affects the repair cost in the decision model), then the certain equivalent gain variation is represented by curve 2 in fig. 6.7. We can see from this example that $\tilde{g}$ is almost constant for the risk-preference decision-maker ($\gamma \in [-1.0, -0.2]$) and is decreasing for a risk-aversion maintenance engineer.

Replacement rules under a semi-Markov deterioration. Consider a complex technical system under a deterioration process. Its performance at any point in time is completely characterized by a finite number of states $0, 1, 2, \ldots, L$. The significance of state description is that previously described. The deterioration process is of a semi-Markov type, where "transitions from state to state are made in accordance with a Markov chain and the time spent in each state before a transition occurs is a random variable depending on the transition" (see Kao [27]).

The model parameters are: a) the probability transition p_{ij} that the system moves to a less desirable state j, ($j = i+1, \ldots, L$), such that $\sum_{j=i+1}^{L} p_{ij} = 1, i = 0, 1, 2,$ $\ldots, L-1$, $p_{ij} = 0$ for $j \leq i$; b) the holding time, defined by $\tau_{ij}(h_{ij}(.) = \Pr\{\tau_{ij} = (.)\}$ for $i = 0, 1, 2, \ldots, L-1$ and $j = i+1, \ldots, L$); c) in the terminal state, in the absence of replacement, $p_{LL} = 1$ and $h_{LL}(1) = 1$. The flow diagram of an unattended system is given in fig. 6.8.

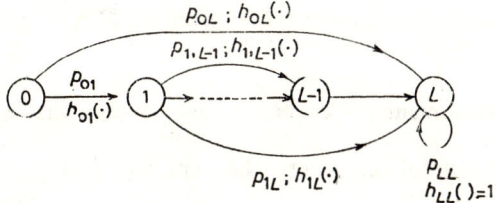

Fig. 6.8

Costs replacement structure. The occupancy cost A_i (monetary units/unit time), ($i = 0, 1, 2, \ldots, L$), is associated with the system when it is holding in state i; A_0 represents the maintenance cost (per unit time) of a new technical system and A_L is the cost associated with the system failure.

The fixed cost R_i, ($i = 0, 1, 2, \ldots, L$), is associated with replacement of the system when it is in state i. Due to real maintenance activities, we can also include a variable cost D_i, such that the total system replacement costs are $R_i + D_i \tau_{i0}$, where τ_{i0} is a finite random variable, integer-valued for the amount of time units required to make a replacement in state i ($h_{i0}(.) = \Pr\{\tau_{i0} = (.)\}$).

We shall use the following notation *.

---

* the symbol $\leq$ ($>$) denotes the corresponding cumulative (complementary cumulative) distribution and the bar symbol denotes the conditional expectation.

N_{ij} represents the maximum possible holding time in state i before moving to state j (for $i = 0, 1, \ldots, L-1$ and $p_{ij} > 0$);

$$\hat{N}_i = \max_j \{N_{ij} | j = 1+i, \ldots, L; \, p_{ij} > 0\};$$

$$\check{N}_i = \min_j \{N_{ij} | j = i+1, \ldots, L; \, p_{ij} > 0\};$$

$$N_{ij}^* = \max \{N_{i,i_1} + \ldots + N_{i_k,j}\}$$

and represents the maximum number of time units required to move from state i to state $j \, (i < j)$;

$$\theta_{0i} = \dot{\sum} p_{0i_1} \ldots, p_{i_k,i} \text{ (e.g. } \theta_{00} = 1); \quad {}^{\leqslant}h_{ij}(n) = \sum_{m=1}^{n} h_{ij}(m); \quad {}^{>}h_{ij}(n) =$$

$$= \sum_{m=n+1}^{N_{ij}} h_{ij}(m); \quad {}^{\geqslant}h_{ij}(n) = \sum_{m=n}^{N_{ij}} h_{ij}(m); \quad \overline{h_{ij}(n)} = \sum_{m=1}^{n} m h_{ij}(m); \quad c_{ij}(m) = p_{ij} h_{ij}(m);$$

$${}^{\leqslant}c_{ij}(n) = p_{ij} {}^{\leqslant}h_{ij}(n); \quad {}^{>}c_{ij}(n) = p_{ij} {}^{>}h_{ij}(n); \quad d_{ij}(n) = p_{ij}\overline{h_{ij}(n)}.$$

The decision of the system engineer to replace the technical system at the time a transition is made to state i is denoted by 1 and the decision "do nothing" by 0. A *state-dependent replacement rule* is given by $\mathbb{R}_s = (k_0, k_1, k_2, \ldots, k_{L-1}, k_L)$ for $k_i = 0, 1, \, i = 0, 1, 2, \ldots, L-1$ and $k_L = 1$.

If $\widetilde{\mathbb{R}}_s$ = class of $\mathbb{R}_s$, then card $\widetilde{\mathbb{R}}_s = 2^L$.

An optimal replacement policy $\mathbb{R}_s^*$ is defined such that $\Gamma_{\mathbb{R}_s^*} = \min_{\mathbb{R}_s \in \widetilde{\mathbb{R}}_s} \{\Gamma_{\mathbb{R}_s}\}$,

where the bracketed symbol represents the expected average cost per unit time.

The parameters below describe the maintenance process formulated as a discrete time semi-Markov decision process.

— Transition probabilities,

$$p_{ij}^0 = \begin{cases} p_{ij} & \text{for } i = 0, 1, 2, \ldots, L-1; \, j = i+1, \ldots, L \\ 0 & \text{otherwise} \end{cases}$$

$$p_{ij}^1 = \begin{cases} 1 & \text{for } i = 0, 1, \ldots, L; \, j = 0 \\ 0 & \text{otherwise} \end{cases}$$

— Mean waiting times,

$$\bar{\tau}_i^0 = \sum_{j=i+1}^{L} p_{ij} \bar{\tau}_{ij}, \quad (i = 0, 1, \ldots, L-1)$$

$$\bar{\tau}_i^1 = \bar{\tau}_{i0}, \quad (i = 0, 1, \ldots, L),$$

where $\bar{\tau}_{ij}$ indicates the mean holding times of the deterioration process, and $\bar{\tau}_{i0}$ is the mean time for replacement when the system is in state i;

— Cost rates,

$$r_i^0 = A_i \bar{\tau}_i^0, \qquad (i = 0, 1, \ldots, L-1)$$

$$r_i^1 = D_i \bar{\tau}_{i0} + R_i, \qquad (i = 0, 1, \ldots, L)$$

$$q_i^k = r_i^k/\bar{\tau}_i^k, \qquad (i = 0, 1, \ldots, L; \; k = 0, 1).$$

Denote i^* to prescribe a control limit decision rule such that when the technical system reaches state i, if $i \geqslant i^*$, "do replacement" and "do nothing" otherwise.

For an optimal state-dependent replacement rule, Kao [27] indicated the following conditions.

Condition 1: $\{r_i^0\}_{i=0}^{L}$ is a non-decreasing bounded sequence and $\{\bar{\tau}_i^0\}_{i=0}^{L}$ is a non-increasing sequence;

Condition 2: $r_i^1 = r^1$ and $\bar{\tau}_i^1 = \bar{\tau}^1 = \bar{\tau}_{i0}$ for $i = 0, 1, 2, \ldots, L$;

Condition 3: $\sum_{j=k}^{L} p_{ij}$ is non-decreasing in i for each $k = 0, 1, 2, \ldots, L$.

LEMMA 6.5 [27]. *In accordance with Condition 3, for each non-decreasing function $\xi(j)$, the function $\sum_{j=0}^{L} p_{ij}\xi(j)$ is also non-decreasing in i.*

LEMMA 6.6 [27]. *Let $V_\beta(i, n)$ be the minimal total expected discounted maintenance cost, given that the technical system is in state i and there are n transitions remaining (β is the discount factor $0 < \beta < 1$). Let $V_\beta(i) = \lim_{n \to \infty} V_\beta(i, n)$, where this represents the minimal total expected discounted cost.*

If Conditions 1−3 hold, then $V_\beta(i)$ is non-decreasing in i.

LEMMA 6.7 [27]. *If conditions 1 − 3 hold, then $V_\beta(i) - V_\beta(0)$ is uniformly bounded for $i = 1, 2, \ldots, L$.*

THEOREM [27]. *If lemma 6.7 holds, then there exists a constant g and a bounded non-decreasing function v_i such that $\mathbb{R}_s$, which minimizes the right-hand side of the equation, is $\mathbb{R}_s^*$ which minimizes the expected average cost:*

$$v_i = \min_{k=0,1} \left\{ r_i^k + \sum_{j=0}^{L} p_{ij}^k v_j - g\bar{\tau}_i^k \right\}, \qquad (i = 0, 1, 2, \ldots, L).$$

THEOREM 6.5 [27]. *Under Conditions 1−3, there exists a control limit $\mathbb{R}_s$ that is optimal over all $\mathbb{R}_s \in \widetilde{\mathbb{R}}_s$.*

The optimal state-dependent control-limit replacement rule implies a replacement action as soon as $i > i^*$ and "do nothing" otherwise, where

$$i^* = \min_{i=0,1,\ldots,L} \left\{ i \,\Big|\, r_i^0 + \sum_{j=0}^{L} p_{ij}^0 v_j - g\bar{\tau}_i^0 > r^1 + v^0 + g\bar{\tau}^1 \right\}.$$

The optimal control limit i^* is calculated *via* the non-decreasing function $\xi(j)$, defined as

$$\xi(j) = \begin{cases} r_i^1/\overline{\tau}_i^1 & \text{for } i = 0 \\ \dfrac{\left[\sum_{j=0}^{j=i-1} \theta_{0j} r_j^0 + \sum_{j=i}^{L} r_j^1 \sum_{k=0}^{k=i-1} \theta_{0k} p_{kj}\right]}{\left[\sum_{j=0}^{j=i-1} \theta_{0j} \overline{\tau}_j^0 + \sum_{j=i}^{j=L} \overline{\tau}_j^1 \sum_{k=0}^{k=i-1} \theta_{0k} p_{kj}\right]} & \text{for } i = 1, 2, \ldots, L. \end{cases}$$

When Conditions 1—3 are fulfilled in the above expression for $\xi(j)$, we have $r_j^1 = r^1$ and $\overline{\tau}_j^1 = \overline{\tau}^1$, and finally

$$i^* = \{i \mid \Gamma_{\mathbb{R}_s^*} = \min_{i=1,\ldots,L} \{\xi(j)\}\}.$$

For solving the maintenance problem concerning optimal state-dependent replacement rule, we can use the policy-iteration method (see Howard [8]).

A numerical example [27]. The performance of a complex technical equipment at any time $t \geq 0$ is characterized by one of the four states 0, 1, 2, 3 (the initial, good, marginal and absorbing/terminal states). The system is considered to be completely observable and the deterioration process of the technical equipment is approximated by a semi-Markov process with the following transition probabilities matrix.

$$P = \begin{bmatrix} 0 & 0.8 & 0.15 & 0.05 \\ 0 & 0 & 0.97 & 0.03 \\ 0 & 0 & 0 & 1.0 \\ 0 & 0 & 0 & 1.0 \end{bmatrix}$$

The holding-time mass functions are given in table 6.13 and the occupancy costs and chosen example in table 6.14. All the necessary information to be used in applying the policy-iteration algorithm are tabulated in table 6.15.

Table 6.13

m	$h_{01}(m)$	$h_{02}(m)$	$h_{03}(m)$	$h_{12}(m)$	$h_{13}(m)$	$h_{23}(m)$
1	0.01	0.15	0.02	0.07	0.04	0.01
2	0.03	0.07	0.03	0.07	0.07	0.02
3	0.05	0.02	0.07	0.12	0.13	0.02
4	0.07	0.02	0.10	0.15	0.76	0.02
5	0.09	0.02	0.13	0.20		0.05
6	0.21	0.02	0.16	0.27		0.88
7	0.54	0.35	0.20	0.10		
8		0.20	0.12	0.02		
9		0.15	0.07			
10			0.05			
11			0.04			
12			0.01			

Table 6.14

Item	State i			
	0	1	2	3
Occupancy cost, A_i (monetary units/unit time)	1	2	3	7
Fixed replacement cost, R_i	5	10	20	30
Variable replacement cost, D_i	2	3	4	5
Mean time for replacement, τ_{i0}	1	1	2	3

Table 6.15

State i	Alternative k	p_{ij}^k				$\bar{\tau}_i^k$	r_i^k	q_i^k
		p_{i0}^k	p_{i1}^k	p_{i2}^k	p_{i3}^k			
0	0	0	0.80	0.15	0.05	6.017	6.017	1
	1	1	0	0	0	1	7	7
1	0	0	0	0.97	0.03	4.6188	9.2376	2
	1	1	0	0	0	1	13	13
2	0	0	0	0	1	5.72	17.16	3
	1	1	0	0	0	2	28	14
15	1	1	0	0	0	3	45	15

The results obtained by the use of the policy iteration method are summarized in table 6.16. Two iterations are required to obtain the optimal solution: $\mathbb{R}_s^* = [0, 1, 1, 1]$ and $\Gamma_{\mathbb{R}_s^*} = 3.1467$.

A different class of replacement rules is for the *state-age-dependent maintenance* case. When the holding-time mass functional probability in a given state is very large for a certain time period, a replacement rule that is based on the underlying state of the system and the amount of time the system has been in that state may result in a lower long-run average cost per unit time.

Table 6.16

a) Policy-improvement phase

State i	Alternative k	Iteration 1 Γ_i^k	Iteration 2 Γ_i^k	Iteration 3 Γ_i^k
0	0	1	4.1256	3.1467
	1	7	7	7
1	0	2	4.1256	4.6563
	1	13	3.56	3.1467
2	0	3	4.1256	5.4219
	1	14	0.9077	3.1467
3	1	15	4.1256	3.1467

b) Policy-evaluation operations

	Iteration 1	Iteration 2
g	4.1256	3.1467
v_0	−32.6232	−35.5599
v_1	−16.0632	−25.7066
v_2	− 6.4385	−13.8533
v_3	0	0

A state-age-dependent replacement rule is indicated by $\mathbb{R}_{sa}$ such that $\mathbb{R}_{sa} = [n_0, n_1, n_2, \ldots, n_L]$ and $0 \leq n_i \leq \hat{N}_i$, $(i = 0, 1, \ldots, L-1)$, and $n_L = 0$, where n_i evaluates the time units the technical system has been in state i, $(i = 0, 1, 2, \ldots, L)$. It is also considered that $h_{ij}(m) > 0$ for all $p_{ij} > 0$ and $m = 1, 2, \ldots, N_i$; $I = 0, 1, \ldots, L-1$, and $n_L = 0$, where n_i evaluates the time units the technical system has been in state i, $(i = 0, 1, 2, \ldots, L)$. It is also considered that $h_{ij}(m) > 0$ for all $p_{ij} > 0$ and $m = 1, 2, \ldots, N_{ij}$. If $\tilde{\mathbb{R}}_{sa}$ = class of $\mathbb{R}_{sa}$, then card $\tilde{\mathbb{R}}_{sa} = \prod_{i=0}^{L-1}(N_i + 1)$.

An optimal state-age-dependent replacement rule is defined such that
$$\Gamma_{\mathbb{R}_{sa}^*} = \min_{\mathbb{R}_{sa} \in \tilde{\mathbb{R}}_{sa}} \{\Gamma_{\mathbb{R}_{sa}}\},$$

where the bracketed symbol represents the expected average cost per unit time associated with the above maintenance rule.

For finding an optimal policy to this problem, Kao [27] presented a solution by enumeration and a solution by the policy iteration method.

We shall deal now with the solution by the policy-iteration method. The systems engineer (decision maker) can select the age of replacement in each state and this generates a semi-Markov decision process with $(\hat{N}_i + 1)$ alternatives in each state i, $(i = 0, 1, \ldots, L-1)$. The policy iteration method can be used with the fixed state space and simple modifications of process parameters *.

For the case in which $0 < n_i \leqslant \check{N}_i$ for all i, $i = 0, 1, 2, \ldots, L-1$, the following model parameters must be considered.

a) Transition probabilities,

$$p_{i0}^* = \begin{cases} \sum_{j=i+1}^{L} {}^{>}c_{ij}(n_i) & \text{for } i = 0, 1, \ldots, L-1 \\ 1 & \text{for } i = L \end{cases}$$

$$p_{ij}^* = \begin{cases} {}^{\leqslant}c_{ij}(n_i) & \text{for } i = 0, 1, \ldots, L-1; \ j = i+1, \ldots, L \\ 0 & \text{otherwise;} \end{cases}$$

b) Mean holding times,

$$\bar{\tau}_{i0}^* = \begin{cases} \bar{\tau}_{i0} + n_i & \text{for } i = 0, 1, \ldots, L-1 \\ \bar{\tau}_{i0} & \text{for } i = L \end{cases}$$

$$\bar{\tau}_{ij}^* = \begin{cases} \overline{h_{ij}(n_i)}/{}^{\leqslant}h_{ij}(n_i) & \text{for } i = 0, 1, \ldots, L-1; \\ & j = i+1, \ldots, L \\ 0 & \text{otherwise;} \end{cases}$$

c) Cost rates

$$q_i^* = \begin{cases} \dfrac{\left[A_i \sum_{j=i+1}^{L} d_{ij}(n_i) + (n_i A_i + D_i \bar{\tau}_{i0} + R_i) \sum_{j=i+1}^{L} {}^{>}c_{ij}(n_i)\right]}{\left[(\bar{\tau}_{i0} + n_i) \sum_{j=i+1}^{L} {}^{>}c_{ij}(n_i) + \sum_{j=i+1}^{L} d_{ij}(n_i)\right]} & \text{for } i = 0, 1, 2, \ldots, L-1 \\ (D_i \bar{\tau}_{i0} + R_i)/\bar{\tau}_{i0} & \text{for } i = L; \end{cases}$$

d) Limiting interval transition probabilities,

$$\Phi_i^* = \begin{cases} (1/U) e_{0i} \left[[\bar{\tau}_{i0} + n_i] \sum_{j=i+1}^{L} {}^{>}c_{ij}(n_i) + \sum_{j=i+1}^{L} d_{ij}(n_i)\right] \\ (1/U) e_{0i} \bar{\tau}_{i0} \qquad \text{for} \quad i = L, \end{cases}$$

where

$$e_{0i} = \begin{cases} \sum \# {}^{\leqslant}c_{0i_1}(n_0) \ldots {}^{\leqslant}c_{i_k}, i(n_{ik}), & \text{for } 1 \leqslant i \leqslant L \\ 1 & \text{for } i = 0 \end{cases}$$

---

* The asterisks are used to denote the process parameters under the present replacement rule.

and $\sum_{\#}$ represents the sum over all possible paths going from state 0 to state i;

$$U = \sum_{i=0}^{L-1} e_{0i} \left[(\bar{\tau}_{i0} + n_i) \sum_{j=i+1}^{L} {}^{>}c_{ij}(n_i) + \sum_{j=i+1}^{L} d_{ij}(n_i) \right] + e_{0L}\bar{\tau}_{L0}.$$

The long-run average cost $\Gamma_{\mathbb{R}_{sa}}$ can be calculated using Φ_i^* and q_i^*.

For the case in which $n_i = 0$ for some or all i, $(i = 0, 1, 2, \ldots, L-1)$, and $0 \leq n_i \leq \check{N}_i$ for all i, $(i = 0, 1, \ldots, L-1)$, if $n_i = 0$, then ${}^{\leq}h_{ij}(0) = 0$, ${}^{>}h_{ij}(0) = 1$, $\bar{h}_{ij}(0) = 0$, $d_{ij}(0) = 0$ for all j; $p_{i0} = 1$ and 0 otherwise. All the results given previously follow Kao [27], who also investigated the case in which $\check{N}_i < n_i \leq N_i$ for some or all i, $(i = 0, 1, \ldots, L-1)$.

The optimal solution for the state-age-dependent replacement rules is the same either using an enumeration or a policy-iteration solution.

A numerical example [27]. For the chosen technical system considering the class of state-age-dependent replacement rules $\tilde{\mathbb{R}}_{sa}$, we have

$$\mathbb{R}_{sa} = [n_0, n_1, n_2, 0]$$

$0 \leq n_0 \leq 12$; $0 \leq n_1 \leq 8$; $0 \leq n_2 \leq 6$.

The data used in finding an optimal state-age-dependent replacement rule for the technical system are given in table 6.17. The results obtained by the use of the policy-iteration method are presented in table 6.18.

After two iterations the optimal solution is $\mathbb{R}_{sa}^* = [5, 0, 0, 0]$ with $\Gamma_{\mathbb{R}_{sa}^*} = 2.5248$.

There are practical cases when the state of the system is unknown except when a replacement occurs or the cost of monitoring the state of the technical system under observation is relatively high. The type of replacement rule when the decision to replace the system depends only on the time elapsed since the completion of the last replacement is referred to as the *age-dependent replacement rule* (see Kao [27]).

Such a replacement rule is denoted by $\mathbb{R}_a = t$, where t represents the age of the technical system at which a replacement takes place. If $\tilde{\mathbb{R}}_a$ is the class of age-dependent replacement rules, then it consists of integers $0, 1, 2, \ldots, N_{0L}^*$.

6.6. Inspection-maintenance-replacement model for technical systems

The model described in this paragraph is due to Klein [23]. The deterioration process is considered to have a stochastic character, the identification of which is possible upon inspection. Decision alternatives for system maintenance are 0) complete replacement of the system, 1) immediate repair of system components and subsystems or 2) rescheduling for a new inspection on the system. A repair or replacement decision on the technical system restores the system to the "as new" state. The deterioration of the system is described by a discrete time, finite Markov chain.

Table 6.17

I. Policy-improvement phase

State	Alternative	Iteration number		
		1	2	3
i	k	Γ_i^k	Γ_i^k	Γ_i^k
0	0	7.0000	7.0000	7.0000
	1	3.9520	4.3352	4.2739
	2	2.0286	3.4373	3.3375
	3	2.4037	3.0192	2.8804
	4	2.0701	2.8294	2.6445
	5	1.8305	2.7620 ←	2.5248 ←
	6	1.5700	2.9410	2.5583
	7	1.0672	3.9042	3.0657
	8	1.0310	4.0165	3.1710
	9	1.0050	4.1050	3.2509
	10	1.0025	4.1148	3.2630
	11	1.0005	4.1234	3.2733
	12	1.0000 ←	4.1256	3.2759
1	0	13.000	−3.5600 ←	2.5248 ←
	1	7.3032	−0.3321	2.7085
	2	5.3920	0.7726	2.8005
	3	4.3063	1.5542	3.0557
	4	3.3337	2.2747	3.4789
	5	2.0282	2.9086	3.8712
	6	2.2784	3.7503	4.3043
	7	2.0462	4.0618	4.7448
	8	2.0000 ←	4.1256	4.7950
2	0	14.0000	0.9077 ←	2.5248 ←
	1	10.3087	1.6314	2.7328
	2	8.4300	2.0163	2.8759
	3	7.3004	2.2483	2.9631
	4	6.5459	2.4036	3.0224
	5	5.9333	2.5591	3.1363
	6	3.0000 ←	4.1256	5.5306
3	0	15.000 ←	4.1256 ←	2.5248 ←

I. Policy-evaluation phase

	Iteration number	
	1	2
g	4.1256	2.5248
v_0	−32.6232	−37.4256
v_1	−16.0632	−26.9504
v_2	−6.4385	−14.4752
v_3	0.0	0.0

Table 6.18

Age of repl.	Conditional expected total occupancy times in each state				Interval transition probabilities				Exp. const. per cycle S_t	Exp. time per cycle M_t	Expected average cost with $R_a = t$
	$E(T_0\|t)$	$E(T_1\|t)$	$E(T_2\|t)$	$E(T_3\|t)$	$\Phi_{00}(t)$	$\Phi_{01}(t)$	$\Phi_{02}(t)$	$\Phi_{03}(t)$			
0	0.0000	0.0000	0.0000	0.0000	1.0000	0.0000	0.0000	0.0000	7.00	1.00	7.00
1	1.0000	0.0000	0.0000	0.0000	0.9685	0.0080	0.0225	0.0010	8.56	2.02	4.23
2	1.9685	0.0080	0.0225	0.0010	0.9325	0.0315	0.0333	0.0327	10.05	3.04	3.31
3	2.9010	0.0394	0.0558	0.0800	0.8800	0.0692	0.0379	0.0069	11.65	4.05	2.87
4	3.7870	0.1086	0.0957	0.8220	0.8220	0.1198	0.0127	0.0127	13.52	5.07	2.67
5	4.6090	0.2285	0.1392	0.7405	0.7405	0.1809	0.0205	0.0205	15.73	6.10	2.58 ⇐
6	5.3495	0.4094	0.1973	0.0437	0.5615	0.3296	0.0778	0.0311	18.86	7.14	2.64
7	5.9110	0.7390	0.2570	0.0748	0.0670	0.7248	0.1445	0.0637	25.54	8.27	3.09
8	5.9780	1.4637	0.4196	0.1386	0.0310	0.6502	0.2349	0.0839	30.16	9.40	3.21
9	6.0090	2.1140	0.6545	0.2224	0.0050	0.5639	0.3321	0.0990	34.88	10.55	3.31
10	6.0140	2.6778	0.9866	0.3214	0.0025	0.4506	0.4266	0.1203	39.81	11.67	3.41
11	6.0165	3.1284	1.4132	0.4417	0.0005	0.3203	0.5272	0.1320	45.38	12.83	3.54
12	6.0170	3.4478	1.9404	0.5938	0.0000	0.1844	0.6323	0.1833	51.24	14.00	3.66
13	6.0170	3.6331	2.5728	0.7770	0.0000	0.0535	0.6727	0.2738	58.29	15.22	3.83
14	6.0170	3.6866	3.2455	1.0508	0.0000	0.0084	0.6141	0.3775	64.77	16.37	3.96
15	6.0170	3.6950	3.8596	1.4282	0.0000	0.0000	0.5139	0.4841	71.21	17.48	4.07
16	6.0170	3.6950	4.3756	1.9123	0.0000	0.0000	0.4088	0.5912	77.97	18.59	4.39
17	6.0170	3.6950	4.7844	2.5035	0.0000	0.0000	0.2923	0.7077	85.32	19.71	4.33
18	6.0170	3.6950	5.0767	3.2112	0.0000	0.0000	0.1651	0.8349	93.31	20.83	4.48
19	6.0170	3.6950	5.2418	4.0461	0.0000	0.0000	0.0475	0.9525	101.65	21.95	4.63
20	6.0170	3.6950	5.2893	4.9986	0.0000	0.0000	0.0074	0.9926	109.14	23.00	4.75
21	6.0170	3.6950	5.2967	5.9912	0.0000	0.0000	0.0000	1.0000	116.24	24.00	4.84

The inspection procedure is capable of identifying the state of the system. A repair activity will put the system in one of the possible operational states. The identification of an inspection-repair-replacement policy will be formulated in linear programming terms. The operational states of the system are classified in accordance with the performance level of the total structure function $\Phi_o^*(X)$.

At any point in time t, the system can be completely characterized by classifying it as in one of the possible states $0, 1, \ldots, L$. (State 0 represents the process before any deterioration and state L represents the terminal state of deterioration — an absorbing state.)

The sequence of successive states of the system forms a discrete Markov chain with stationary transition probabilities $p_{ij}(i, j = 0, 1, \ldots, L)$, with the following supplementary conditions. $\lim_{t\to\infty} p_{iL}(t) = 1$, $(i = 0, 1, \ldots, L-1)$ and $p_{LL} = 1$, where $p_{ij}(t)$ represents the t-step transition probability from state i to state j, $t \geq 1$. It is also assumed that $p_{iL} > 0$, $(i = 0, 1, \ldots, L-1)$.

Let $d_{s:k}(i)$ denote a general maintenance decision to translate the system from its present state i to that described by state s and to schedule the next inspection k periods from the present, where

$$i = 0, 1, \ldots, L, L(1), \ldots, L(k-1);$$

$$k = 1, 2, \ldots, K; \; s \in S_i \text{ and}$$

$$S_i = \{0, 1, \ldots, L-1\} \cap \{i\}.$$

In the above notation, $L(m)$ indicates that the technical system has been inoperative for m time periods. It is assumed in this model that inspection repairs and replacements are made almost instantaneously; next transitions will take place at later time periods.

All possible randomized decision rules are of the form $D_{is:k} = \Pr\{d_{s:k}(i)\}$ and $\sum_{s:k; s \in S_i} D_{is:k} = 1$, $(i = 0, 1, \ldots, L(k-1))$.

The controlled process is finally expressed by the stationary transition probabilities q_{ij}, $(i, j = 0, 1, \ldots, D(k-1))$, where $q_{ij} = \sum_{s:k} D_{is:k} v_{s:kj}$, $(i, j = 0, 1, \ldots, L(k-1))$ and

$$v_{s:kj} = \begin{cases} p_{sj}^k, (j = 0, 1, \ldots, L; \; s = 0, 1, \ldots, L-1; \; k = 1, 2, \ldots, k); \\ q_{sL}^{k-n}, (s = 0, 1, \ldots, L-1; \; j = L(n); \; n = 1, 2, \ldots, K-1); \; (k-n) > 0; \\ 1, \; s = L(m), \; (m = 0, 1, \ldots, K-2; \; j = L(n); n = 1, 2, \ldots, k-1; \\ \quad k = (n-m) > 0; \; j = s+k; \; k = 1, 2, \ldots, L(k-1)-s); \\ 0, \quad \text{otherwise.} \end{cases}$$

Costs relevant to the management of deteriorating technical systems consist of maintenance costs (i.e. repair and replacement costs), inspection and penalty costs related to the length of time between failure of the system and its identification through inspection.

The following notation will be used in the context of the present model: $a_{s:kj}$ represents the cost of inspection in state j, given that k periods in the past the state was s ($a_{s:kj} = 0$ if $s = L, L(1), \ldots, L(k-1)$); $b_{s:kj}$ represents the penalty cost associated with state j, given that k periods earlier the system was in state s.

The structure of penalty costs is

$$b_{s:kj} = \begin{cases} 0 & \text{if } j = 0, 1, \ldots, L-1 \\ b_j \geq 0 & \text{if } s = 0, 1, \ldots, L-1; \\ & j = L, L(1), \ldots, L(K-1) \\ b_j - b_s \geq 0 & \text{if } s = L, L(1), \ldots, L(K-2) \\ & \text{and } j = s + k. \end{cases}$$

For convenience, we can combine the inspection and penalty costs as given by $t_{s:kj}$.

If $c_{is:k}$ represents the cost associated with the decision $d_{s:k}(i)$, then it consists of m_{is}, i.e. the maintenance cost associated with the management decision to change from state i to state s (m_{i0} represents the cost of replacement before failure if $i \leq (L-1)$, and after failure if $i > (L-1)$, and m_{is} is the cost of repair from state i to state s if $0 < s \leq L-1$ or zero if $i = s \neq L(K-1)$).

The expected cost per inspection of the technical system is given by

$$\bar{C} = \sum_i \sum_{s:k} \sum_j \pi_i D_{is:k} v_{s:kj} t_{s:kj} + \sum_i \sum_{s:k} \pi_i D_{is:k} c_{is:k},$$

where all π_i, ($i = 0, 1, \ldots, L(K-1)$), represent the equilibrium probabilities of the different states of the control maintenance process with the additional constraints

$$\pi_j = \sum_i \pi_i p_{ij}, \ (j = 0, 1, \ldots, L(K-1)); \ \sum_j \pi_j = 1.$$

The decision problem can be formulated as a linear program $\min \varphi = \sum_i \dfrac{c_i}{d_i} z_i^2$ with the following constraints. $\sum_i z_i^2 = 1, \ \sum_i \dfrac{a_{ij}}{d_i} z_i^2 = 0, \ (i = 1, 2, \ldots, n); \ z_i^2 \geq 0$, ($i = 1, 2, \ldots, n$), where $z_i = y_i/t$; $y_i = +\sqrt{d_i x_i}$; $x_i = \pi_i D_i$. In order to simplify the form of the linear program, the notation was altered to single subscripts.

6.7. Maintenance processes where "cannibalization" is the only repair activity

There are many practical cases when the only decision alternative concerning the maintenance of a technical system is that of replacing failed components, i.e. modules, by others which are in a good state of operation. These last components could be taken from the other system just being repaired or in a standby reserve.

In the literature on systems engineering, such maintenance strategies are known under the name of cannibalization (cannibalization as a repair activity). For more details, see Hirsch et al. [43] and Rolfe [117]. Application of such mathematical models refers to aircraft and submarine repair as well as other transportation means which have to continue their mission or to be available at the requested moment. The model presented in this paragraph uses Markov processes for a system with a very general coherent structure and cannibalization is the only repair activity at hand at a given moment (see Rolfe [117]).

Let us consider a group of S identical aircraft which have to operate under the following conditions:

1) Any aircraft includes in its coherent structure K parts or components. The probability distribution function for the time-to-failure of the i-th part subsystem is exponentially distributed such that it is of the form $1 - e^{-\lambda_i t}$. For practical reasons, we consider that k parts in the model are assumed to be non-critical. This means that irrespective of the number of the K assemblies which fail during a given mission, in principle, it is always possible for an aircraft to return to the base.

2) Any failed equipment/subsystem part can neither be repaired nor resupplied. It is considered further that there is enough repair capability to cannibalize unfailed parts from other failed aircraft.

3) The fly mission for the aircraft is of a fixed length T. It is also assumed that there is enough time between consecutive mission takeoff times for all aircraft to return from the previous mission.

First, the model has to provide the expected number of good aircraft after t missions, $t \geq 1$, which we denote by $E[A(t)]$. The analysis is performed using completely observable Markov processes.

Let $n_i(t)$ be the number of copies of part i which will survive t missions, such that $\mathbf{n}(t) = [n_1(t), n_2(t), \ldots, n_k(t)]$ is a vector which represents the number of parts of each type which survive t missions. The Markov process used has $(S+1)^K$ states for each t, where $n_i(t)$ is an integer ($0 \leq n_i(t) \leq S$; $i = 1, 2, \ldots, K$).

In order to decrease the dimensionality of the model, we lump all states with one or more $n_i(t) = 0$ into a single "zero" state. The new cardinality of the states of the process becomes $S^K + 1$. Using a Markov model, the evolution from mission t to $t+1$ is given by the one-step transition matrix P of dimension $((S^K+1) \times (S^K; 1))$. For the case of lexicographically ordered states, P becomes a lower triangular matrix with zero above the principal diagonal.

We shall use the following notation: $d_1 =$ (zero state); $d_2 = [1, 1, \ldots, 1]; \ldots$; $d_{S^K+1} = [S, S, \ldots, S]$.

For $d_i = [i_1, \ldots, i_k]$, $d_j = [j_1, \ldots, j_k]$ and $M_i = \min_{l=1,\ldots,K} i_l$, the p_{ij} elements of the transition matrix P become

1) $\prod_{l=1}^{K} \binom{M_i}{j_l - i_l + M_i} \cdot (e^{-\lambda_l T})^{j_l - i_l + M_i} \cdot (1 - e^{-\lambda_l T})^{i_l - j_l},$

if $i, j > 1$ and $j_l \leq i_l \leq j_l + M_i$ for $l = 1, 2, \ldots, K$;

2) 0 if $i, j > 1$ and $i_l < j_l$ or $i_l > j_l + M_i$ for at least one $l = 1, 2, \ldots, K$;

3) 1 if $i = j = 1$;
4) 0 if $i > 1, j > 1$;
5) $1 - \sum_{d=2}^{i} p_{id}$ if $i > 1, j = 1$.

If $\pi(t)$ is a row vector, then from the general theory of Markov chains,

$$\pi(t) = \pi(0) P^t; \quad t \geq 1, \quad \text{where} \quad \pi(0) = [0, 0, \ldots, 0, 1],$$

because the process starts in state $(S, S, \ldots, S)$. The states of $\pi(\cdot)$ and P are lexicographically ordered.

The first two moments of the number of good aircrafts after t missions are given by the following relations.

$$E[A(t)] = \sum_{j=1}^{S^K+1} \pi_j(t) M_j \quad \text{and} \quad E[A(t)^2] = \sum_{j=1}^{S^K+1} \pi_j(t) M_j^2,$$

where
$$M_j = \min_{l=1,2,\ldots,K} \{j_l | d_j = (j_1, \ldots, j_K)\}.$$

It is assumed that M_j is independent of t for each j and as soon as $\pi(t)$ is determined, the value for $E[\cdot]$ can be easily calculated.

Example 1 [117]. Consider the case $K = 2$ when only two cannibalizable parts per aircraft are allowed. Let $q = (e^{-\lambda_1 T}, e^{-\lambda_2 T})$ be the vector whose components represent the probabilities of non-failure for each part on one mission of length T.

Similarly, S represents the number of aircraft under repair or inspection. Computational results for a fixed q and S variable are given in table 6.19 and for q variable and S fixed in table 6.20 and the quantity of interest is the expected number of good aircraft after t missions ($E[A(t)]$). Entries in table 6.20 are $E[A(t)]/S$; $q = (0.9; 0.8)$ and for table 6.21, $E[A(t)]/E[A(t-1)]$; $S = 10$.

Table 6.19

t \ S	2	5	10	13
1	0.734	0.759	0.777	0.782
2	0.554	0.598	0.620	0.626
3	0.423	0.473	0.495	0.501
4	0.323	0.374	0.396	0.401
5	0.247	0.296	0.316	0.320
6	0.187	0.233	0.252	0.256
7	0.142	0.184	0.201	0.204
8	0.106	0.144	0.160	0.163
9	0.080	0.113	0.127	0.130
10	0.059	0.088	0.101	0.103
11	0.044	0.068	0.080	0.082
12	0.032	0.052	0.063	0.065
13	0.024	0.040	0.049	0.052
14	0.017	0.030	0.039	0.041
15	0.013	0.023	0.030	0.032

Table 6.20

q \ t	(0.9, 0.8)	(0.8, 0.8)	(0,7, 0.7)	(0.6, 0.6)	(0.8, 0.5)	(0.5, 0.5)	(0.9, 0.4)	(0.5, 0.4)
1	0.777	0.731	0.620	0.514	0.494	0.412	0.399	0.353
2	0.804	0.768	0.661	0.558	0.506	0.456	0.400	0.391
3	0.803	0.773	0.667	0.563	0.504	0.456	0.401	0.389
4	0.806	0.775	0.668	0.558	0.503	0.450	0.400	0.382
5	0.805	0.777	0.668	0.556	0.503	—	—	—
6	0.802	0.776	0.667	0.550	—	—	—	—
8	0.803	0.776	0.660	—	—	—	—	—
10	0.804	0.775	—	—	—	—	—	—
12	0.804	0.773	—	—	—	—	—	—
14	0.804	—	—	—	—	—	—	—

For the case when no cannibalization is allowed, $E[A(t)]/S = (0.72)^t$ results from the fact that 0.72 represents the probability that any aircraft survives one mission and a sequence of binomial trials are allowed.

From table 6.21, we can see the behaviour of successive values of $E[A(t)]$ for a number of ten aircraft and for different values for q. It is clear that for $t \geq 2$, the ratio ρ ($\rho = E[A(t)]/E[A(t-1)]$) is a constant. This is equivalent with $E[A(t-1)]/S = \rho^{t-1}$, $t \geq 2$ which leads to the relations $E[A(t)]/S = \rho_0(\rho)^{t-1}$, $t \geq 1$ and $\rho_0 = E[A(1)]/S$.

Table 6.21

S \ t	Any S, no cannibalization $(0.72)\,t$	2	5	10	13
1	0.720	0.734	0.762	0.777	0.783
2	0.518	0.557	0.600	0.622	0.628
3	0.373	0.431	0.480	0.500	0.504
4	0.268	0.336	0.384	0.402	0.405
5	0.193	0.263	0.308	0.323	0.326
6	0.139	0.206	0.247	0.260	0.261
7	0.100	0.161	0.198	0.209	0.210
8	0.072	0.125	0.159	0.168	0.169
9	0.054	0.097	0.128	0.135	0.136
10	0.037	0.075	0.102	0.108	0.109
11	0.027	0.058	0.082	0.087	0.087
12	0.019	0.044	0.066	0.070	0.070
13	0.014	0.034	0.053	0.056	0.056
14	0.010	0.026	0.042	0.045	0.045
15	0.007	0.019	0.033	0.036	0.036

An approximative solution for $K > 2$. According to Rolfe [117], the value for $E[A(t)]$ with $q = [q_1, q_2, \ldots, q_k]$ and $\alpha = \min_i q_i$ is given by $E[A(t)]/S^i = \alpha^t$; $t \geq 0$.

The lower bounds for $E[A(t)]/S$ and $E[A(t)]$ are $E[A(t)]/S \geq \beta^t$, $t \geq 0$, and $\beta = \prod_{i=1}^{k} q_i$, or considering that every surviving part is tested on every mission then, from ref. [117],

$$E[A(t)]/S \geq \frac{1}{S} \sum_{k=1}^{S} \prod_{i=1}^{K} \sum_{j=k}^{S} \binom{S}{j} q_i^{t\,j}(1 - q_i^t)^{S-j}.$$

Applying the above relation to the data given in table 6.19, we obtain the results in table 6.21.

A computational method for calculating the generating function of state probabilities. If $N = S^{K+1}$ and $\{p_{ij};\ i,j = 1, 2, \ldots, N\}$ are elements of the transition matrix P and if $\pi(t) = [\pi_1(t), \pi_2(t), \ldots, \pi_N(t)]$ is the state probability vector, then the generating functions are defined as below [117]. $\tilde{\pi}_i(z) = \sum_{t=0}^{\infty} \pi_i(t) z^t$, $(i = 1, 2, \ldots, N)$.

According to the Chapman-Kolmogorov equation for Markov processes, we have $\pi(t) = \pi(t-1)\,P$.

Using the above relations and the fact that $p_{ij} = 0, j > 1$, we can write

$$\tilde{\pi}_N(z) = p_{NN} z \tilde{\pi}_N(z) + 1$$

$$\tilde{\pi}_i(z) = \sum_{j=i}^{N} p_{ji} z \tilde{\pi}_j(z); \ i = N-1, N-2, \ldots, 1.$$

Similarly,

1) $\quad \tilde{\pi}_N(z) = 1/(1 - p_{NN} z)$ \hfill (6.65)

2) $\quad \tilde{\pi}_i(z) = \left(z \sum_{j=i+1}^{N} p_{ji} \tilde{\pi}_j(z)/(1 - p_{ji} z) \right); \ i = N-1, N-2, \ldots, 1.$ \hfill (6.66)

Using an inverse-transformation, Eqn. (6.65) becomes $\pi_N(t) = (p_{NN})^t$, $t \geq 0$.

Example 2 [117]. Consider a technical system in a repair environment with $S = 2$, $K = 2$, $q = (0.5;\ 0.4)$. Apply the above relations and, hence,

$$P = \begin{bmatrix} 1 & 0 & 0 & 0 & 0 \\ 0.8 & 0.2 & 0 & 0 & 0 \\ 0.5 & 0.3 & 0.2 & 0 & 0 \\ 0.6 & 0.2 & 0 & 0.2 & 0 \\ 0.52 & 0.24 & 0.08 & 0.12 & 0.04 \end{bmatrix}.$$

Equations (6.65) and (6.66) generate

$$\tilde{\pi}_5(z) = 1/(1 - 0.04z),$$

$$\tilde{\pi}_4(z) = (0.12z)/(1 - 0.2z)(1 - 0.04z),$$

$$\tilde{\pi}_3(z) = (0.8z)/(1 - 0.2z)(1 - 0.04z),$$

$$\tilde{\pi}_2(z) = (0.24z)/(1-0.2z)^2(1-0.04z),$$

$$\tilde{\pi}_1(z) = (0.0016z)(5+z)(65-z)/(1-z)(1-0.2z)^2 \cdot (1-0.04z).$$

By using methods to invert the generation functions, we have $E[A(t)] = 0\{\pi_1(t)\} + 1\{\pi_2(t) + \pi_3(t) + \pi_4(t)\} + 2\{\pi_5(t)\}$ or: $E[A(t)] = \dfrac{3}{2} t (0.2)^t + \dfrac{7}{8} (0.2)^t + \dfrac{9}{8} (0.04)^t;\ t \geqslant 0.$

6.8. Generalized cannibalization policies in multicomponent systems

The study of the consequence of interchanging components within a generalized coherent structure (see Hochberg [212]) — previously defined as cannibalization — is highly important in systems engineering. In this section the general theory of cannibalization is extended to the case of multichotomic behaviour of each component of the system. A representation theorem developed in [212] is given, which expresses the state of a system as a function of the number of working parts at each level.

Following the general framework of coherent structures, we introduce the abstract set $\Lambda = \{\lambda_1, \lambda_2, \ldots, \lambda_n\}$, which represents the loci in the technical structure. We consider that each locus can be in k possible states such that if it is the highest operational state, then $a_1 = 1$, and further assume one of the $k-2$ intermediate operational values $\{a_2, a_3, a_4, \ldots, a_{k-1}\}$ such that $1 = a_1 > a_2 > a_3 > \ldots > a_{k-1} > 0$.

If $a_k = 0$, then the locus fails to contain an operational part. Recalling previous results, at any fixed moment of time, the states of the structure are described by a mapping $v: \Lambda \to \{a_1, a_2, a_3, \ldots, a_{k-1}, a_k\}$ and the state of λ_i is denoted by $x_i = v(\lambda_i)$.

The total number of states of the structure is K^n such that $K^n = \{(z_1, z_2, \ldots, z_n);\ z_i = a_j,\ i = 1, 2, \ldots, n;\ j = 1, 2, \ldots, k\}$.

For the case of cannibalization, we shall assume that two subsystems/parts O_1 and O_2 can be interchanged if and only if (1) they are capable of functioning in precisely the same set of loci and (2) when O_1 is installed and operating at a given level in a given locus λ, then its "contribution" to the level of performance of an engineering structure is exactly the same as that of O_2 when O_2 is installed and operating at the same given level in λ.

The performance of the structure is measured by a set of possible performance levels $\mathcal{A} = \{0, 1, \ldots, M\}$; the state 0 represents a total failure and state M denotes a perfect performance of the system. For the structure function Φ, we have the mapping $\Phi: K^n \to \mathcal{A}$ which has a monotonic behaviour.

With respect to the interchangeability of a coherent structure, we introduce the set $\Gamma = \{\gamma_1, \gamma_2, \ldots, \gamma_N\}$, where γ_i, $(i = 1, 2, \ldots, N)$, represent the part types in the structure. The collection of all possible subsets of Λ is given by 2^Λ and $Q: \Gamma \to 2^\Lambda$.

Values $Q(\gamma_i)$ represent the set of all loci in which parts of type γ_i were installed initially. Interchanges are allowed only within $Q(\gamma_i)$ but not between a locus $\lambda \in Q(\gamma_i)$ and a locus $\lambda' \notin (\gamma_i)$. It is also assumed that $Q(\gamma_i) \neq \Phi$, $(i = 1, 2, \ldots, N)$, $Q(\gamma_i) \cap Q(\gamma_j) = \Phi$ if $i \neq j$, $\bigcup_{i=1}^{N} Q(\gamma_i) = \Lambda$.

For each integer i, $i \in [1, n]$, there exists a unique $j = \delta(i)$ that satisfies the relation $\lambda_i \in Q(\gamma_j)$ such that each part type used at each locus is identified by δ.

DEFINITION 6.4 [212]. For a given set F, the indicator function I_F is defined by

$$I_F(v) = \begin{cases} 1 & \text{if } v \in F \\ 0 & \text{if } v \notin F. \end{cases}$$

DEFINITION 6.5 [212]. For each integer l, $l \in [1, N]$, we set $w_l = (w_{l_1}, w_{l_2}, \ldots, w_{l, k-1})$, where $w_{l_q}: K^n \to \{0, 1, \ldots, n\}$ is the map defined by

$$w_{l_q}(v) = \sum_{\{i : \lambda_i \in Q(\gamma_l)\}} I_{\{(v)_i = a_q\}}(v); \quad (q = 1, 2, \ldots, k-1).$$

The number of operational parts of type γ_l operating at level a_q, $l_q = 1, 2, \ldots, k-1$, is denoted by w_{l_q}.

Property 1. If and only if all $\lambda_i \in Q(\gamma_l)$ are operating at some positive level of performance, then

$$0 \leqslant \sum_{q=1}^{k-1} w_{l_q}(v) \leqslant |Q(\gamma_l)|.$$

Property 2. Under the assumptions given in Property 1,

$$\sum_{q=1}^{k-1} w_{l_q}(v) = |Q(\gamma_l)|.$$

Property 3. $w_{l_1}(v) = |Q(\gamma_l)|$ if and only if $(v)_i = 1$ for all i such that $\lambda_i \in Q(\gamma_l)$.

DEFINITION 6.6. Two vertices $v \in K^n$ and $v' \in K^n$ are w-equivalent in symbols $(v \stackrel{w_l}{\sim} v')$, if and only if $w_l(v) = w_l(v')$.

DEFINITION 6.7. The vertex v is said to be equivalent to the vertex $v'(v \sim v')$ if and only if for each integer j, $j \in [1, N]$ we can write $v \stackrel{w_l}{\sim} v'$.

We shall consider that the class $[v]$ consists of all locus-vector states v' for which the number of operational parts of type γ_l, $l \in [1, N]$ operating at level a_i, $(i = 1, 2, \ldots, k-1)$, is the same for both v and v'.

DEFINITION 6.8. A cannibalization maintenance procedure is any transformation $T: K^n \to K^n$ such that for all $v \in K^n$, we have $Tv \in [v]$.

The class of all cannibalizations are denoted by $\mathcal{T}$.

DEFINITION 6.9. A cannibalization structure function Φ^T, given a cannibalization T, is defined by $\Phi^T(v) = \phi T(v) = \phi(Tv)$.

An admissible cannibalization T is given by $\Phi^T \geq \Phi^{T'}$ for all $T' \in \mathcal{T}$ or, more precisely, by

$$\Phi^T(v) = \max_{v' \in [v]} \Phi(v'), \qquad v \in K^n.$$

The class of admissible cannibalization is expressed by $\mathcal{T}^*$.

DEFINITION 6.10. A maximum point of the restriction $(\Phi|[u])$ of Φ to $[u]$ is a point $\bar{u}$ in $[u]$ such that for all $u' \in [u]$ we have $\Phi(\bar{u}) \geq \Phi(u')$.

The following theorems will be given without proofs. For further details, see Hochberg [212].

THEOREM 6.6. *If Φ is a monotone (coherent) structure function, then Φ^* is also a monotone structure function, where Φ^* is a cannibalized structure function induced by $T \in \mathcal{T}^*$.*

The performance of an engineering system depends individually on each part type.

If $\Pi_i: K^n \to K^n$ is the mapping defined by

$$(\Pi_i u)_j = \begin{cases} (u)_j & \text{if } \lambda_j \in Q(\gamma_i), \\ 1 & \text{otherwise,} \end{cases}$$

then "the effect of Π_i on a locus-vector state is to transform it into one in which all the loci occupied by parts other than type γ_i are operational in state 1 and in which the states of the loci corresponding to part type γ_i are left unchanged" [212].

The above mapping Π_i is non-decreasing and order preserving, and if the definition of a cannibalization T is used, then

$$T\Pi_i v \sim \Pi_i v \stackrel{w_i}{\sim} v.$$

A structure function relative to an integer i ($i \in [1, N]$) is defined by

$$\Phi_i = \Phi \Pi_i,$$

and describes how the structure would perform without cannibalization when an infinite number of spares are available.

DEFINITION 6.11. A cannibalized structure function Φ_i^* relative to i is given by

$$\Phi_i^*(v) = \Phi^*(\Pi_i v) = \Phi(T \Pi_i v)$$

and if $v \stackrel{w_i}{\sim} v'$, then

$$\Phi_i^*(v) = \Phi^*(\Pi_i v) = \Phi^*(\Pi_i v') = \Phi_i^*(v').$$

There exists a function

$g_i: \{0, 1, \ldots, |Q(\gamma_i)|\}^{k-1} \to S$ such that for all $v \in K^n$, $\Phi_i^*(v) = g_i w_i(v)$.

THEOREM 6.7. *If Φ^* is a monotone structure function, then g_i is non-decreasing in each variable.*

DEFINITION 6.12. A structure function Φ is said to satisfy the minimum condition if $\Phi = \min_{1 \leqslant l \leqslant N} \Phi_l$, and it states that "the value that the structure function assumes at any vertex is determined by one particular part type, in the sense that if all other part types were to be made fully operational, the value of the structure function remains unchanged" [212].

THEOREM 6.8. If $\Phi = \min_{1 \leqslant l \leqslant N} \Phi_l$, then $\Phi^* = \min_{1 \leqslant l \leqslant N} \Phi_l^*$.

THEOREM 6.9. *Let Φ be any structure function that induces Φ^*. The relation $\Phi^* = \min_{1 \leqslant l \leqslant N} \Phi_l^*$ holds if and only if to each maximum point $\bar{v}$ of $\Phi[v]$, there corresponds an integer i_0 depending on $\bar{v}$ such that $\Pi_{i_0}\bar{v}$ is a maximum point of $\Phi[\Pi_{i_0}\bar{v}]$ and $\Phi(\bar{v}) = \Phi(\Pi_{i_0}\bar{v})$.*

THEOREM 6.10. If $\Phi^* = \min_{1 \leqslant l \leqslant N} \Phi_l^*$, then

$$\min_{1 \leqslant j \leqslant N} \Phi T\Pi_j = \min_{1 \leqslant j \leqslant N} \Phi \Pi_j T.$$

To give a representation of the cannibalized structure function Φ^*, we may recall that $\Phi_i^* = g_i w_i$. If $n_i(q)$, $q \in [0, M]$ is the set of minimal points of A_i, q^1, then we have the following

THEOREM 6.11. *For all integers i, q, $i \in [1, N]$, $q \in [0, M+1]$ and $v \in K^n$, $g_i w_i(v) \geqslant q$ if and only if $w_i(v) \geqslant n_i(q)$.*

THEOREM 6.12. *If $q < q'$, then $n_i(q) \leqslant n_i(q')$.*

We give now the representation theorem for generalized cannibalization structures.

THEOREM 6.13. *If $\Phi^* = \min_{1 \leqslant l \leqslant N} \Phi_l^*$, then $\Phi^* = \sum_{k=1}^{M} \prod_{i=1}^{N} I_{\{w_i \geqslant n_i(k)\}} = \sum_{k=1}^{M} I_{\bigcap_{i=1}^{N} \{w_i \geqslant n_i(k)\}}$.*

For a given moment of time we shall determine the probability that the system is operating at level k, $(k = 0, 1, \ldots, M)$.

In compliance with [212], we shall make the following assumptions.

a) the technical system is governed by a monotone structure function Φ;

b) at time $t = 0$, there are s_i spares available of part type γ_i, $i = 1, 2, \ldots, N$;

c) the maintenance policy is that a spare is installed only when its locus in the structure has zero operational value;

d) a failure of a given part type is serviced by performing an admissible cannibalization.

Let us define a stochastic process of the form $\{V^T(t) = (V_1^{(T)}(t), V_2^{(T)}(t), \ldots, V_n^{(T)}(t), t \geqslant 0\}$. The random variable $V_i^{(T)}$ for each fixed t may have the values a_j, $(j = 1, 2, \ldots, k)$.

---

[1] $A_{i,q}$, $i \in [1, N]$, $q \in [0, M+1]$ denote the set of $(k-1)$-tuples in the domain of g_i, for which g_i takes a value at least as large as q.

DEFINITION 6.13. $V_i^T(t)$ represents the state of the part in locus λ_i at time t if cannibalization T is used.

Let $\varphi^*(t)$ represent the state of the cannibalized system $\varphi^*(t) = \Phi^* V^T(t)$, at time t and let $W_i^T(t)$ be the random number of operating parts of type γ_i at time t such that $W_i^T(t) = w_i V^T(t)$.

From the representation theorem, we can write

$$\varphi^*(t) = \sum_{k=1}^{M} \prod_{i=1}^{N} I_{\{W_i^T(t) \geq n_i(k)\}} = \sum_{k=1}^{M} I_{\bigcap_{i=1}^{N} \{W_i^T(t) \geq n_i(k)\}}.$$

Hochberg [212] proved that the expected state of the cannibalized system at time t is given by

$$E[\varphi^*(t)] = \sum_{k=1}^{M} \prod_{i=1}^{N} E[I_{\{W_i(t) \geq n_i(k)\}}] = \sum_{k=1}^{M} \prod_{i=1}^{N} \Pr\{W_i(t) \geq n_i(k)\}.$$

For

$$I_k = I_{\bigcap_{i=1}^{N} \{W_i(t) \geq n_i(k)\}}, \quad k = 1, 2, \ldots, M; \ n_i(1) \leq \ldots \leq n_i(M),$$

we have

$$I_1 \geq I_2 \geq \ldots \geq I_M,$$

and

$$\Pr\{\varphi^*(t) \geq j\} = \sum_{i=1}^{N} \Pr\{W_i(t) \geq n_i(j)\}.$$

By computing $\Pr\{W_i(t) \geq n_{i(j)}\}$ for $i = 1, 2, \ldots, N$, $j = 1, 2, \ldots, M$, we can evaluate the probability distribution and the expected value of $\varphi^*(t)$.

If $\zeta_v(t)$ denotes the number of operational parts at level a_v at time t, $|Q|$ are the installed parts, F are the distributed random variables with a common distribution function for $z_v(O_j)$, $(v = 1; j = 1, 2, \ldots, |Q|)$, the random variable representing the lifetime of part O_j in state a_v, G a common distribution function for $Z_v(O_j)$, $v = 2, j = 1, 2, \ldots, |Q|$, then for $s = 0$ (s represents the spares available $|Q|$) we have

$$\Pr\{\xi_1(t) = j, \ \xi_2(t) = k\} = (j, k, |Q| - j - k) \cdot [1 - F(t)]^j =$$
$$= [F(t) - F * G(t)]^k [F * G(t)]^{|Q|-j-k},$$

where $t \geq 0$ and $*$ denotes a congruent multiplication.

For $s > 0$ and exponential lifetimes in both states (operating and failure) we have the relations

$$\Pr\{\xi_1(t) = j \text{ and } \xi_2(t) = k\} = \int_0^t K_t(y) \, dF_s(y),$$

where

$$K_t(y) = \sum_{v=1}^{|Q|} \sum_{q=0}^{k} \binom{v}{j,q,v-j-q} [C(t-y)]^j [B(t-y)]^q [A(t-y)]^{v-j-q}$$

$$\cdot \binom{|Q|-v}{k-q} [D(t-y)]^{k-q} [E(t-y)]^{|Q|-v-k+q} \binom{|Q|-1}{v-1} [\mu_1(y)]^{v-1} [\mu_2(y)]^{|Q|-v};$$

$$D(t) = \Pr\{z_2(O_j) \geq t\} = \begin{cases} e^{-\lambda_2 t}, & t \geq 0 \\ 0; & t < 0 \end{cases}$$

$$E(t) = \Pr\{z_2(O_j) < t\} = \begin{cases} 1 - e^{-\lambda_2 t}; & t \geq 0 \\ 0, & t < 0 \end{cases}$$

$$A(t) = 1 - e^{-\lambda_1 t} - \frac{\lambda_1}{\lambda_1 - \lambda_2} [e^{-\lambda_2 t} - e^{-\lambda_1 t}]; \quad t \geq 0$$

$$B(t) = \frac{\lambda_1}{\lambda_1 - \lambda_2} [e^{-\lambda_2 t} - e^{-\lambda_1 t}]; \quad t \geq 0.$$

$$C(t) = 1 - [1 - e^{-\lambda_1 t}] = e^{-\lambda_1 t}, \quad t \geq 0$$

$$\mu_1(t) = \frac{\lambda_2}{\lambda_1 + \lambda_2} + \frac{\lambda_1}{\lambda_1 + \lambda_2} e^{-(\lambda_1 + \lambda_2) t}$$

$$\mu_2(t) = 1 - \mu_1(t),$$

and $F_s(y)$ denotes the probability that the time of occurrence of the s-th complete failure precedes y.

The results given above can be generalized to the case of k levels of performance with exponential lives in each state.

An alternative relation for $\Pr\{\xi_1, \xi_2\}$ is

$$\Pr\{\xi_1(t) = j, \xi_2(t) = |Q| - j\} = \binom{|Q|}{j} [\mu_1(t)]^j [\mu_2(t)]^{|Q|-j} \cdot [1 - F_s(t)] +$$

$$+ \left[\int_0^t L_t(y) \, dF_s(y)\right] F_s(t),$$

where

$$L_t(y) = \sum_{v=1}^{|Q|} \binom{v}{j} [C(t-y)]^j [B(t-y)]^{v-j} [D(t-y)]^{|Q|-v} \binom{|Q|-1}{v-1}.$$

$$\cdot [\mu_1(y)]^{v-1} [\mu_2(y)]^{|Q|-v}.$$

6.9. A Markov decision model for multicomponent system maintenance

For many technical systems, the solution of multicomponent inspection and repair problems is found by using a stochastic model consisting of a number of independent Markov chains [113]. The failure and wear processes in individual components of the system are considered to be mutually independent. For practical and model reasons, the maintenance cost of equipments are expressed in terms of the nature of failure and maintenance decisions (e.g. repair, replacement, inspection) for the individual parts/subsystems. For general cases, the maintenance cost functional is highly nonlinear.

In the following model, each individual component is modelled by a single Markov chain. As was noted in [113], "the mutual independence of the Markov chains for the individual components is reflected in these constraints, which can be decomposed into a number of independent blocks corresponding to the system components".

For technical systems with a series structure configuration, the objective function is concave. The solution of the above problem is given by applying concave programming techniques. The approach is valid even for series-parallel structures by identifying a set of components such that the failure of all components in the set induces the failure of the system itself.

We shall deal now with systems where each of the components accepts two modes of operation: "failed" and "not failed". By applying the model we have to find the maintenance schedule that minimizes the expected cost under the assumption that the inspection and repair procedures are available for each component of the system.

Technical system with boolean behaviour. As was already mentioned, the components within the system accept a dichotomic behaviour. The Markov chain for this type of equipment contains only two operational states. The operating time of the system is given by the sequence of discrete time points $n = 0, 1, 2, \ldots, N$. The planning period for the system's maintenance is given by $n \in [0, N]$. The model refers only to the case of systems with a finite horizon planning period [113]. It is assumed that if a part subsystem is inspected and found to be in a failed state, then it is a mandatary decision to repair it immediately. The failed unit will always be in an operating state after inspection (e.g. the state after inspection is determined in advance). In the sequel, we shall define the following probabilities: $p(n) = \Pr\{\text{the component failed at time } (n+1) | \text{ operating at time } n\}$, $q(n) = \Pr\{\text{the component operating at time } (n+1) | \text{ operating at time } n\}$ such that $p(n) + q(n) = 1$.

The probability $x(n+1)$ is defined as $x(n+1) = \Pr\{\text{the component failed at time } (n+1)\}$. After some manipulations we can note (see [113]) that $x(n+1) = q(n)x(n) + p(n)$. After an inspection followed by a repair decision at time n, the failed component is reset into an operating state.

Owing to the dichotomic state description of components' behaviour, the transition relation in a probability form is $x(n+1) = p(n)[1 - x(n)] + p(n) x(n)$ or $x(n+1) = p(n)$.

In view of including the effect of maintenance actions, we can use the following equation. $x(n+1) = [1 - \alpha(n)][q(n) x(n) + p(n)] + \alpha(n) p(n) = q(n)[1 - \alpha(n)] x(n) + p(n)$, where $\alpha(n) = \Pr\{\text{the component is inspected and repaired if failed at time } n\}$.

If $u(n) = x(n) \alpha(n)$ represents the probability for inspection and repair at time n and component failed at time n, then the above equation becomes $x(n+1) = q(n)[x(n) - u(n)] + p(n)$, and because $0 \leq \alpha(n) \leq 1$ the control variables $u(n)$ satisfy the inequalities $0 \leq u(n) \leq x(n)$, $n = 0, 1, 2, \ldots, (N-1)$.

The objective function in the adopted model includes failure and maintenance costs. Failure costs are calculated by expressing the failure probability of the system in terms of the failure probability of the individual components such that $\Pr\{\text{system failed at time } n\} = 1 - \prod_{j=1}^{M} [1 - x^j(n)]$, where $x^j(n)$ is the failure probability of the j-th component at time n and M represents the number of components in the technical system. Usually, $x^j(n) \ll 1$ ($x^j < 10^{-3}$) for $j = 1, 2, \ldots, M$ and then we can use the linear approximation $\Pr\{\text{system failed at time } n\} \cong \sum_{j=1}^{M} x^j(n)$.

The penalty cost, c, of the system failure is given by the functional equation $C[x^1(n), \ldots, x^j(n), \ldots, x^M(n)] = c \sum_{j=1}^{M} x^j(n)$.

For a multicomponent system, the total maintenance cost is

$$C = \sum_{l=1}^{L} K_l \Delta[\sum_{j \in J_l} u^j(n)] + \sum_{j=1}^{M} a^j u^j(n),$$

where $\{J_l | l = 1, 2, \ldots, L\}$ are sets of components for which a separate setup cost is incurred and K_l is the setup inspection cost for the l-th set J_l. The expected value of the repair cost for the j-th component at time n is $a^j u^j(n)$, where a^j is the cost of repairing component j. For only one component, the total maintenance cost (inspection and repair costs for the component) is $k\Delta[u^j(n)] + a^j u^j(n)$, where k is the setup cost for inspecting the component and $\Delta[\cdot]$ is the unit step function.

The general optimization problem of inspection and repair in a dichotomic multicomponent system has the form

$$\min: \sum_{n=0}^{N} \left\{ c \sum_{j=1}^{M} x^j(n) + \sum_{j=1}^{M} a^j u^j(n) + \sum_{l=1}^{L} K_l \Delta[\sum_{j \in J_l} u^j(n)] \right\}$$

subject to

$$x^j(0) = x_0^j, \quad (j = 1, 2, \ldots, M)$$
$$x^j(n+1) = q^j(n)[x^j(n) - u^j(n)] + p^j(n), \quad (n = 0, 1, \ldots, N-1)$$
$$0 \leq u^j(n) \leq x^j(n), \quad (n = 0, 1, \ldots, N-1).$$

The above programming optimization model can be solved by using Hoang's algorithm [133] for convex mathematical programming. As was mentioned in [113], by applying the algorithm "during the search, a list of suboptimal schedules can be generated and ordered according to decreasing cost. This feature is advantageous in parametric investigations and in cases where interest is focused not primarily on the optimal schedule but rather on investigating the range of possibilities for improving current maintenance procedures".

Technical systems with multichotomic components behaviour. For technical systems with multichotomic components behaviour the previous constraints must be modified. Components which require different repair procedures will allow different types of failure modes.

We introduce the following notation. $[P_r(n)| r = 0, 1]$ is a transition matrix for "no repair" ($r = 0$) and "repair" ($r = 1$), respectively, for the associated Markov process; $P_{ijr}(n)$ represents the ij-th element of matrix $P_r(n)$; L is the number of states in the Markov process;

$\alpha_{ir}(n) = \Pr\{$repair action r at time $n|$ state i at time $n\}$; $x_i(n) = \Pr\{$state i at time $n\}$; $u_{ir}(n) = x_i(n)\, \alpha_{ir}(n) = \Pr\{$repair r and state i at time $n\}$.

Using these notation and definitions we can write

$$x_i(n+1) = \sum_{k=0}^{L} \sum_{r=1}^{L} P_{kir}(n)\, \alpha_{kr}(n)\, x_k(n), \quad (i = 1, 2, \ldots, L).$$

The transition mechanism of the multicomponent technical system can be expressed by means of the expression

$$\sum_{k=0}^{1} u_{ik}(n+1) = \sum_{k=0}^{1} \sum_{r=1}^{L} P_{rik}(n)\, u_{rk}(n), \quad (i = 1, 2, \ldots, L),$$

if

$$0 \leqslant u_{ik}(n) \text{ for } k = 0, 1 \text{ and } n = 0, 1, \ldots, (N-1)$$

and

$$\sum_{k=0}^{1} u_{ik}(n) = x_i(n) \text{ for } i = 1, 2, \ldots, L \text{ and } n = 0, 1, \ldots, (N-1).$$

The formulation of the maintenance, inspection and repair problem for multichotomic multicomponent systems is analogous to that presented in the previous section. We have to minimize a concave functional subject to linear equality and inequality constraints. Compared with the previous model, "in a multistate component, however, the repair need not be performed immediately following an inspection since several operating states may exist in which a repair is neither needed nor advantageous" [113]. For the latter case (multicotomic behaviour) the state of a component is not necessarily known before an inspection has been made. The suboptimal maintenance policies obtained from the model have important practical value to the maintenance system engineer.

Chapter 7

Inspection, diagnosis, reliability and maintenance policies for coherent structures

As we have seen before, the complexity of coherent structures allows a detailed modelling by means of Markov models. In this chapter, we shall consider that the system is allowed a pertinent observation space, and cause-effect models are appropriate for a probabilistic description of the underylying process regarding the system behaviour. Generally speaking, any well-defined (coherent) system has been designed for some technological reasons; engineering "judgement" has been incorporated into the system. However, cause-effect models could be built for any particular structure. The model given in this chapter can also be used for fault isolation in complex systems (see Gheorghe [84]), when the dynamic behaviour of the system is described by a Markov process. The states of the internal (core) process are not directly observable by the decision-maker (controller). However, a finite set of cause-effect models are available to an observer after some inputs which are probabilistically related to the internal state of the process have been observed. For instance, let us take a BWR system; information about the performance level of subsystems or devices from the reactor can be obtained by performing a set of observations (i.e. pertinent measurements in the control room).

Models described in the literature for fault detection and isolation are either statistical (see Ransom [85]) or stochastic, such as the kinetic fault tree model given by Vesely [16]. However, models like those given in [85] refer to a set of observations which are available for fault detection, but the system has a statistical representation. Vesely uses "kinetic" models for the above purposes, but he does not use the explicit concept of observation.

7.1. Cause-effect models and Markovian decision processes

We shall give first several simple relationships in order to understand cause-effect models. (For a more sophisticated analysis, see Rousseau [50].) In what follows, the notions of cause and effect are assumed to be sufficiently defined for our practical maintenance reasons.

If we suppose that there exist two events A and B as those given in fig. 7.1, and if the arrow indicates the direction of causality, then the description of the relationship between these two events is given by the statement "A always causes B".

Symbol I indicates the "only" causality. However, if causality is not certain, then we need to describe it in a probabilistic fashion. In fig. 7.2, we give a probabilistic cause-effect model, where the appropriate statement for describing this model is "A causes B with probability p".

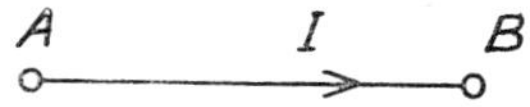

$$A \xrightarrow{I} B \qquad A \xrightarrow{p} B$$

Fig. 7.1 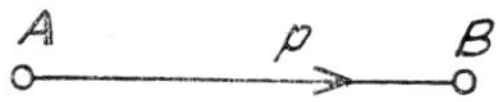 Fig. 7.2

This can be qualified by the relations $\Pr\{B|A, \mathscr{E}\} = p$ and $\Pr\{A|B, \mathscr{E}\} = 1$, where, for fig. 7.2, we can write the relations $\Pr\{B|A, \mathscr{E}\} = 1$ and $\Pr\{A|B, \mathscr{E}\} = 1$.

By introducing a boolean operator, more complicated relationships between events can be described. We shall use "logical gates" such as AND, OR to show the connection of different primary events to produce a secondary event (i.e. one of the different primary events produces a secondary event). This implies that one of the events has more than one cause, which is realistic in the description of large and complex systems. In the model presented in fig. 7.3, the appropriate logical gate for system description is OR. The meaning of the notation can be interpreted probabilistically as $\Pr\{B/A, \text{NOT } C, \mathscr{E}\} = p$, $\Pr\{B/\text{NOT } A, C, \mathscr{E}\} = q$, $\Pr\{B/A, C, \mathscr{E}\} = p + q - pq$, where the last relation derives from the basic properties of probabilities and can be proved using a Venn diagram. A collection of such simple models connected by boolean rules is known in the literature under the name of "fault tree" or event-tree models (see Haasl [79], Vesely [16] and Fussell [46]). A fault tree gives information about the probabilistic behaviour of any structure function $\Phi_\lambda(X)$, $(\lambda = 1, 2, \ldots, \omega)$, or a set of such functions when we can well define primary events belonging to the components state vector X, or an appropriate modular decomposition and boolean connections such as AND, OR gates. A *fault tree* is a static boolean representation of the possible behaviour of the system and *does not* incorporate any "dynamics". Attempts have been made to define *kinetic fault trees* [46]; a large set of applications of these techniques have been found for the safety analysis of nuclear reactors. It is not the purpose of this chapter to show the limits of such techniques for an overall analysis in large systems. In this section, we shall develop a methodology to show that it is always possible to couple cause-effect models,

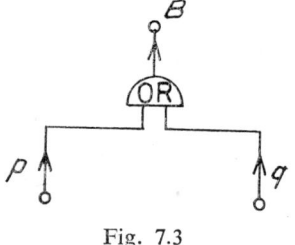

Fig. 7.3

as those described before, with the core process of a Markov process. The state space of the core process is defined by an index for the structure function performance (the system deterioration dynamics). By a set of maintenance decision alternatives, such as inspection, repair or replacement, we can obtain more powerful models for systems analysis at the price of more difficult computational problems.

As we have pointed out, in a cause-effect model events occur simultaneously, and for the example given in fig. 7.4, we have IF $(A: \text{TRUE}) \Rightarrow (B: p_{AB})$, $\Pr\{A|B, \mathscr{E}\} = p_{AB}$.

In a dynamic probabilistic model (e.g. Markov or semi-Markov), events (the pertinent states of the deteriorating system) are thought of as occurring one at a time. In this case, it appears to be practically possible and theoretically acceptable to include into the dynamic probabilistic model a cause-effect representation of the underlying process which, in principle, does not affect the dynamic mechanism. However, more complicated logical models could have TRUE and FALSE logical functions; this means that some primary events could happen or not at any time instant. These assumptions do not change the general structure of the model, but introduce computational complications for the process [50].

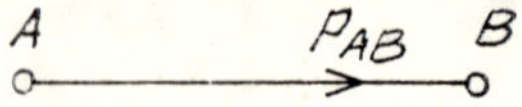

Fig. 7.4

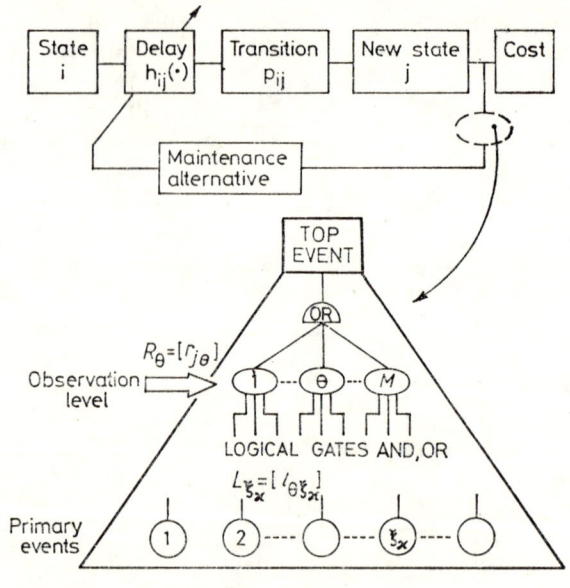

Fig. 7.5

With all these in mind, it is quite clear that we can change the configuration of the state space of the process at each transition t, rather than changing the mechanism of evolution of the core process itself. We refer to fig. 7.5, where a representation of such models for the most general case of a semi-Markov process can be seen.

7.2. Model formulation

For the model presented in this chapter, we make the following assumptions:
— the system has n components,
— a fault tree analysis gives information about the probabilistic behaviour of any structure function $\Phi_\lambda(X)$, $(\lambda = 1, 2, \ldots, \omega)$. We have to consider the dichotomic behaviour of each component of the system $x_i = 0, 1; i = 1, 2, \ldots, n$;
— at the level of the "core process" we can consider the performance index for the total structure function $\Phi_\omega^*(X)$.

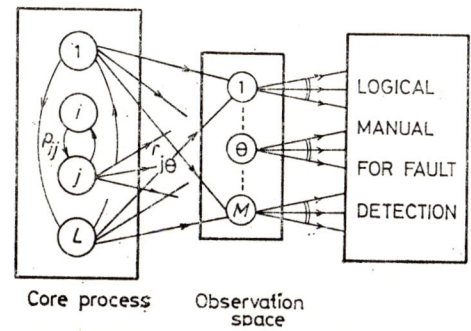

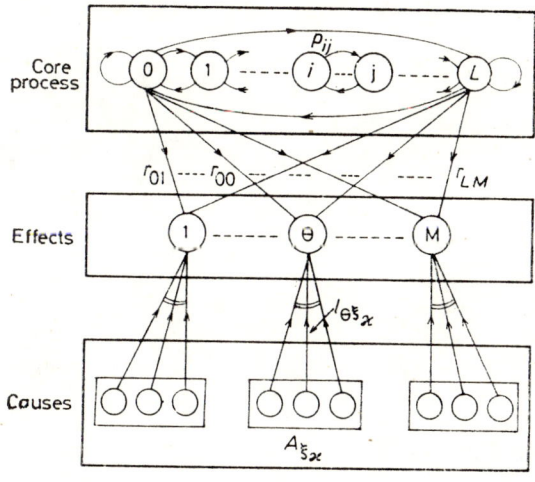

Fig. 7.6

It is clear that the core process describes the dynamic behaviour of the system given by the total structure function; on the other hand, a fault tree technique is limited to a boolean representation for component performance ($x_i = 0, 1$; $i = 1, 2, \ldots, n$). In the case of partially observable processes ([13], [31]), the first order Bayesian inference must be considered for correlating the internal states of the deterioration process with the observation space after an inspection has been performed. If we have some observable outputs which eventually can be found in the description of the fault tree, then we can build a "lottery" from the cause-effect models described above. A "second order" Bayesian inference is given by

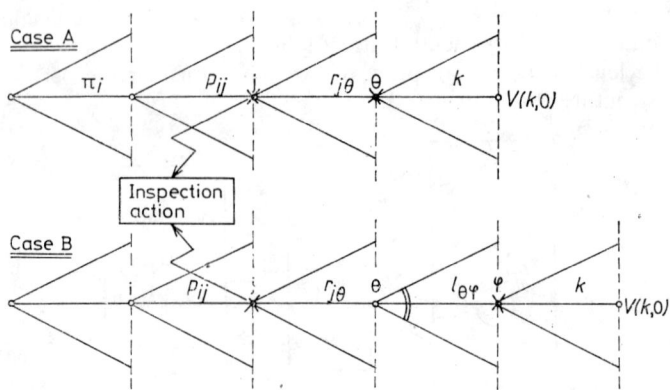

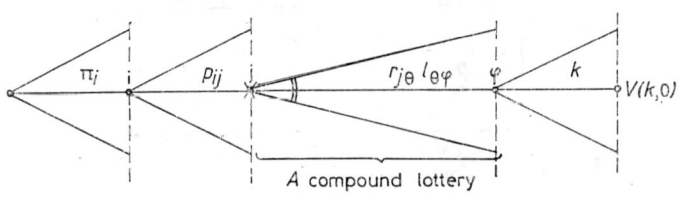

Fig. 7.7

the logical structure of the fault tree. Figure 7.6 shows a complete representation of such a process. Figure 7.7 gives a decision tree representation for the Markov process with logical conditions where the difference between partially observable Markov processes and the one proposed in this chapter is emphasized.

7.2.1. The core process

Our assumption is that the system has a Markov behaviour [7]. The system dynamics is given by (see fig. 7.5)

$$p_{ij} = \Pr\{S(t+1) = j|\ S(t) = i,\ S(t-1) = 1,\ldots,\ S(0) = m,$$
$$\mathscr{E}(t)\} = \Pr\{S(t+1) = j|\ S(t) = i,\ \mathscr{E}(t)\},\quad (i,j = 0, 1,\ldots, L), \qquad (7.1)$$

and
$$\sum_{j=0}^{L} p_{ij} = 1, \quad (i = 0, 1, 2, \ldots, L). \tag{7.2}$$

The above probabilistic measure gives sufficient information on the system's behaviour and shows that the present state of the system depends only on the immediate past transition and our knowledge $\mathscr{E}(t)$ about the system evolution. For a time-dependent Markov process, vector π can be written as

$$\pi(t) = [\pi_i(t), \ i = 0, 1, \ldots, L] = [\Pr\{S(t) = i| \ \mathscr{E}(t)\}]. \tag{7.3}$$

We can define the set of all possible initial state vectors as

$$\mathbf{\Pi} = \text{set}\left[\pi_i \colon \pi_i \geq 0; \ \sum_{i=0}^{L} \pi_i = 1\right]. \tag{7.4}$$

7.2.2. The observation space

Independent observation outputs $\theta = 1, 2, \ldots, M$ are available after the observer has performed a stochastic inspection. Following an observation, we need to take advantage of a "diagnosis manual". Our new state of knowledge about the process at the $(t+1)$-st transition is given by $\mathscr{E}(t+1) = [\mathscr{E}(t), Z(t+1) = \theta, Y(t+1) = \xi_n]$. The random inspection is given by the probability

$$r_{j\theta} = \Pr\{Z(t+1) = \theta| \ S(t+1) = j, \ \mathscr{E}(t)\},$$
and
$$(j = 0, 1, \ldots, L, \ \theta = 1, 2, \ldots, M), \tag{7.5}$$

$$0 \leq r_{j\theta} \leq 1, \ \sum_{j=0}^{L} r_{j\theta} = 1, \quad (\theta = 1, 2, \ldots, M). \tag{7.6}$$

Subsequently, a cause-effect model is characterized by

$$l_{\theta \xi_n} = \Pr\{Y(t+1) = \xi_n| \ Z(t+1) = \theta, \ S(t+1) = j, \ \mathscr{E}(t)\},$$
$$1 \geq l_{\theta \xi_n} \geq 0 \tag{7.7}$$
$$l_{\theta \xi_n} + \bar{l}_{\theta \xi_n} = 1; \ \theta = 1, 2, \ldots, M; \ \xi_n = 1, 2, \ldots, N_{A_n}.$$

However, a cause-effect model contains logical gates such as AND, OR, TRUE, FALSE.

7.3. Computing probabilities in cause-effect models and overall system dynamics

It is necessary to have methods for calculating the probabilities $l_{\theta \xi_n}$ in cause-effect models which involve logical combinations such as AND/OR gates. Examples of calculating these probabilities for simple cause-effect models are given in table 7.1. More extensive results can be found in ref. [50].

The state space of the deterioration-inspection-diagnosis process for the complex system is no longer Markov; its form is even more complicated than that given in [13] and [31]. The dynamics of the new state-space are described below.

Let

$$X^k = (R_\theta^k \vee L_{\xi_n}^k), \qquad (7.8)$$

where $\vee$ is in fact a logical multiplication operator. Therefore, X^k measures the probability of a particular cause accounting for a specific performance state which was achieved as a result of maintenance decision alternative k.

Table 7.1

	Cause-effect model	Probabilities evaluation
1	$A \to p$, $B \to q$, OR $\to C$	$\Pr\{B\|A\} = p$; $\Pr\{B\|C\} = q$ $\Pr\{B\|A, \text{NOT } C\} = p$ $\Pr\{B\|\text{NOT } A, C\} = q$ $\Pr\{B\|\text{NOT } A, \text{NOT } C\} = 0$ $\Pr\{B\|A, C\} = p + q - pq$ $B = (A: p) \text{ OR } (C: q)$
2	$A \to p$, $B \to q$, AND $\to C$	$\Pr\{B\|A\} = 0$; $\Pr\{B\|C\} = 0$ $\Pr\{B\|A, \text{NOT } C\} = 0$ $\Pr\{B\|\text{NOT } A, C\} = 0$ $\Pr\{B\|\text{NOT } A, \text{NOT } B\} = 0$ $\Pr\{B\|A, C\} = p \cdot q$ $B = (A: p) \text{ AND } (C: q)$
3	(two equivalent diagrams with A, B, C, D, OR gates)	$D = (A: p) \text{ OR } (B: q) \text{ OR } (C: r)$ $D = \text{OR}((A: p), (B: q), (C: r))$ $D = ((A: p) \text{ OR } (B: q)) \text{ OR } (C: r)$

For an example in which the total structure function could take the values $\Phi_\omega^*(X) = 0, 1, 2$, two observations and two causes are considered. Then X^k has 12 elements. Considering the system in the performance state j as a result of maintenance alternative k,

$$X_j^k = (R_j^k \vee L_{\xi_n}^k) = [R_{j1}^k \; L_{11}^k \; R_{j1}^k \; L_{12}^k \; R_{j2}^k \; L_{21}^k \; R_{j2}^k \; L_{22}^k], \qquad (7.9)$$

where L_{uv} refers to observation u and the v-th element in the L-vector, and R_{jw} refers to the w-th element in the row.

Let d be an index for the cause-effect model. In the example with two causes and two observations, d has four elements corresponding to each combination of cause and observation.

If X_d^k is defined as diag $[X^k]$ for constant d, then
$$\{d/\pi, k\} = \pi P^k X_d^k \mathbf{1}, \qquad (7.10)$$
where
$$\mathbf{1} = [\underbrace{1 \ldots 1 \ldots 1}_{N}]^T.$$

In a matrix formulation, the dynamics of the new state space becomes
$$T(\pi| d, k) = \frac{\pi P^k X_d^k}{\{d| \pi, k\}}, \qquad (7.11)$$

where k represents a maintenance decision alternative (repair, replace, do nothing) at time t in the triplex $D = \{\mathcal{S}, \mathcal{O}, \mathcal{D}\}$ such that $k \in \mathcal{D}$ and $d \in \mathcal{O}$.

7.3.1. State-space dynamics for Markov processes with logical conditions

Consider a system which makes a transition between two successive times t and $(t + 1)$. The overall system dynamics may be computed using the Bayes' theorem. A decision tree approach is given in fig. 7.8. A cause-effect model at the $(t + 1)$th transition may be described by the pair $d = (\theta, \xi_n)$ of the observation after an inspection (effect) and primary event (cause) such that the model may be designated $\mathcal{M}(t + 1)$. Given the decision maker's state of knowledge at the t-th transition,

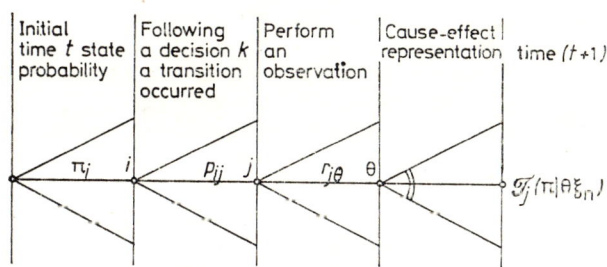

Fig. 7.8

the process dynamics and a complete description for cause-effect models, and in virtue of Bayes' theorem,

$$\Pr\{S(t + 1) = j| \mathcal{E}(t + 1) = [\mathcal{E}(t), \mathcal{M}(t + 1) = d]\} = \Pr\{S(t + 1) = j,$$
$$\mathcal{M}(t + 1) = d| \mathcal{E}(t)\}/\Pr\{\mathcal{M}(t + 1)| \mathcal{E}(t)\}. \Pr\{\mathcal{M}(t + 1) = d| \mathcal{E}(t)\}^{-1} \sum_i \Pr\{S(t) = i,$$
$$S(t + 1) = j, \mathcal{M}(t + 1) = d| \mathcal{E}(t)\} = \Pr\{\mathcal{M}(t + 1) = d| \mathcal{E}(t)\}^{-1} \sum_i \Pr\{S(t) =$$
$$= i| \mathcal{E}(t)\} \cdot \Pr\{S(t + 1) = j| S(t) = i, \mathcal{E}(t)\} \cdot \Pr\{\mathcal{M}(t + 1) = d| S(t + 1) = j,$$

$$S(t) = i, \mathscr{E}(t)\} = \Pr\{\mathscr{M}(t+1) = d| \mathscr{E}(t)\}^{-1} \cdot \sum_i \pi_i(t) \, p_{ij}(t+1) \Pr\{\mathscr{M}(t+1) =$$
$$= d| S(t+1) = j, S(t) = i, \mathscr{E}(t)\} = \Pr\{\mathscr{M}(t+1) = d| \mathscr{E}(t)\}^{-1} \sum_i \pi_i(t) \, p_{ij}(t+1)$$
$$\cdot \Pr\{Z(t+1) = 0, Y(t+1) = \xi_n| S(t+1) = j, S(t) = i, \mathscr{E}(t)\} = \Pr\{\mathscr{M}(t+1) = d| \mathscr{E}(t)\}^{-1} \cdot \sum_{i=0} \pi_i(t) \, p_{ij}(t+1) \cdot \Pr\{Z(t+1) = 0| S(t+1), S(t) = i,$$
$$\mathscr{E}(t)\} \cdot \Pr\{Y(t+1) = \xi_n| Z(t+1) = 0, \mathscr{E}(t)\} = \Pr\{\mathscr{M}(t+1) = d| \mathscr{E}(t)\}^{-1} \cdot$$
$$\cdot \sum_{i=0} \pi_i(t) \, p_{ij}(t+1) \, r_{j\theta}(t+1) \, 1_{\theta\xi_n}(t+1) = \mathscr{T}_j(\pi| \theta, \xi_n) = \mathscr{T}_j(\pi| d). \quad (7.12)$$

From the above relation, it can be readily seen that completely and partially observable Markov processes (Howard [7], Sondik [13]) are particular cases of the present one. For a simpler notation, we can use the operator

$$\{d| \pi\} = \Pr\{\mathscr{M}(t+1) = d| \mathscr{E}(t)\} = \sum_{i,j,\theta} \pi_i(t) \, p_{ij}(t+1) \, r_{j\theta}(t+1) \, l_{\theta\xi_n}(t+1).$$
$$(7.13)$$

Thus, we have

$$\pi'_j(t+1) = \mathscr{T}_j(\pi| d) = \frac{\sum_i \pi_i(t) \, p_{ij}(t+1) \, r_{j\theta}(t+1) \, l_{\theta\xi_n}(t+1)}{\sum_{i,j,\theta} \pi_i(t) \, p_{ij}(t+1) \, r_{j\theta}(t+1) \, l_{\theta\xi_n}(t+1)}. \quad (7.14)$$

A matrix form of eqn. (7.14) is given by eqn. (7.11).

$T(\cdot|\cdot)$ specifies the dynamics of the new state vector π, given a particular combination of cause and observation as a result of having carried out maintenance k. Dimensionally, $\{d/\pi, k\}$ involves the matrix multiplication $(1 \times N)(N \times N)(N \times N)(N \times 1)$ leaving a scalar quantity. The elements of $T(\cdot|\cdot)$ are therefore normalized by dividing by the index d, so that the sum of the probabilities constituting the elements of T are equal to unity.

In a finite horizon model, we may consider the TRUE or FALSE statements at each time period over the horizon plan for the primary events in the cause-effect model. In fig. 7.8 we give a representation of the system dynamics for a process where the core process is characterized by $L = 3$ states. However, the model presented here could also be used for a real time (on-line) process implementation. The cut sets for the structure could be found using the fault tree representation [16]. If we can observe an event θ, this can be associated with the elements in a cut set, eventually a minimal cut set.

7.3.2. The value of a complete analysis for maintenance policies

An interesting analysis of the economic value of modelling has been given by Matheson [89]. We shall follow his approach to prove that an MDP with logical conditions could follow the pattern given in ref. [89].

Let us consider that Pr $\{o|\mathscr{E}\}$ represents our prior knowledge about the partially observable process. After a primary analysis has been performed, a more complete one is suggested. A partially observable process would give information about the state of the process and the output from observations (the inspection process).

We need to structure the process of outcome generation in finer detail and also to use the best experts' judgement (eventually measurement) to obtain data for each primary event. However, we can imagine that by so doing we could obtain a narrow distribution of outcomes that would in turn lead to better decisions such that, finally, the MDP would have the minimum expected maintenance cost.

If d represents the index for a cause-effect model over possible distribution of outcomes resulting from the complete analysis, then Pr $\{o/d, \mathscr{E}\}$ represents the density function of outcome provided that the value of the index d is known and our prior knowledge about the process $\mathscr{E}$ is available. The complete analysis is viewed as the determination of the value of the index d.

We have calculated the probability distribution of the occurrence of each possible cause-effect model $\{d/\mathscr{E}\}$ and, hence,

$$\Pr\{o|\mathscr{E}\} = \int_d \Pr\{o|d, \mathscr{E}\} \cdot \{d|\mathscr{E}\}. \tag{7.15}$$

Let a set of maintenance decisions in each state of the process be given. We denote this set by $\mathscr{D}$.

Let $v(k, o)$ be the value over MDP ($k \in D$, $o \in \mathcal{O}$). The decision equation is concerned with choosing the best action $k^*(v/d, \mathscr{E})$ such that finally we need to minimize the expected maintenance cost

$$\langle v| d, k, \mathscr{E} \rangle = \int_d \langle v| d, k, o, \mathscr{E} \rangle \cdot \Pr\{o|d, k, \mathscr{E}\}. \tag{7.16}$$

We shall assume that

$$\Pr\{o|d, k, \mathscr{E}\} = \Pr\{o|d, \mathscr{E}\}, \tag{7.17}$$

and hence,

$$\langle v| d, k, \mathscr{E} \rangle = \int_o V(k, o) \cdot \Pr\{o|d, \mathscr{E}\} \cdot \langle v| k^*(v|d, \mathscr{E}), d, \mathscr{E}\rangle =$$

$$= \langle v| \hat{k}(v|d, \mathscr{E}), d, \mathscr{E}\rangle = \min_k \int_o V(k, o) \cdot \Pr\{o|d, \mathscr{E}\}, \tag{7.18}$$

and $\hat{k}(\cdot|\cdot)$ is the value of k that minimizes $\langle v/k, \mathscr{R}\rangle$ (see ref. [89]), where $\mathscr{R}$ is the state of information on which probability assignments will be made.

We can see that the expected value of the decision process before a complete analysis has been performed is

$$\langle v|\ k^*(v|\ d,\ \mathscr{E}),\ \mathscr{E}\rangle = \int_d \{d|\ \mathscr{E}\}\ \langle v|\ \hat{k}(v|\ d,\ \mathscr{E}),\ d,\ \mathscr{E}\rangle =$$

$$= \int_d \{d|\ \mathscr{E}\}\ \min_d \int_o V(k,\ o)\ \Pr\{o|\ d,\ \mathscr{E}\}, \quad (7.19)$$

under the assumption that

$$\Pr\{d|\ k^*(v|\ d,\ \mathscr{E}),\ \mathscr{E}\} = \{d|\ \mathscr{E}\}.$$

Finaly, we have

$$\langle v|\ k^*(v|\ d,\ \mathscr{E}),\ \mathscr{E}\rangle = \min_k \int_k \int_o \{d|\ \pi,\ \mathscr{E}\}\ V(k,\ o). \quad (7.20)$$

The decision alternatives k are concerned with "do nothing", "inspect and repair", "inspect and replace".

7.4. Inspection, diagnosis and maintenance policies

Over the planning horizon, the maintenance engineer is faced with repair/replacement decisions for the multicomponent system, after a set of pertinent inspections has been performed. However, in our approach, we do not have to inspect each of the n components. After an observation has been performed from the structure topology *via* the fault tree analysis, we have information about the state of the components.

In this section, we shall consider the analysis of optimal maintenance policies after an inspection-diagnosis process has been performed for the case of finite and infinite horizon planning period. The models given in this chapter could also be used in computer-aided medical diagnosis or dynamic fault diagnosis in computer operation.

7.4.1. A finite-horizon maintenance planning model

We shall consider first that the maintenance decision-maker has to take decisions over a finite horizon time period. The cost of the system transition and inspection is $W_{ij\theta}$ (see Eckles [71]) and the further detection cost (a diagnosis process) is $C_{\theta\xi_n}$. The expected value $v_\pi(t)$ of the process when t units time are left for the system to run is given by

$$v_\pi(t) = \sum_{A_n=1}^{M} \sum_{i=0}^{L} \pi_i(t) \sum_{j=0}^{L} p_{ij} \sum_{\theta=1}^{M} r_{j\theta} \sum_{\xi_n=1}^{N_{A_n}} l_{\theta\xi_n} \cdot (W_{ij\theta} + C_{\theta\xi_n} + V_T(\pi|\ \theta,\ \xi_n)(t-1)),\ t \geqslant 1. \quad (7.21)$$

For the case when maintenance alternative k is appropriate (do nothing, do inspection with repair/replacement), the expected inspection-diagnosis-maintenance cost can be written as

$$V_\pi(t) = \min_k \left[\sum_{A_n=1}^{M} \sum_{i=0}^{L} \pi_i(t) \sum_{j=0}^{L} p_{ij}^k \sum_{\theta=1}^{M} r_{j\theta}^k \cdot \sum_{\xi_n=1}^{N_{A_n}} l_{\theta\xi_n}(W_{ij\theta}^k + C_{\theta\xi_n} + V_{T(.|.)}(t-1)) \right] t \geq 1. \quad (7.22)$$

The immediate expected return from the maintenance process is given by

$$q_i^k = \sum_{A_n, j, \theta, \xi_n} p_{ij}^k r_{j\theta}^k l_{\theta\xi_n}^k t_{ij\theta\xi_n}^k, \quad (7.23)$$

where $t_{ij\theta\xi_n}^k = W_{ij\theta}^k + C_{\theta\xi_n}^k$.

Using the above results, we can write

$$V_\pi(t) = \min_k \left[\sum_{i=0}^{L} \pi_i q_i^k + \sum_{i,j,\theta,\xi_n, A_n} \pi_i p_{ij}^k r_{j\theta}^k l_{\theta\xi_n}^k V_{T(.|.)}(t-1) \right], \quad (7.24)$$

with the initial condition that

$$V_\pi(0) = \left[\sum_{i=0}^{L} \pi_i q_i^0 \right] = \pi q^0. \quad (7.25)$$

The state space dynamics is given by

$$\pi_j' = \frac{\sum_i \pi_i p_{ij}^k r_{j\theta}^k l_{\theta\xi_n}^k}{\sum_{i,j,\theta} \pi_i p_{ij}^k r_{j\theta}^k l_{\theta\xi_n}^k} \quad (7.26)$$

or in a vector form $\pi' = T(\pi | k, d)$, where $v_\pi(t)$ is again the minimum expected maintenance cost of the system during its lifetime, if the current information vector is π, and t control intervals remain before the process terminates. In order to simplify the notation, we shall use the index d for the cause-effect models. With the above assumptions and using the previous results, we can write

$$V_\pi(t) = \min_k \left[\sum_{i=0}^{L} \pi_i \sum_{j=0}^{L} p_{ij}^k \sum_{d=1}^{M} x_{jd}^k \cdot (t_{ijd}^k + V_{T(\pi|k,d)}(t-1)) \right], \quad t \geq 1, \quad (7.27)$$

where $\mathscr{M}$ is the combination of all possible pairs (θ, ξ_n); in other words, it is the collection of all possible cause-effect models.

An alternative form of the above relation is

$$V_\pi(t) = \min_k [\pi q^k + \sum_d \{d | \pi, k\} \cdot V_{T(\pi|d,k)}(t-1)], \quad t \geq 1, \quad (7.28)$$

where $\{d/\pi, k\} = \{d/\pi\}$ by using the triplex $(R_\theta^k, P^k, L_{\xi_n}^k)$.

In the remainder of this section, we shall show that the problem has a solution given by the "one pass algorithm" (see refs. [13] and [31]). First, we prove some properties for the decision process.

THEOREM 7.1. $V_\pi(t)$ is piecewise linear and concave for all t.

Proof. We have shown that $V_\pi(0) = \pi q^0$. From this relation, it is clear that $V_\pi(0)$ is not only linear in π but also piecewise linear and concave.

Let us suppose now that $V_\pi(t-1)$ is linear and concave.

LEMMA 7.1. $T(\pi/k, d)$ preserves straight lines if $0 \leq \rho \leq 1$, $\bar{\rho} = (1-\rho)$; then for $\pi^1, \pi^2 \in \Pi$, and every $\theta = $ constant, $T(\rho\pi^1 + \bar{\rho}\pi^2) = \mu e^1 + \bar{\mu} e^2$, where

$$\mu = \frac{\rho \cdot \Pr\{d|\pi^1, k, \theta = \text{ct.}\}}{\rho \Pr\{d|\pi^1, k, \theta = \text{ct.}\} + \bar{\rho} \Pr\{d|\pi^2, k, \theta) = \text{ct.}\}} \quad \bar{\mu} = 1 - \mu$$

and $e^i = T(\pi^i|d, k) \in \Pi$.

Proof.

$$T(\pi|d, k) = \frac{\pi P^k X_d^k}{\pi P^k X_d^k \mathbf{1}},$$

for $d = 1, 2, \ldots$, and then we can write

$$T(\rho\pi^1 + \bar{\rho}\pi^2 | d, k) = \frac{(\rho\pi^1 + \bar{\rho}\pi^2) Q_d}{(\rho\pi^1 + \bar{\rho}\pi^2) Q_d \mathbf{1}} = \frac{\rho\pi^1 Q_d}{(\rho\pi^1 + \bar{\rho}\pi^2) Q_d \mathbf{1}} +$$

$$+ \frac{\bar{\rho}\pi^2 Q_d}{(\rho\pi^1 + \bar{\rho}\pi^2) Q_d \mathbf{1}} = \frac{1}{(\rho\pi^1 + \bar{\rho}\pi^2) Q_d \mathbf{1}} [\rho e^1(\pi^1 Q_d \mathbf{1}) + \bar{\rho} e^2(\pi^2 Q_d \mathbf{1})] =$$

$$= \frac{1}{(\rho\pi^1 + \bar{\rho}\pi^2) Q_d \mathbf{1}} [e^1 \rho \Pr\{d|\pi^1, k\} + e^2 \bar{\rho} \Pr\{d|\pi^2, k\}]$$

$$(\rho\pi^1 + \bar{\rho}\pi^2) Q_d \mathbf{1} = \rho \Pr\{d|\pi^1, k\} + \bar{\rho} \Pr\{d|\pi^2, k\}$$

$$T(\rho\pi^1 + \bar{\rho}\pi^2|d, k) = \mu e^1 + \bar{\mu} e^2$$

$$T(\rho\pi^1 + \bar{\rho}\pi^2|d, k) = \mu T(\pi^1|d, k) + \bar{\mu} T(\pi^2|d, k).$$

The minimum of concave piecewise linear functions is also concave and linear and the cost functional is

$$f(\pi) = \Pr\{d|\pi, k\} V_{T(\pi|d,k)}(t-1). \tag{7.29}$$

By a linear combination for vector π, $\pi\rho = \pi^1\rho + \pi^2\bar{\rho}$, the cost functional is immediate.

$$f(\pi\rho) = \Pr\{d|\pi\rho, k\} V_{T(\pi|d,k)}(t-1) = \Pr\{d|\pi\rho, k\} V^{t-1}[\mu T(\pi^1|d, k) +$$

$$+ \bar{\mu} T(\pi^2|d, k)] \geq \Pr\{d|\pi\rho, k\} \cdot [\mu V_{T(\pi^1|d,k)}(t-1) + \bar{\mu} V_{T(\pi^2|d,k)}(t-1)] =$$

$$= [\rho \Pr\{d|\pi^1, k\} + \bar{\rho} \Pr\{d|\pi^2, k\}] \cdot [\mu V_{T(\pi^1|d,k)}(t-1) + \bar{\mu} V_{T(\pi^2|d,k)}(t-1)] =$$

$$= \rho \Pr\{d|\pi^1, k\} V_{T(\pi^1|d,k)}(t-1) + \bar{\rho} \Pr\{d|\pi^2, k\} V_{T(\pi^2|d,k)}(t-1) = \rho f(\pi^1) + \bar{\rho} f(\pi^2). \tag{7.30}$$

From the above demonstration, we can see that $V_{(\cdot)}(t-1)$ is concave. In order to prove that $V_\pi(t-1)$ is piecewise linear with $\alpha(\pi)$ being piecewise constant over Π, we can write

$$f(\pi) = \Pr\{d|\pi, k\}\, V_{T(\pi|d,k)}(t-1) = \Pr\{d|\pi, k\} \frac{\pi P^k X_d^k}{\{d|\pi, k\}} \cdot \alpha(T(\cdot|\cdot)) =$$
$$= \pi P^k X_d^k \alpha(T(\cdot|\cdot)) = \pi \xi(\pi), \tag{7.31}$$

where $\xi(\cdot)$ is a piecewise constant over the state space Π of the system.

It has been proved before (Theorem 7.1) that the cost functional $v_\pi(t)$ is piecewise linear and concave for the general case when the pair of points (θ, ξ_n) is in the information space of the system dynamics (see fig. 7.9). Accordingly, we can write

$$V_\pi(t) = \min_k \left[\sum_{i=}^{L} \alpha_i^k(t)\, \pi_i \right], \tag{7.32}$$

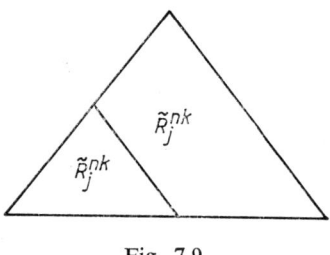

Fig. 7.9

where $\alpha_i^k(t)$ is a component of the vector

$$\alpha^k(t) = [\alpha_0^k(t), \alpha_1^k(t), \ldots, \alpha_L^k(t)]. \tag{7.33}$$

A partition of the information vector, given by the above cost functional, can be written as

$$\tilde{R}_j^t = [\pi : V_\pi(t) = \pi \alpha_j^t], \tag{7.34}$$

where the set union

$$\tilde{R}^t = \bigcup_j \tilde{R}_j^t. \tag{7.35}$$

Let
$$A^n = [\alpha_0^t, \alpha_1^t, \ldots] \tag{7.36}$$

be a set defined as a collection of all "α-vectors" in the information space Π. Subsequently, we can write the following functional relations.

$$V_\pi(t) = \min_k [\pi \alpha_j(t)], \quad j \geqslant 1$$

$$V_\pi(t) = \min_k [\pi q^k + \sum_d \min_j \pi P^k X_d^k \alpha_j(t-1)]$$

$$V_\pi(t) = \pi \alpha_j(t) = \min_k [\pi q^k + \sum_d \min_a \pi P^k X_d^k \alpha_a(t-1)] =$$
$$= \pi [q^{k_j} + \sum_d P^{k_j} X_d^{k_j} \alpha_{l(\pi|k,d)}(t-1)]. \tag{7.37}$$

Then, for $\alpha_j(t)$, we have the following value

$$\alpha_j(t) = q^{k_j} + \sum_d P^{k_j} X_d^{k_j} \alpha_{l(\pi,d)}(t-1). \tag{7.38}$$

An algorithm for calculating "α-vectors" at time t under the assumption that these vectors are known at time $(t-1)$ was developed in ref. [13].

We shall use this method for our solution to the above maintenance-diagnosis process. From the above results, we can write

$$V_\pi^k(t) = [\pi q^k + \sum_d \{d|\pi, k\} V_{T(\pi|d,k)}(t-1)]$$

$$V_\pi^k(t) = \pi \alpha_\pi^k(t), \tag{7.39}$$

where

$$\alpha_\pi^k(t) = q^k + \sum_d P^k X_d^k \alpha_{T(\pi|d,k)}(t-1) \tag{7.40}$$

and

$$\alpha_j^k(t) = q^k + \sum_d P^k X_d^k \alpha_{l(\pi|d,k)}(t-1)$$

$$V_\pi^k(t) = \pi q^k + \sum_d \min_{a_d} \pi P^k X_d^k \alpha a_d(t-1) = \pi \alpha_j^k(t). \tag{7.41}$$

In general, two regions $\tilde{R}_j^{t-k}$ and $\tilde{R}_a^{tk}$ (see fig. 7.9) share a common border, if the set $[\pi: \min_l \pi \alpha^k(t) = \pi \alpha_j^k(t) = \pi \alpha_a^k(t)]$ is nonempty.

In the remainder of this section, we shall follow [13]; the "one-pass algorithm" is given as the optimization solution to the maintenance-diagnosis MDP.

LEMMA 7.2. *The set of points π or $\tilde{R}_j^{tk}$ that lie on a boundary of $\tilde{R}_j^{tk}$, resulting from the transformation of a boundary of $\tilde{R}^{t-1}$, satisfies the equation below*

$$\pi P^k X_d^k(\alpha_a(t-1) - \alpha_l(t-1)) = 0.$$

Proof.

$$\tilde{R}_a^{t-1} \cap \tilde{R}_1^{t-1} = [\pi \in \Pi: \min_m \pi \alpha_m(t-1) = \pi \alpha_a(t-1) = \pi \alpha_l(t-1)]$$

$$[\pi \in \Pi: (\pi \alpha_a(t-1) - \pi \alpha_l(t-1) = 0]$$

$$T(\pi|d, k)(\alpha_a(t-1) - \alpha_l(t-1)) = 0$$

$$\frac{\pi P^k X_d^k}{\{d|\pi, k\}} (\alpha_a(t-1) - \alpha_l(t-1)) = 0$$

$$\pi P^k X_d^k(\alpha_a(t-1) - \alpha_l(t-1)) = 0$$

$$\pi \in \Pi.$$

The next theorems will be given without proof (for more theoretical details, see refs. [13] and [31]).

THEOREM 7.2. $\tilde{R}_j^{tk} = [\pi \in \Pi: \pi P^k X_d^k(\alpha_{l(\pi|d, k)}(t-1) - \alpha_a(t-1)) \leq 0: \forall d, \forall a]$.

THEOREM 7.3. *Let $\alpha_j^k(t)$ be a value of $\alpha_\pi^k(t)$ with g-representation.*
$g_j^k(t) = [g_{jd}^k(t), \ldots]$.
Suppose that the constraint $\pi P^k X_d^k[\alpha_{g_{jd}^k(t)}^k(t-1) - \alpha_l(t-1)] \leq 0$, *where $g_{jd}^k(t)$ is given for each pair corresponding to $d = (\theta, \xi_n)$, forms a boundary of $\tilde{R}_j^{tk}$. Then the cost functional $V_\pi^k(t)$ over some region in Π is given by $V_\pi^k(t) = \pi \alpha_m^k(t)$, where the g-representation of $\alpha_m^k(t)$ is*

$$g_m^k(t) = [g_{m_1}^k(t) = g_{j_1}^k(t), \ldots, g_{md}^k(t) = l, g_{mM}^k(t) = g_{jM}^k(t)].$$

Suppose now that a set of constraints $\pi f_j \leq 0, 0 \leq j \leq L$ are given on the space Π. These constraints are to be used in order to find that

$$R = [\pi \in \Pi : \pi f_j \leq 0, \forall j].$$

Then

$$\max \pi f_1$$

s.t.

$$\pi f_j \leq 0; \ 0 \leq j \leq L; \ \pi \in \Pi.$$

From Theorems 7.2 and 7.3, and from the above remarks, we can build a "one-pass algorithm" similar to that given in ref. [31].

This algorithm is carried out in four steps for calculating the optimal control policy for the t horizon case assuming that the α-vectors for the $(t-1)$ horizon case are known.

Step 1: Pick an initial state vector π and calculate the optimal control and its corresponding α vector k^* and $\alpha^*(t)$.

Step 2: Construct the complete list of inequalities for the region using

(1) $V_\pi(t) = \min_k \left[\sum_{i=0}^{L} \pi_i(q_i^k + \sum_{A_n, \xi_n, \theta, j} p_{ij}^k r_{j\theta}^k l_{\theta \xi_n}^k \alpha_{l(\pi|k, \theta, \xi_n)}(t-1)) \right]$

(2) $\pi P^k X_d^k(\alpha_{l(\pi|d,k)}(t-1) - \alpha_d(t-1)) \leq 0; \ \forall d, \forall a$

(3) $\pi(\alpha^*(t) - \alpha_k(t)) \leq 0; \ \forall k$

(4) $\sum_i \pi_i = 1.$

Step 3: Use the linear programming for the convex hull of $\tilde{R}_j^{tk}$ in order to calculate the maximum set of inequalities that define the region for $\alpha^*(t)$. From each boundary of the region construct, using (1) from Step 2, a new α-vector; store the following in a list for each:
— the vector, its corresponding optimal control alternative and an information vector for which it is the α-vector.

Step 4: Store the indices of the α-vectors that are neighbours to the region under consideration in order to limit the number of inequalities of the type in Step 2, during the $(t+1)$ horizon calculations. If there exist any α-vectors on the list whose region has not been calculated, pick a new α-vector and return to Step 2. Otherwise, the complete specification of the optimal control policy has been calculated.

7.4.2. An infinite-horizon maintenance planning model

Costly technical systems, e.g. nuclear reactors and computer units, are designed for a long period of operation. The system maintenance activities (i.e. repair) are very small compared with its life time. Then we can consider the maintenance process over an infinite-time horizon. In the following, we shall give a Branch and Bound solution for the inspection-diagnosis-maintenance process. When the "core process" is described by a stationary Markov process, we need to take advantage of cause-effect models with MDP. However, some other assumptions are required in order to formulate the optimization process in a Branch and Bound form. The technique used has some limitations for large-scale systems, but this is a general difficulty for a Markovian class of models. A general survey for Branch and Bound (B & B) algorithms was given by Lawler and Wood [90]; an approach to B & B for decision tree analysis is due to Chen and Patton [91]. Satia and Lave [34] used for the first time B & B (and the implicit enumeration algorithm) for partially observable MDP. In the remainder of this section we give a B & B algorithm for the inspection-diagnosis-maintenance problem. Computational aspects will be emphasized.

We shall use several results from a general formulation of B & B methods. A complete presentation was given by Mitten [86]:

> "In essence Branch and Bound methods are enumerative schemes for solving optimisation problems. The utility of the method derives from the fact that in general only a small fraction of the possible solutions need actually be enumerated, the remaining solutions being eliminated from consideration through the application of bounds that establish that such solutions cannot be optimal".

The value of information from the inspection-diagnosis process. In the case under consideration, the observer conveys in different modes information regarding the state of the process. We are faced with a problem of the *value of information* for the underlying component state vector X, by performing a set of observations for the system performance index $\Phi_\omega^*(X)$. We can see that when the behaviour of a system is described by its performance index, then the only possibility to resolve the uncertainty about the component state of operation is to use partially observable models (eventually Markovian models for the case of dynamic probabilistic systems) or models with a complete cause-effect description.

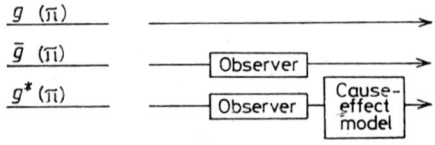

Fig. 7.10

We assume that $g(\pi)$ is the expected return in the case of a completely observable process; $\bar{g}(\pi)$ is the expected return for a partially observable process (a probabilistic observation for the system states through a "noisy channel"); $g^*(\pi)$ is the expected return for a probabilistic observation of the system states and cause-effect models as auxiliary.

A representation of such models is given in fig. 7.10.

Then the following relations

$$g(\pi) \leqslant \bar{g}(\pi) \tag{7.42}$$

$$\bar{g}(\pi) \leqslant g^*(\pi) \tag{7.42'}$$

are immediate. Satia [34] proved that relation (7.42) holds for Markovian models. A complete demonstration for (7.42) and (7.43) is given in

PROPOSITION 7.1. $g(\pi) \leqslant \bar{g}(\pi) \leqslant g^*(\pi)$.

Proof. Let the cost structure for a completely observable process be given by c_{ij}^k, when the system is in state i and the next visited state is j under a maintenance decision alternative k. Then the expected cost is given by

$$C_{\pi}^k = \sum_j \pi_i p_{ij}^k c_{ij}^k, \tag{7.43}$$

which is convex in π for every k. Subsequently, we can write

$$g(\pi) = g(\sum_j \sum_i \pi_i r_{ij}(\pi'|j)) \leqslant \sum_j \sum_i \pi_i r_{ij} g(\pi'|j) \leqslant \bar{g}(\pi), \tag{7.44}$$

where $r_{ij} = \Pr\{\text{output state} = j/\text{System State} = l\}$ and (π'/j) is the new state of knowledge about the process after a transition in the "core process", when the observable output was j.

By introducing a cause-effect model as an interface between the observation process and the maintenance decision process we can write

$$g(\pi) \leqslant \bar{g}(\pi) \leqslant \sum_m \sum_j \sum_i \pi_i r_{ij} l_{jm} g(\pi''|m), \tag{7.45}$$

where $l_{jm} = \Pr\{\text{primary event} = m/\text{observable state} = j\}$ and $g(\cdot|\cdot)$ is a convex function and, finally, $g(\pi) \leqslant \bar{g}(\pi) \leqslant g^*(\pi)$, which proves the above proposition.

As we have previously pointed out, the problem of the "value of information" arises when the system is allowed a probabilistic inspection. We have to pay for obtaining observations and, alternatively, for cause-effect models. A *perfect observer* is one who can identify the state of the process (the performance index for the total structure function) deterministically; in other words, $R_\theta = I$, where I is the identity matrix. It is obvious that for this case $g(\pi) = \bar{g}(\pi)$ as one has to pay to an observer and, hence, the maintenance decision-maker might change the subsequent state decisions owing to his better information. The value of observation (VO) for the above case is given by the difference between the value that can be attained with or without it.

$$\text{VO} = \bar{g}(\pi) - g(\pi). \tag{7.46}$$

The decision-maker who makes decisions on the basis of the expected cost (an expected value decision-maker) would measure the value of observation by the decrease in expected cost that it brings.

For the case of Markovian processes with cause-effect models as auxiliary, we can use the concept of "clairvoyant" (we use this concept in an acceptation different from Howard's [14]). The clairvoyant can predict the possible primary events that induce the deterioration of the structure function index. In this case, the clairvoyant partially eliminates the uncertainty about the state of the components instead of simply resolving it in advance. The value (VC) of the clairvoyant is

$$\text{VC} = g^*(\pi) - g(\pi), \tag{7.47}$$

and we can easily derive that $g^*(\pi) \geq \bar{g}(\pi)$, which proves again the results given previously.

In the remainder of this chapter we shall use the general B&B method presented by Mitten [86].

Branch and Bound Algorithm. The Branch and Bound method deals with (see ref. [86])

"*Branching* which consists of dividing collections of sets of solutions into subsets. *Bounding* which consists of establishing bounds on the value of the objective function over the subset of solutions".

The practical implementation of the method for the problem formulated above implies the following [86]:

"The B&B procedure involves recursive application of the branching and bounding operations with provision made for deleting subsets known not to contain an optimal solution".

The mathematical model presented below extends the results given in ref. [34] and uses as a value function branching criterion either

(a) DIFF = $(\beta)^{\text{level}} \times $ (UPB(I) − LOB(I)) × probability that the node is maximum for that branch Prob (I),

where DIFF is simply a branching criterion, β is the discount factor and level refers to the level of iteration in the B&B algorithm. Prob (I) is the probability that the decision alternative for the node is the optimum for that branch, or

(b) the lower bound for decision k is greater than the upper bound for any decision $j \neq k$ for that particular branch.

A Fibonnacci search method could be used for further branching in the decision tree representation for the above MDP.

It can be shown that an upper bound of the cost functional for the maintenance strategy exists and is given by

$$g(\pi) \leq \sum_{i, \xi_n} \pi_i v_i l_{\xi_n}, \qquad (7.48)$$

where

$$v_i(t) = \min_k \left\{ \sum_{j=0}^{L} p_{ij}^k (c_{ij}^k + \beta v_j(t-1)) \right\}, \; (i = 0, 1, \ldots, L). \qquad (7.49)$$

c_{ij}^k is the maintenance cost structure for a completely observable process which makes the transition from state i to state j, $v_i(t)$ is the minimum expected cost of the process terminating in state i as a result of decision alternative k: $v_i = \lim_{t \to \infty} v_i(t)$, (i.e. the steady state value or infinite horizon).

It can be shown that a lower bound of the cost functional exists and is given by

$$g(\pi) \geq \min [I - \beta P^{k-1}] c^k, \qquad (7.50)$$

where the expected maintenance cost for one transition is $c_i^k = \sum_j p_{ij}^k c_{ij}^k$ for each $i = 0, 1, \ldots, L$.

Proofs of (7.48) and (7.50) are given in the sequel.

PROPOSITION 7.2. *An upper bound of the cost functional exists and is given by*

$$g(\pi) \leq \sum_{i, \xi_n} \pi_i v_i l_{\xi_n},$$

where $v_i(t) = \min_k \left\{ \sum_{j=0}^{L} p_{ij}^k (c_{ij}^k + \beta v_j(t-1)) \right\}$, $(i = 0, 1, \ldots, L)$.

Proof. The above proposition is proved by induction.
Let $g(\pi, 1)$ be the cost functional associated with state vector at time 1. Then

$$g(\pi, 1) = \min_k \left\{ \sum_i \sum_j \sum_{\xi_n} \pi_i p_{ij}^k c_{ij}^k l_{\xi_n} \right\} \leq \sum_{\xi_n} l_{\xi_n} \sum_i \pi_i \min_k \left\{ \sum_j p_{ij}^k c_{ij}^k \right\} \leq$$

$$\leq \sum_{\xi_n} l_{\xi_n} \sum_i \pi_i v_i(1) \leq \sum \pi_i l_{\xi_n} v_i(1).$$

Suppose that

$$g(\pi, t) \leq \sum_{i, \xi_n} \pi_i l_{\xi_n} v_i(t).$$

By induction, we need to prove that

$$g(\pi, t+1) = \min_k \left\{ \sum_{\xi_n} \sum_i \sum_j \pi_i p_{ij}^k c_{ij}^k l_{\xi_n} + \beta \sum_{\xi_n} \sum_\theta \sum_i \sum_j \pi_i p_{ij}^k r_{j\theta}^k l_{\theta \xi_n} g(T(\pi \theta, \xi_n, k), t) \right\} \leq$$

$$\leq \sum_{\xi_n} \sum_i \pi_i l_{\xi_n} \min_k \left\{ \sum_j p_{ij}^k c_{ij}^k + \beta \sum_\theta \sum_j p_{ij}^k r_{j\theta}^k v_j(t) \right\} \leq \sum_{\xi_n} \sum_i \pi_i l_{\xi_n} \min_k \left\{ \sum_j p_{ij}^k c_{ij}^k + \right.$$

$$\left. + \beta \sum_j p_{ij}^k v_j(t) \right\} \leq \sum_{i, \xi_n} \pi_i l_{\xi_n} v_i(t+1).$$

For large t, we can write $\lim_{t \to \infty} g(\pi, t) \leq \lim_{t \to \infty} \sum_{i, \xi_n} \pi_i l_{\xi_n} v_i(t) \leq \sum_{i, \xi_n} \pi_i l_{\xi_n} \lim_{t \to \infty} v_i(t) \leq$

$$\leq \sum_{i, \xi_n} \pi_i l_{\xi_n} v_i.$$

PROPOSITION 7.3 (Satia [34]). *There exists a lower bound of the cost functional that is given by*

$$g(\pi) \geq \min_k \pi[I - \beta P^k]^{-1} c^k,$$

where $c_i^k = \sum_j p_{ij}^k c_{ij}^k$, $(i = 0, 1, \ldots, L)$.

Proof. Suppose that we can take a reasonable policy $A, k \in A$, for each stage of the process. Then

$$g(\pi) \geq \pi c^k + \beta \pi P^k c^k + \ldots \geq \pi[I - P^k]^{-1} c^k \geq \min_k \pi[I - P^k]^{-1} c^k.$$

In Proposition 7.1 it was shown that $g^*(\pi) > g(\pi)$ and so we can write

$$g^*(\pi) \geq \min_k \pi[I - \beta P^k]^{-1} c^k,$$

which is the lower bound for the problem.

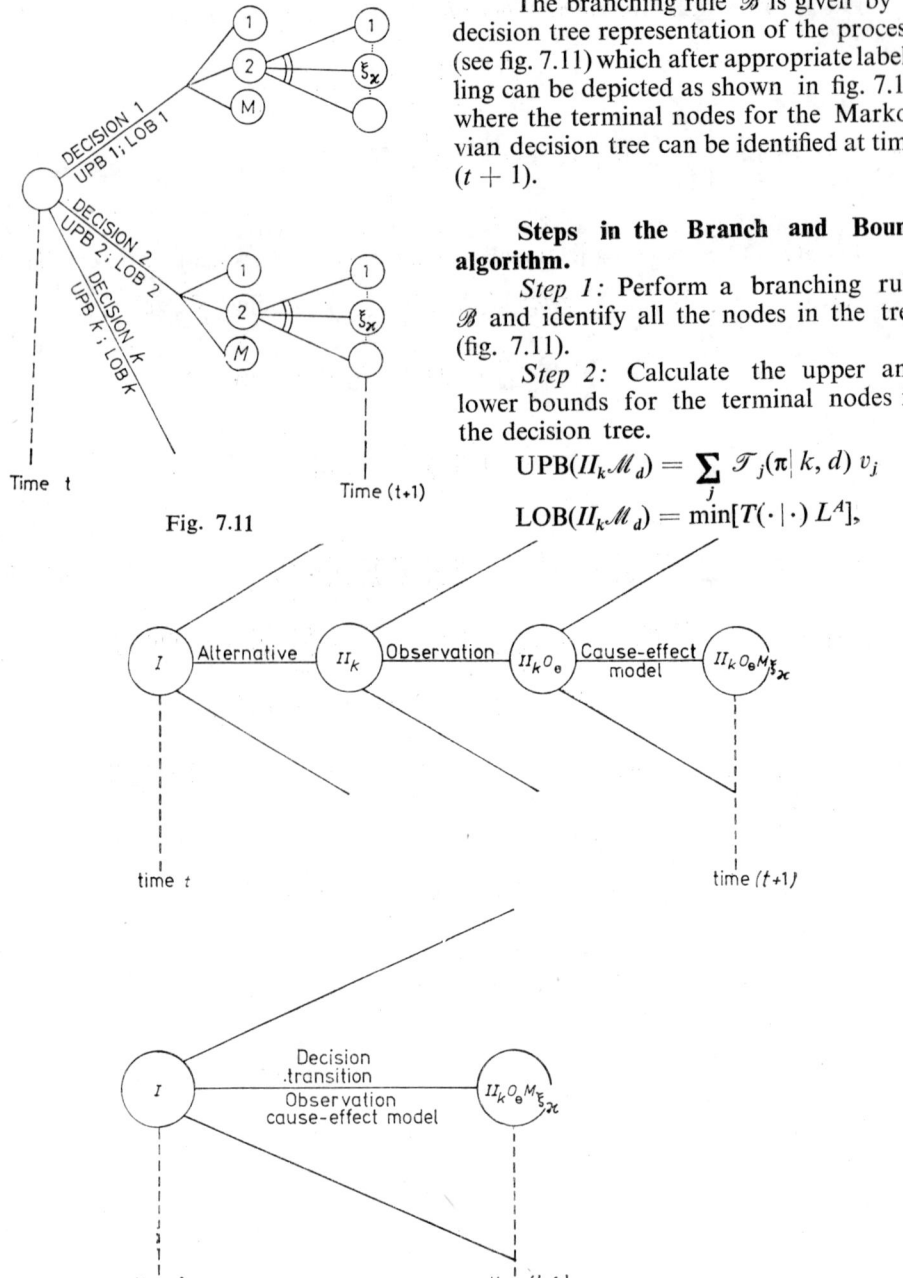

Fig. 7.11

Fig. 7.12

The branching rule $\mathscr{B}$ is given by a decision tree representation of the process (see fig. 7.11) which after appropriate labelling can be depicted as shown in fig. 7.12 where the terminal nodes for the Markovian decision tree can be identified at time $(t + 1)$.

Steps in the Branch and Bound algorithm.

Step 1: Perform a branching rule $\mathscr{B}$ and identify all the nodes in the tree (fig. 7.11).

Step 2: Calculate the upper and lower bounds for the terminal nodes in the decision tree.

$$\text{UPB}(\Pi_k \mathscr{M}_d) = \sum_j \mathscr{T}_j(\pi | k, d) v_j$$

$$\text{LOB}(\Pi_k \mathscr{M}_d) = \min[T(\cdot | \cdot) L^A],$$

where $(II_k \mathcal{M}_d)$ is an information vector in the B&B algorithm. Node $II_k \mathcal{M}_d$ indicates that the system is in state I, taking decision alternative k, and at the end of the transition a cause-effect model is observed.

$$T(\pi|\cdot) = [\mathcal{T}_j(\pi|\cdot)].$$

$\mathcal{T}_j$ above is the j-th component of the vector corresponding to the new state of knowledge after one transition where the original state vector was π, given that decision alternative k was chosen and the cause-effect model observed. L^A is the lower bound cost vector following some policy A.

Step 3: For any initial node which has other branches emanating from it, calculate the bounds for nodes I (the number of initial nodes I is equal to the number of states of the internal process characterized by the Markov model).

$$\text{UPB}(I) = \min_k \{ \sum_{\mathcal{M}_d} \Pr\{\text{Cause effect model} = \mathcal{M}_d | \mathcal{E}\} \cdot \text{UPB}(II_k \mathcal{M}_d) + \sum_j p_{ij}^k c_{ij}^k \},$$

$$\text{LOB}(I) = \min_k \{ \sum_{\mathcal{M}_d} \Pr\{\text{Cause effect model} = \mathcal{M}_d | \mathcal{E}\} \cdot \text{LOB}(II_k \mathcal{M}_d) + \sum_j p_{ij}^k c_{ij}^k \},$$

where $\sum_j p_{ij}^k c_{ij}^k$ is the expected immediate cost resulting from the transition.

Step 4: Select nodes for further branching in accordance with the criteria given above.

Step 5: Scan all the nodes in the decision tree starting from the terminal nodes (which correspond now to the highest level in the tree).

Step 6: If at a given level l, a dominating decision exists, (LOB $(a, l) >$ > UPB (b, l), $a \neq b$, where a and b are maintenance decision alternatives), then go to Step 7. If not, repeat the procedure with the new tree and go to Step 2.

Step 7: Stop and print the results.

Using the theoretical results of Balaș [87], the algorithm finds an optimal solution to the problem in a finite number of iterations. We give below a numerical example for the above algorithm applied to a rather different (medical diagnosis) problem.

An example: A Markovian decision model for clinical diagnosis and treatment as applied to the respiratory system. The Markovian models given in this chapter are widely used in diagnosing dynamic systems. However, we shall introduce a problem of medical diagnosis in order to present a numerical example and the appropriate Branch and Bound solution just explained. The only change in the algorithm is that the process return must be maximized by applying a set of treatment decisions to the respiratory system. (For an extensive approach, see Gheorghe [33].)

Clinical diagnosis. A problem of system diagnosis. The complexity of medical diagnosis has become an acknowledged fact, depending on the experience and judgement of the practising clinician. Recently, the techniques of systems science have been applied to medical problems and it is relevant and timely to consider the transfer of systems ideas to the diagnostic situation.

Decision theory is well established and a number of decision models have been applied to medical problems. For a good survey on the practical problems in the use of computers in medical diagnosis, see Ledley [51]. Several clinically applicable decision models have already been reported. Ginsberg and Offensend [49] proposed a medical diagnosis-treatment model using decision theory. A specific problem and its solution are described in terms of "what course of action in the form of diagnostic tasks and /or treatment should be taken". Rousseau [50] described a diagnostic procedure applicable to heart disease. A cause-effect model representation is adopted with explicit incorporation of AND/OR logical gates. This general approach combining probabilistic and boolean concepts develops procedures which allow the computation of probabilities even in reasonably large medical problems. These two approaches ([49],[50]), however, do not contain any dynamic concept of the process state, that is the patient's state of health. Sondik [13] presented clinical diagnosis-treatment problems in dynamic probabilistic (Markovian) terms and treated them in terms of partially observable Markov processes.

In this unified approach to clinical decision making, it is assumed that the clinician wishes to choose an action from a finite set of treatment alternatives. The states of the patient are considered to follow Markovian dynamics (Howard [7], [8]), describable in terms of a performance index based upon deterioration of the system with time or due to a fault, such as infection. The system is complex and not completely observable, so any change of state can only be assessed on the basis of clinical observations which may be cheap or expensive to perform. It is on the basis of these observations that decisions are made.

A numerical example for the respiratory system. Like all physiological processes, the respiratory system exhibits complex dynamic behaviour. In simplistic terms, the system can be regarded as a gas exchange plant in which blood acts as the carrier of gases from lung to tissues and *vice versa*. In the lung, blood becomes oxygenated and this extra oxygen is conveyed to the tissues which, in exchange, give up CO_2 to the blood which is then returned to the lung for further oxygenation. The process of releasing this CO_2 to the atmosphere and the take-in of fresh oxygen is achieved by breathing (Talbot & Gessner [47]; Bali et al. [48]). A pictorial representation is given in fig. 7.13.

In carrying out the diagnosis, the clinician aims to ensure that the outcome of his decision is satisfactory both to the patient and himself. During this process, three stages have to be considered:

- taking measurements (e.g. pulse rate, temperature);
- reaching a conclusion as to the state of the process, i.e. whether it is a case of heart disease, lung disease, lack of blood flow, poor environment, etc.;
- taking a decision regarding any course of action (e.g. do nothing, apply medication, change environment).

For instance, if as a result of "doing nothing" the patient is in state 2, there is 0.9 probability that his next transition is to state 2.
Initial state vector

$$\pi = [0.3 \quad 0.2 \quad 0.5]$$

Random inspection (clinical observation) probabilities

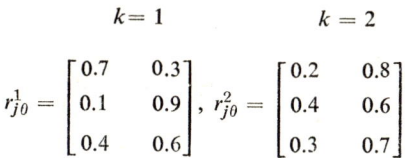

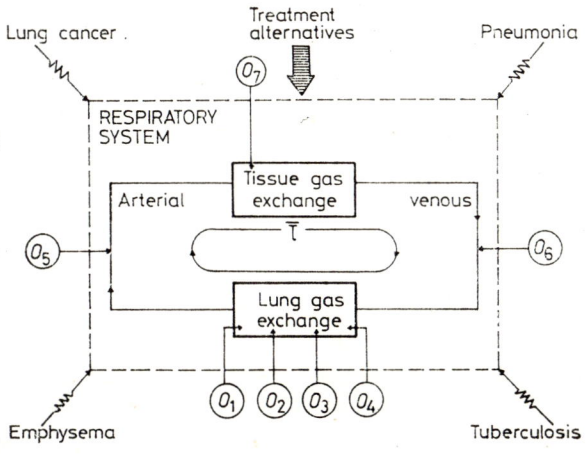

O_i — an observable output; O_1 — rate of breathing;
O_2 — tidal volume; O_3 — heart rate;
O_4 — ventilation/perfusion ratio; O_5 — arterial gas content;
O_6 — venous gas content; O_7 — tissue function;
T — blood circulation time in respiratory system.

Fig. 7.13

For instance, the probability of achieving observation $\theta = 1$, given that state 2 resulted from treatment alternative $k = 1$, is 0.1.

Cause-effect model probabilities

$$k=1 \qquad k=2$$

$$L_1^1 = \begin{bmatrix} 0.6 \\ 0.5 \end{bmatrix}, \quad L_1^2 = \begin{bmatrix} 0.6 \\ 0.5 \end{bmatrix}$$

$$L_2^1 = \begin{bmatrix} 0.7 \\ 0.6 \end{bmatrix}, \quad L_2^2 = \begin{bmatrix} 0.7 \\ 0.2 \end{bmatrix}$$

For instance, there is a probability of 0.7 that observation $\theta = 2$ results from cause 1 and a probability of 0.2 that the same observation results from cause 2, where in both cases the patient was in his present state as a result of treatment alternative $k = 1$.

Return on termination of the process

$$k=1 \qquad k=2$$

$$v^1 = \begin{bmatrix} 6 \\ 4 \\ 5 \end{bmatrix}, \qquad v^2 = \begin{bmatrix} 7 \\ 9 \\ 6 \end{bmatrix}$$

For the respiratory system, the measurement space may consist of heart rate, respiratory rate, temperature, tidal volume, noise level in lungs and oxygen and carbon dioxide concentrations in venous and arterial blood (these last observations being time consuming and expensive). Any set of observations could result from a number of different malfunctions. For example, a high ventilation rate could be due to low perfusion ventilation or a slow rate of blood flow through the system. The clinician has also to decide at what point to stop making further observations and start making decisions about medication or some alternative treatment. So decision taking concerns not only final treatment but also whether a given set of observations reveals the cause of the ailment with any degree of certainty.

Markovian decision modelling techniques like those given in this chapter are now applied to the respiratory system. In order to reduce computational complexity, only a small model is considered. The data used fall within physiological limits and are based on current clinical knowledge. However, the methods can be readily applied to more complex situations as well.

Let us consider the respiratory system with $N = 3$ states (ill, satisfactory and good) and $M = 2$ observable outputs (low peak ventilation i.e. less than normal but greater than 100 l/min, and very low peak ventilation, i.e. less than 100 l/min.). Let us suppose that there are two primary events (causes) for each observation as shown in figs. 7.14, 7.15 and 7.16.

Observation $\theta = 1$ is very low peak ventilation; $\theta = 2$ is low peak ventilation. Cause 1 = = emphysema; Cause 2 = pneumonia. The other process parameters are listed below. Consider that the number of treatment alternatives $k = 1, 2$. Alternative 1 ($k = 1$) is to apply medication and alternative 2 is to "do nothing". The following expressions represent the state transition probabilities.

Decision alternative 1 \qquad Decision alternative 2

$$p_{ij}^1 = \begin{bmatrix} 0.1 & 0.1 & 0.8 \\ 0.2 & 0.5 & 0.3 \\ 0.7 & 0.1 & 0.2 \end{bmatrix}, \quad p_{ij}^2 = \begin{bmatrix} 0.1 & 0.8 & 0.1 \\ 0.7 & 0.1 & 0.2 \\ 0.1 & 0.9 & 0 \end{bmatrix},$$

where state 1 = ill, 2 = satisfactory and 3 = good.

Reward function for change of state

$$k = 1 \qquad k = 2$$

$$c^1 = \begin{bmatrix} 3 & 5 & 7 \\ 2 & 4 & 3 \\ 6 & 4 & 2 \end{bmatrix}, \quad c^2 = \begin{bmatrix} 2 & 4 & 6 \\ 3 & 5 & 7 \\ 2 & 2 & 5 \end{bmatrix}$$

COHERENT STRUCTURES

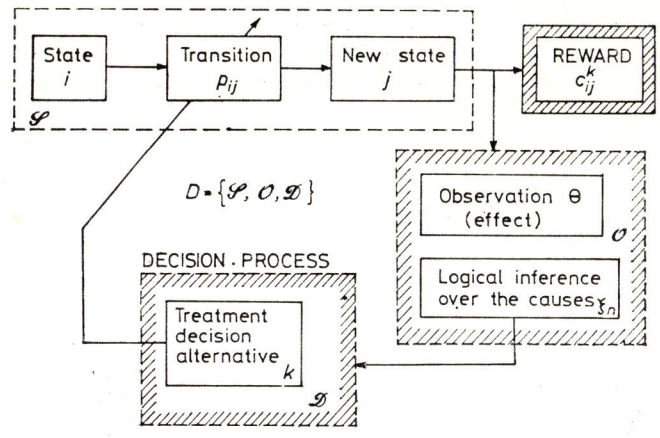

Markovian model representation for clinical diagnosis and treatment

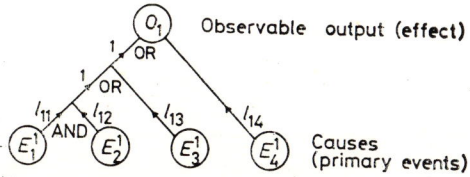

O_1 — peak gas flow rate abnormal
E_1^1 — pneumonia E_3^1 — lung cancer
E_2^1 — emphysema E_4^1 — tuberculosis

Fig. 7.14

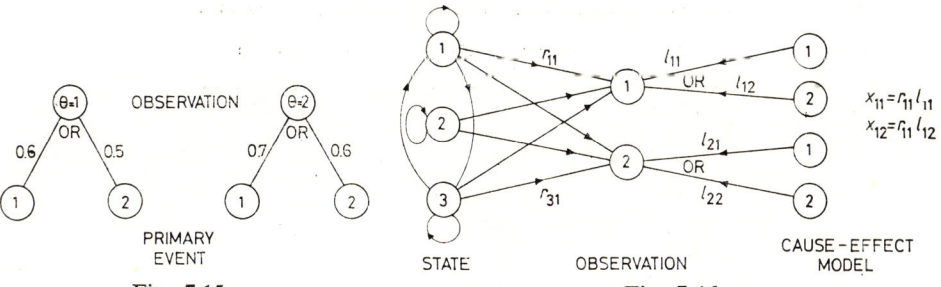

Fig. 7.15   Fig. 7.16

Using the cause-effect model representation and the relations derived above, causes may be related to states as is shown below.

$$x^1 = \begin{bmatrix} 0.42 & 0.35 & 0.21 & 0.18 \\ 0.06 & 0.05 & 0.63 & 0.54 \\ 0.24 & 0.20 & 0.42 & 0.36 \end{bmatrix}, \quad x^2 = \begin{bmatrix} 0.12 & 0.10 & 0.56 & 0.16 \\ 0.24 & 0.20 & 0.42 & 0.12 \\ 0.18 & 0.15 & 0.49 & 0.14 \end{bmatrix}$$

This "logical multiplication" is depicted in fig. 7.16. The dynamics of the new state space can be calculated using eqn. (7.26) from this chapter. The results for the two decision alternatives are shown in table 7.2. We apply now the Branch and Bound algorithm and obtain the results

Table 7.2

a) *First iteration*

Decision alternative $k = 1$

$\mathcal{T}_j(\pi/d, 1)$	State j		
	1	2	3
$d = 1$	0.30	0.14	0.56
$d = 2$	0.295	0.14	0.565
$d = 3$	0.056	0.57	0.374
$d = 4$	0.056	0.57	0.374

b) *First iteration*

Decision alternative $k = 2$

$\mathcal{T}_j(\pi/d, 2)$	State j		
	1	2	3
$d = 1$	0.429	0.571	0.0
$d = 2$	0.428	0.572	0.0
$d = 3$	0.66	0.34	0.0
$d = 4$	0.36	0.64	0.0

Table 7.3

Nodes	Upper bound	Lower bound
1, 1, 1	5.16	5.16
1, 1, 2	5.155	5.155
1, 1, 3	4.486	4.486
1, 1, 4	4.486	4.486
1, 2, 1	8.142	4.856
1, 2, 2	8.144	4.856
1, 2, 3	7.68	5.32
1, 2, 4	8.28	4.72
1, 1	4.7209	4.7209
1, 2	7.9854	5.0146
1	11.9854	9.0146

shown in table 7.3. The node distribution is that shown in fig. 7.17, with the nodes starting from the value corresponding to state 1 (ill). The general node nomenclature follows the pattern (Π_k, M_d). Thus, for instance, node (1, 2, 3) corresponds to being in state 1, taking decision alternative 2 and observing cause-effect model 3. The bounding procedure is carried out once a new

COHERENT STRUCTURES

branching has been implemented, using the branching rule $\mathscr{B}$. Adopting a discount factor $\beta = 0.5$ and $\Pr\{I\} = 0.8$ (probability of realizing a segment history of event such as states, observations, decisions and cause-effect models), the value of the branching criterion DIFF can be evaluated.

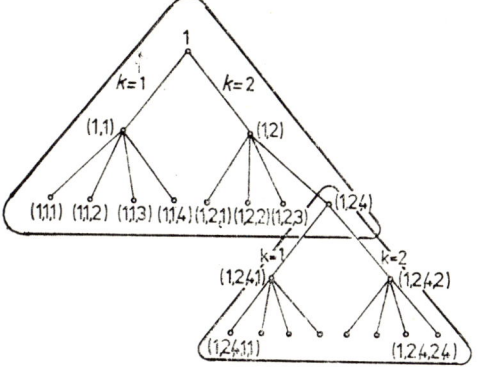

Table 7.4
Second iteration

NODES	DIFF Value
*	1.18832
1, 2, 1	0.52544
1, 2, 2	0.6576
1, 2, 3	0.472
1, 2, 4	0.712

Fig. 7.17

For instance, the value corresponding to node 1 is DIFF $= 1.18832$ (see table 7.4). Similarly, the values of DIFF are calculated for the eight terminal nodes corresponding to $II_k \cdot \mathscr{M}_d$ in the first iteration (see table 7.2 and fig. 7.18). Out of these terminal nodes, node $(1, 2, 4)$ yields the value of DIFF which is nearest to that calculated for the initial node 1.

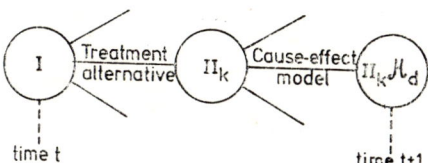

Fig. 7.18

Hence, node $(1, 2, 4)$ becomes node 1 for the second iteration which is expanded in a manner identical to that carried out in the first iteration. The bounds for this second iteration are listed in table 7.3. (Node $(1, 1)$ in table 7.3 corresponds to node $(1, 2, 4, 1)$ in fig. 7.17).

Table 7.5

Node	Upper bound	Lower bound
1, 2	2.18	2.1805
1, 2	1.063	0.6688

Since LOB $(1, 1)$ is greater than UPB $(1, 2)$, decision 1 dominates decision 2 and therefore decision 1 is optimal (see table 7.5). That is to say that for a patient who is currently in state 1, the optimal strategy is to apply medication ($k = 1$). In a similar manner, optimal decision strategies can be calculated for the other patient states.

A practical application could be given for the case of fault detection and repair/replacement in a nuclear reactor. In fig. 7.19 we give a modern scheme for fault detection in a nuclear reactor [88], when one can easily adopt the method presented in this chapter for reliability prediction and optimal maintenance decisions regarding the maintenance policies.

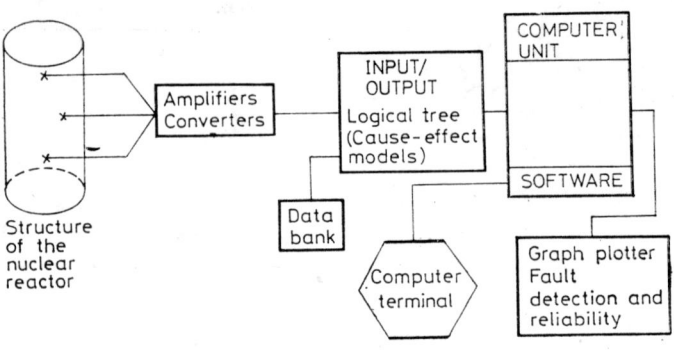

Fig. 7.19

For the type of the above mentioned application, the decision-maker could use this model as a buffer software programme, to find the best decision alternatives for the system and also for reliability prediction of the components or subsystems of the nuclear reactor.

Chapter 8

Semi-Markov maintenance and safety models for polyfunctional systems

It was already shown that alternative forms of Markov decision models can be used for maintenance strategies in technical systems. However, there are many realistic situations when the system is better described by a semi-Markov process. We shall use a system decomposition similar to that given in Chapter 2. We introduce first a semi-Markov population model for maintenance planning and safety analysis of technical systems. This model is an integrated approach between the underlying semi-Markov process of deterioration-maintenance for each structure function $\Phi_\lambda(X)$, $(\lambda = 1, 2, \ldots, \omega)$, and the stochastic process of "failure arrival" for components i, $(i = 1, 2, \ldots, n)$, with all the consequences for coherent structure (as defined in Chapter 2). Second, a semi-Markov decision model for optimal and suboptimal maintenance strategies will be formulated. The results obtained will be used in a hierarchical approach to management-maintenance decisions using $0-1$ mathematical programming ([143], [254]).

8.1. A semi-Markov population model for maintenance planning

In the design of a maintenance planning system, it is important to study the dynamics of component deterioration and repair activities. The repair activities for polyfunctional systems are generally related to the state vector X, which in turn defines a specific set of maintenance requirements. The underlying population model consists of a stochastic process which approximates the improvement of operation level for function λ by maintenance strategies, and another stochastic process describing the "failure arrival" over the components i, $(i = 1, 2, \ldots, n)$, which in turn will decrease the performance index $\Phi_\lambda(X)$, $(\lambda = 1, 2, \ldots, \omega)$ (see fig. 8.1). A decrease in the performance level $\Phi_\lambda(X)$ can be caused by a set of initiating events (e.g. mechanical failure, natural causes, human errors), which in turn will lead the system (i.e. the state vector X) to some less desirable operational state. After some initiating events have happened, we can classify the failed components belonging to different functions λ in classes of accidents (e.g. mechanical, electrical, electronic or a combination of them) in accordance with the maintenance staff available. Let us suppose that there exist two finite numbers E and F of initiating events and classes of accidents, respectively. From Chapter 2, we have $\omega' = \sum_{\lambda=1}^{\omega} M_\lambda$ equivalent functions for the (polyfunctional) system, which represents a mode of failure for the system.

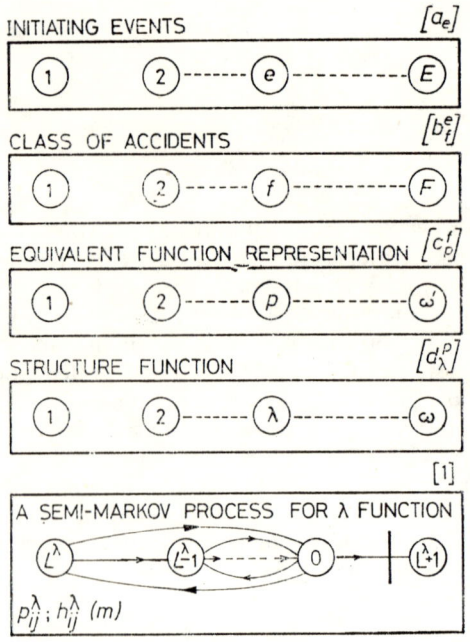

Fig. 8.1

The mode of failure p, $(p = 1, 2, \ldots, \omega')$, is a coordinator which probabilistically leads function λ to one of the states i, $(i = 0, 1, 2, \ldots, L^\lambda)$, of the underlying semi-Markov maintenance decision process. A pictorial representation for the above process is given in figs. 8.1 and 8.2. We recall the assumption (given in Chapter 2) that the function's behaviours are independent of one another from the deterioration/repair standpoint.

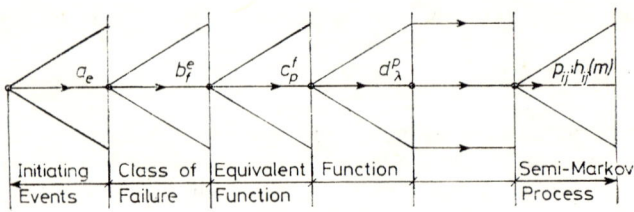

Fig. 8.2

In the remainder of this section, we shall introduce a semi-Markov process for modelling maintenance strategies over X for distinct λ-functions. We shall derive several results to be used in estimating the expected number of functions in operation at any time and the corresponding variance, as well as the probability of a number of functions to operate at a given operational level at some future time. An explicit concept — hypercorrelation — for polyfunctional systems will be introduced in the sequel.

8.2. Semi-Markov process statistics for maintenance modelling

In this paragraph we shall use results given in ref. [8] for modelling the improvement in structure function performance $\Phi_\lambda(X)$, $(\lambda = 1, 2, \ldots, \omega)$, by maintenance actions. Consider the function λ whose performance index at any point in time can be characterized by classifying the behaviour of the function on a finite number of states $0, 1, \ldots, L^\lambda$. State 0 represents the process before a deterioration in X such that $\Phi_\lambda(X) = M$, i.e. it is a fully operational state for function λ. State L^λ represents a terminal state for function λ, i.e. the state for which X leads the system to $\Phi_\lambda(X) = 0$. The intermediate states are ordered in accordance with the lexicographical ordering over X. We assume that the process is completely observable, i.e. at any time instant we have certain knowledge about the operational state for function λ. For technical reasons, in this paragraph we introduce one more state $(L^\lambda + 1)$ — a dummy state — which indicates, for instance, that after a maintenance procedure, the components of X belonging to function λ leave the maintenance department.

The underlying dynamic mechanism of function λ deterioration has the following characteristics.

(a) The probability that the process moves from state i to state j is p_{ij}^λ which satisfies the condition that

$$\sum_{j=0}^{L^\lambda + 1} p_{ij}^\lambda = 1, \quad (i = 0, 1, \ldots, L^\lambda + 1; \; \lambda = 1, 2, \ldots, \omega)$$

$$p_{ij}^\lambda \geq 0, \quad (i, j = 0, 1, \ldots, L^\lambda + 1; \; \lambda = 1, 2, \ldots, \omega).$$

(b) Before making any real transition out of state i to state j, the holding time in i is τ_{ij}^λ — an integer-valued random variable, positive and finite, which randomly transforms the time scale governed by a holding-time mass function $h_{ij}^\lambda(\cdot)$.

(c) Since the dummy state (discharge state as regards the maintenance department) is absorbing, then $p_{L^\lambda+1, L^\lambda+1}^\lambda = 1$; $h_{L^\lambda+1, L^\lambda+1}^\lambda(1)$, and $p_{ii}^\lambda = 0, (i = 0, 1, 2, \ldots, L^\lambda)$ and this clearly shows that a self-transition for the system is not possible.

(d) A prior probability state vector for function λ is given by $\Pi^\lambda(t) = [\pi_0^\lambda(t), \pi_1^\lambda(t), \ldots, \pi_{L^\lambda+1}^\lambda(t)]$ (a starting state vector is $\Pi^\lambda(0)$ with components $\pi_i^\lambda(0)$ for all states i).

By performing a set of maintenance strategies, function λ will eventually improve its performance index. We introduce now the notion of interval transition probabilities (see [8, p. 585]). The interval transition probability is denoted by $\varphi_{ij}^\lambda(t)$ and is the probability that function λ is in state j at time t given that it entered state i on time zero. The mathematical equation is

$$\varphi_{ij}^\lambda(t) = \delta_{ij} {}^{>}w_i^\lambda(t) + \sum_{k=0}^{L^\lambda+1} p_{ik}^\lambda \sum_{m=0}^{t} h_{ik}^\lambda(m) \varphi_{kj}^\lambda(t-m),$$

$$(i, j = 0, 1, \ldots, L^\lambda + 1, \; t = 0, 1, \ldots,) \qquad (8.1)$$

$$\varphi_{ij}^\lambda(0) = \begin{cases} 1 & \text{if } i = j \\ 0 & \text{otherwise,} \end{cases}$$

where δ_{ij} is the Kronecker symbol and

$${}^{>}w_i^\lambda(t) = \sum_{k=0}^{L^\lambda+1} p_{ij}^\lambda \left(1 - \sum_{m=0}^{t} h_{ij}^\lambda(m)\right) \qquad (8.2)$$

$$(i = 0, 1, 2, \ldots, L^\lambda + 1; \; t = 0, 1, 2, \ldots),$$

is the complementary cumulative probability distribution for the waiting times τ_i^λ in state i. To calculate the length-of-stay distribution in any state of the function, we introduce the first passage time probabilities $f_{ij}^\lambda(t)$ as the probability that first passage from state i to state j requires t units of time (see [8, p. 634]). The mathematical formulation is given by

$$f_{ij}^\lambda(t) = \sum_{\substack{k=1 \\ k \neq j}}^{L^\lambda+1} \sum_{m=0}^{t} p_{ij}^\lambda h_{ik}^\lambda(m) f_{kj}^\lambda(t-m) + p_{ij}^\lambda h_{ij}^\lambda(t), \qquad (8.3)$$

$$(i, j = 0, 1, \ldots, L^\lambda + 1; \; t = 0, 1, 2, \ldots),$$

$$f_{ij}(0) = 0, \; (i, j = 0, 1, \ldots, L^\lambda + 1).$$

LEMMA 8.1. *In a polyfunctional system, for function λ under a semi-Markov behaviour, having known constant failure and repair rates (that is the matrices $P^\lambda = [p_{ij}^\lambda]$ and $H^\lambda(m) = [h_{ij}^\lambda(m)]$, are known), the length-of-stay distribution is given by*

$$\mathcal{L}^\lambda(t) = \begin{cases} \pi_{L^\lambda+1}^\lambda(0), & t = 0 \\ \sum_{i=0}^{L^\lambda+1} \pi_i^\lambda(0) f_{i,L^\lambda+1}^\lambda(t), & (t = 1, 2, \ldots). \end{cases} \qquad (8.4)$$

Proof. The proof is straightforward from the above relation on first time passage probabilities and the definition of the prior probability state vector (see characteristic (d) above).

The following results are due to Kao [52], and use the probability figures given above. The probability $\gamma_{ik}^k(s/r)$ represents the probability that the components belonging to function λ are in state i for s units time if they were for r units time in state k

$$\gamma_{ik}^\lambda(s \mid r) = \frac{p_{ki}^\lambda h_{ki}^\lambda(s+r)}{{}^> w_k^\lambda(r)}, \quad (i, k = 0, 1, \ldots, L^\lambda + 1). \tag{8.5}$$

The next lemma is given without proof as it is straightforward from the definition of $\varphi_i^\lambda(t/E_k^\lambda(r))$, i.e. the probability that at time t function λ will be in state i of operation given that it was in state k for r units time.

LEMMA 8.2. *In a polyfunctional system, for function λ under a semi-Markov behaviour, having known matrices P^λ and $H^\lambda(m)$, the conditional maintenance state probability vector is*

$$\varphi_i^\lambda(t/E_k^\lambda(r)) = \sum_{j=0}^{L^\lambda+1} \sum_{s=1}^{t} \gamma_{kj}^\lambda(s/r)\, \varphi_{kj}^\lambda(t-s) + \delta_{ki} \sum_{j=0}^{L^\lambda+1} \left(1 - \sum_{s=0}^{t} \gamma_{kj}^\lambda(s/r)\right), \tag{8.6}$$

$$(i = 0, 1, \ldots, L^\lambda + 1),$$

where $E_k^\lambda(r)$ is the event for the components belonging to function λ to be in state k for r units time.

A "perfect inspection" at time zero will classify the performance index for function λ as given by state i of the underlying semi-Markov process. After t units time, by performing a new perfect inspection on the components belonging to function λ, the performance index is classified for state j. We introduce now the destination probabilities [8, p. 617],

$$\gamma_{ijq}^\lambda(t) = \Pr\{S(t+1) = q \mid S(t) = j,\, S(0) = i,\, \tau_i = m,\, \mathcal{E}(t)\}.$$

The mathematical equation for destination probabilities is given by

$$\gamma_{ijq}^\lambda(t) = \sum_{m=0}^{t} e_{ij}^\lambda(m)\, p_{jq}^\lambda\, {}^> h_{jq}(t-m), \tag{8.7}$$

$$(i, j, q = 0, 1, 2, \ldots, L^\lambda + 1,\ t = 0, 1, 2, \ldots)$$

and (see ref. [8, p. 614])

$$e_{ij}^\lambda(t) = \delta_{ij}\delta(t) + \sum_{k=0}^{L^\lambda+1} \sum_{m=0}^{t} p_{ik}^\lambda h_{ik}^\lambda(m)\, e_{kj}^\lambda(t-m), \tag{8.8}$$

where

$$\delta(t) = \begin{cases} 1 & \text{if } t = 0 \\ 0 & \text{otherwise,} \end{cases}$$

$$e_{ij}^\lambda(0) = \begin{cases} 1 & \text{if } i = j \\ 0 & \text{otherwise.} \end{cases}$$

The destination probabilities give information about scheduling new maintenance activities for future transition over the states of function λ.

Let us define the following events.

A_s^λ — the components belonging to function λ have been for maintenance for s units time,

C_p^λ — function λ failed through some minimal path p, $(p = 1, 2, \ldots, \omega')$,

B_f — failed components belonging to X could be classified into f, $(f = 1, 2, \ldots, F)$, classes of accidents,

G_e — the primary event e, $(e = 1, 2, \ldots, E)$, leads vector X to some failed states,

D_λ — function λ has to be maintained.

The probability values which describe the stochastic process for the "failure arrival" are given by (a) the prior probability for the initiating events G_e which in a vector form is

$$A = [a_e] = [\Pr\{G_e \mid \mathscr{E}\}]; \quad e = 1, 2, \ldots, E. \tag{8.9}$$

$\sum_{e=1}^{E} a_e = 1$; (b) the conditional probability that a failed component is classified in f, $(f = 1, 2, \ldots, F)$, classes of failures, given the initiating event G_e. The matrix representation is given by

$$B = [b_f^e] = [\Pr\{B_f \mid G_e, \mathscr{E}\}], \tag{8.10}$$

$$(e = 1, 2, \ldots, E; f = 1, 2, \ldots, F)$$

$\sum_{f=1}^{F} b_f^e = 1$; $e = 1, 2, \ldots, E$; (c) in a matrix formulation, the conditional probability that the class of failure f affects the structure by its equivalent function representation p, $(p = 1, 2, \ldots, \omega')$, is

$$C = [c_p^f] = [\Pr\{C_p \mid B_f, G_e, \mathscr{E}\}] \tag{8.11}$$

$$(f = 1, 2, \ldots, F, \; p = 1, 2, \ldots, \omega')$$

$\sum_{p=1}^{\omega'} c_p^f = 1$: $f = 1, 2, \ldots, F$; (d) in a matrix formulation, the conditional probability that the equivalent function p in the polyfunctional structure affects function λ, $(\lambda = 1, 2, \ldots, \omega)$, is

$$D = [d_\lambda^p] = [\Pr\{D_\lambda \mid C_p, B_f, G_e, \mathscr{E}\}], \tag{8.12}$$

$$(p = 1, 2, \ldots, \omega', \; \lambda = 1, 2, \ldots, \omega),$$

$$\sum_{\lambda=1}^{\omega} d_\lambda^p = 1, \; (p = 1, 2, \ldots, \omega').$$

Let us define $N_{efpskrq}$ to be the index specifying the number of failed components from X that follow from primary event e, the class of failure f and the equivalent function representation p. In the maintenance system, the components belonging

SEMI-MARKOV MAINTENANCE AND SAFETY MODELS

to X will spend s units time in a particular state k, corresponding to the performance index $\Phi_\lambda(X)$, the process holds for r units time, and q is the predicted state of the process in the maintenance system. A vector $\mathscr{I}$ which provides information about the components movement through the maintenance system is given by

$$\mathscr{I} = (N_{1111111}, \ldots, N_{efpskrq}, \ldots, N_{EF\,\omega'\,SL^\lambda M_{L^\lambda} L^\lambda}). \tag{8.13}$$

The maximum time spent in each state for the components belonging to function λ is given by M_i^λ, $(i = 1, 2, \ldots, L^\lambda; \lambda = 1, 2, \ldots, \omega)$.

THEOREM 8.1. *In a polyfunctional system, for function λ described by a semi-Markov process which approximates the maintenance strategies (this implies knowledge of matrices P^λ and $H^\lambda(m)$) and by the stochastic process describing the "failure arrival" (that is to know matrices A, B, C, D), the probabilities that the components belonging to function λ follow the pattern $N_{efpskrq}$ are given in matrix form by*

(a) $\quad \Gamma_e^q(\lambda \mid E_k^\lambda(r), p, f, s, \mathscr{E}) = \dfrac{A\,B\,C\,D\,\boldsymbol{y}^\lambda \,{>}\boldsymbol{f}}{\{x \mid \lambda\}}, \tag{8.14}$

where $\{x \mid \lambda\} = ABCD\,\boldsymbol{y}^\lambda \,{>}\boldsymbol{f}\,\mathbf{1}$, and

(b) $\quad \Delta_e(p \mid E_k^\lambda(r), \lambda, f, s, \mathscr{E}) = \dfrac{ABCD\,\boldsymbol{z}}{\{y \mid \lambda\}}, \tag{8.15}$

where $\{y \mid \lambda\} = ABCD\,\boldsymbol{z}\,\mathbf{1}$.

Proof. First, we shall prove part (a) of the theorem. By using Bayes' theorem, we write for the above process

$\Pr\{D_\lambda \mid E_k^\lambda(r), C_p, B_f, G_e, A_s^\lambda, \mathscr{E}\} = \Pr\{E_k^\lambda(r) \mid D_\lambda, C_p, B_f, G_e, A_s^\lambda, \mathscr{E}\} \cdot$

$\cdot \Pr\{A_s^\lambda \mid D_\lambda, C_p, B_f, G_e, \mathscr{E}\} \cdot \Pr\{D_\lambda \mid C_p, B_f, G_e, \mathscr{E}\} \cdot \Pr\{C_p \mid B_f, G_e, \mathscr{E}\} \cdot \Pr\{G_e \mid \mathscr{E}\} /$

$\displaystyle\sum_{\lambda=1}^{\omega} \Pr\{E_k^\lambda(r) \mid D_\lambda, C_p, B_f, G_e, A_s^\lambda, \mathscr{E}\} \cdot \Pr\{A_s^\lambda \mid D_\lambda, C_p, B_f, G_e, \mathscr{E}\} \cdot \Pr\{D_\lambda \mid C_p, B_f$

$G_e, \mathscr{E}\} \cdot \Pr\{C_p \mid B_f, G_e, \mathscr{E}\} \cdot \Pr\{B_f \mid G_e, \mathscr{E}\} \cdot \Pr\{G_e \mid \mathscr{E}\}. \tag{8.16}$

If $\eta_e^q(\lambda \mid E_k^\lambda(r), p, f, s, \mathscr{E}) = \Pr\{D_\lambda \mid E_k^\lambda(r), C_p, B_f, G_e, A_s^\lambda, \mathscr{E}\}$ and

$y_q^\lambda(E_k^\lambda(r) \mid p, f, e, s, \mathscr{E}) = \Pr\{E_k^\lambda(r) \mid D_\lambda, C_p, B_f, G_e, A_s^\lambda, \mathscr{E}\}$

then

$$\eta_e^q(\cdot \mid \cdot) = \dfrac{y_q^\lambda(\cdot \mid \cdot) \,{>}f_{L^\lambda+1}^\lambda(s) \cdot d_\lambda^p \cdot c_p^f \cdot b_f^e \cdot a_e}{\displaystyle\sum_{\theta=1}^{\omega} y_q^\theta(\cdot \mid \cdot) \,{>}f_{L^\theta+1}^\theta(s) \cdot d_\theta^p \cdot c_p^f \cdot b_f^e \cdot a_e}, \tag{8.17}$$

where

$$y_q^\lambda(\cdot|\cdot) = \sum_{i=0}^{L^\lambda+1} \pi_i^\lambda(0) \cdot \gamma_{ikq}^\lambda(s-r) \cdot {}^>w_k^\lambda(r), \qquad (8.18)$$

and

$$ {}^>f_{L^\lambda+1}^\lambda(\cdot) = \sum_{i=0}^{L^\lambda+1} \pi_i^\lambda(0) \left(1 - \sum_{m=0}^{s} f_{i,L^\lambda+1}^\lambda(m) \right). \qquad (8.19)$$

In matrix form (8.19) is given by (8.14). For relation (b), by using Bayes' theorem, we can write

$$\Pr\{C_p \mid E_k^\lambda(r), D_\lambda, B_f, G_e, A_s^\lambda, \mathscr{E}\} = \Pr\{D_\lambda \mid C_p, B_f, G_e, A_s^\lambda, \mathscr{E}\} \cdot$$
$$\cdot \Pr\{C_p \mid B_f, G_e, \mathscr{E}\} \cdot \Pr\{B_f \mid G_e, \mathscr{E}\} \cdot \Pr\{G_e \mid \mathscr{E}\} / \qquad (8.20)$$

$$\sum_{\varphi=1}^{\omega} \Pr\{D_\varphi \mid C_p, B_f, G_e, A_s^\varphi, \mathscr{E}\} \cdot \Pr\{C_p \mid B_f, G_e, \mathscr{E}\} \cdot \Pr\{B_f \mid G_e, \mathscr{E}\} \cdot \Pr\{G_e \mid \mathscr{E}\}.$$

If

$$\rho_e(p \mid E_k^\lambda(r), \lambda, f, s, \mathscr{E}) = \Pr\{C_p \mid E_k^\lambda(r), D_\lambda, B_f, G_e, A_s^\lambda, \mathscr{E}\}$$

and

$$z^p(\lambda \mid f, e, s, \mathscr{E}) = \Pr\{D_\lambda \mid C_p, B_f, G_e, A_s^\lambda, \mathscr{E}\},$$

then

$$\rho_e(\cdot|\cdot) = \frac{z^p(\cdot|\cdot) \cdot c_p^f \cdot b_f^e \cdot a_e}{\sum_{\gamma=1}^{\omega'} z^\gamma(\cdot|\cdot) \cdot c_\gamma^f \cdot b_f^e \cdot a_e}, \qquad (8.21)$$

where

$$z^p(\cdot|\cdot) = \sum_{i=0}^{L^\lambda+1} \pi_i M_i^\lambda. \qquad (8.22)$$

The matrix formulation of (8.22) is straightforward from (8.15).

From Theorem 8.1, we have information about which function λ is most likely to fail and from the second part of the theorem we have information about which configuration of components given by the equivalent representation p, ($p = 1, 2, \ldots, \omega'$), is more likely to fail. However, using the above probabilistic figures, we can increase system safety by using better design and maintenance strategies.

THEOREM 8.2. *In a polyfunctional system with all identical functions, if at time zero (beginning of the observation) all ω-functions were operating, given that any function λ was in state k for r units time, the probability that g or more* *functions operate in state i, $(i = 1, 0, \ldots, L^\lambda)$, t units time from now onwards, is given by*

$$\Pr\{g \text{ functions operate in state } i\} =$$

$$= \psi_i^g(t \mid E_k^\lambda(r), \omega, \mathscr{E}) = \sum_{l=0}^{\omega-q} \binom{\omega}{g}\binom{\omega-g}{l}(-1)^l \prod_{\lambda=1}^{\omega-l} \varphi_i^\lambda(t) \mid E_k^\lambda(r) \quad , \quad (8.23)$$

$\Pr\{\text{more then } g \text{ functions operate in state } i\} = \psi_i^{>g}(t \mid E_k^\lambda(r), \omega, \mathscr{E}) =$

$$= \sum_{l=0}^{\omega-g} \frac{g}{\omega-l}\binom{\omega-g}{l}\binom{\omega}{g}(-1)^{\omega-g-l} \cdot \prod_{\lambda=1}^{\omega-l} \varphi_i^\lambda(t) \mid E_k^\lambda(r), \quad (8.24)$$

where $\varphi_i^\lambda(\cdot \mid \cdot)$ is given in Lemma 8.2, for $i, k = 0, 1, \ldots, L^\lambda$, $g = 1, 2, \ldots, \omega$, $t = 0, 1, 2, \ldots$.

Proof. The proof is given by using Lemmas 3.1.1 and 3.1.2 from O'Brien [26, p. 31-33] and the above definitions for the underlying semi-Markov process for the polyfunctional system.

THEOREM 8.3. *In a polyfunctional system, under the conditions given in Theorem 8.2, the probability that the first ε functions operate in a state i, t units time from now onwards is given by*

$$\Pr\{\text{the first } \varepsilon \text{ functions operate}\} =$$

$$= \psi_i^{(1,2,\ldots,\varepsilon)} \quad (t \mid E_k^\lambda(r), \omega, \mathscr{E})$$

$$= \prod_{g^j:\, 1 \in g^j} \Pr\{\psi_{g^j}(X) = 1\} \cdot \prod_{g^j:\, 2 \in g^j,\, 1 \notin g^j} \Pr\{\psi_{g^j}(X) = 1\} \cdot$$

$$\cdots \cdot \prod_{g^j:\, \varepsilon \in g^j,\, 1,\ldots,\varepsilon-1 \notin g^j} \Pr\{\psi_{g^j}(X) = 1\} \cdot \sum_{\lambda=1}^{\varepsilon} \left\{ \sum_{j=0}^{L^\lambda} \sum_{s=0}^{t} \gamma_{kj}(s \mid r) \varphi_{ji}(t-s) + \right.$$

$$\left. + \delta_{kj} \sum_{j=0}^{L^\lambda} \left(1 - \sum_{s=0}^{t} \gamma_{kj}(s \mid r)\right)\right\}, \quad (8.25)$$

$$\lambda = 1, 2, \ldots, \varepsilon; \quad t = 0, 1, 2, \ldots,$$

$$\varepsilon = 1, 2, \ldots, \omega, \quad i = 0, 1, \ldots, L^\lambda.$$

Proof. The proof is derived by using Lemma 3.13 from ref. [26, pp. 33-36] and the above definitions given for the underlying semi-Markov process for the polyfunctional system.

The results given by Theorems 8.2 and 8.3 can be used for systems analysis in order to increase design performance (in the design phase) or to perform better maintenance strategies (during the operational phase). However, one may be interested which is the probability that the first two functions of a BWR safety system are operating at any given time t in the future under some repair/replacement policies.

8.3. An invariant index for polyfunctional systems

We shall introduce here a statistically invariant index to characterize a polyfunctional system for which input-output relations are available. This is the hypercorrelation index which is a generalization of the concept of statistical correlation between two populations, and was introduced in ref. [53].

A "system statistic" takes the statistical interactions of the system homogeneously by considering all the variables at the same time. In the system statistic, the strength of interactions is of prime interest and is independent of the evenness or oddness of the number of subsystems, structure functions, modules, etc. The minimum value of any index which measures the strength of interactions of the structure functions need not be zero, because by grouping a few of them in forming the total structure function $\Phi_\omega^*(X)$, the zero value (any well defined structure has a total structure function) is excluded. When functions λ are identical, the index should have the maximum value of unity.

Let Y_ω^θ be a polyfunctional system, and some observable θ, ($\theta = 1, 2, \ldots, R$), outputs are available at any time instant. For any structure function $\Phi_\lambda(X)$, ($\lambda = 1, 2, \ldots, \omega$), we can have a value for its expected performance level given by $E[\Phi_\lambda(X)]$. In this section, we shall introduce an invariant index when the system has been running for a long period of time ($t \to \infty$).

DEFINITION 8.1. The hypercovariance measure of a polyfunctional system Y_ω^θ is given by

$$\varepsilon_s = E\left[\prod_{\lambda=1}^{\omega} |(\Phi_\lambda(X) - E[\Phi_\lambda(X)])|\right], \tag{8.26}$$

or, alternatively,

$$\varepsilon_s = \left\{E\left[\prod_{\lambda=1}^{\omega} (\Phi_\lambda(X) - E[\Phi_\lambda(X)])\right]\right\}^{1/2k} \tag{8.27}$$

where k is a positive integer.

DEFINITION 8.2. The hypermoment of function λ of a polyfunctional system Y_ω^θ is the system deviation in which every function β, ($\beta = 1, 2, \ldots, \omega$), $\beta \neq \lambda$, is replaced by function λ.

$$\varepsilon_{s\lambda} = E[|(\Phi_\lambda(X) - E[\Phi_\lambda(X)])|^\omega], \tag{8.28}$$

or, alternatively,

$$\varepsilon_{s\lambda} = \{E[(\Phi_\lambda(X) - E[\Phi_\lambda(X)]^{2k\omega}]\}^{1/2k}, \tag{8.29}$$

where k is a positive integer.

DEFINITION 8.3. The hypercorrelation coefficient is defined as

$$\rho_\omega^\theta = \frac{\varepsilon_s}{\left(\prod_{\lambda=1}^{\omega} \varepsilon_{s\lambda}\right)^{1/\omega}}. \tag{8.30}$$

The hypercorrelation coefficient defined in (8.30) gives a numerical measure of the strength of interdependence of any subset $Y_r \subseteq Y_\omega$ of the functions of the set $[1, 2, \ldots, \omega]$.

The following theorem, which was adopted from ref. [53], proves that in any well-defined ω-function system, the hypercorrelation index is less than unity.

THEOREM 8.4. *In a polyfunctional system, the hypercorrelation coefficient cannot exceed the unity ($\rho_\omega^\theta \leq 1$) for any finite number of functions $\lambda = 1, 2, \ldots, \omega$ of the system.*

Proof. By Holder's inequality,

$$a_1 + a_2 + \ldots + a_\omega = 1, \tag{8.31}$$

and if $a_1, a_2, \ldots, a_\omega; a_{11}, a_{2\theta}, \ldots, a_{\omega\theta}$ are all positive,

$$\sum_{\theta=1}^{R} (a_{1\theta}^{a_1} a_{2\theta}^{a_2} \ldots a_{\omega\theta}^{a_\omega}) \leq (\sum_\theta a_{1\theta})^{a_1} (\sum_\theta a_{2\theta})^{a_2} \ldots (\sum_\theta a_{\omega\theta})^{a_\omega}. \tag{8.32}$$

Substituting

$$a_1 = a_2 = \ldots = a_\omega = \frac{1}{\omega}, \tag{8.33}$$

and

$$a_{\lambda\theta} = |\Phi_\lambda^\theta(X) - \overline{\Phi}_\lambda(X)|^\omega, \tag{8.34}$$

where $\overline{\Phi}_\lambda(X)$ is the mean of $\{\Phi_\lambda^\theta(X) \mid \theta = 1, 2, \ldots, R\}$, we can write relations

$$\sum_\theta \left[\prod_{\lambda=1}^{\omega} |(\Phi_\lambda^\theta(X) - \overline{\Phi}_\lambda(X))| \right] \leq \prod_{\lambda=1}^{\omega} \left[\sum_\theta |(\Phi_\lambda^\theta(X) - \overline{\Phi}_\lambda(X)|^\omega \right]^{1/\omega}. \tag{8.35}$$

Hence,

$$\frac{\sum_\theta [\prod_\lambda |(\Phi_\lambda^\theta(X) - \overline{\Phi}_\lambda(X))|]}{\prod_\lambda [\sum_\theta |(\Phi_\lambda^\theta(X) - \overline{\Phi}_\lambda(X))|^\omega]^{1/\omega}} \leq 1. \tag{8.36}$$

We can see that the left-hand side of (8.36) is an alternative definition of the hypercorrelation coefficient ρ_ω^θ in the discrete case.

The index ρ_ω^g provides a fair numerical measure of the correlation of any subset of the functions of the systems.

We may conclude that hypercorrelation coefficient is an index for the system organization measure and may be used in the design of complex systems (e.g. nuclear safety systems).

8.4. The vector semi-Markov process

Let us consider a polyfunctional system in which the behaviour of each function λ, ($\lambda = 1, 2, \ldots, \omega$), is governed by a semi-Markov process, and all functions are acted upon independently. The stochastic process of the failure arrival on the component state vector X leads the system functions to some state i of the underlying semi-Markov process. We shall define a vector semi-Markov process as the process which indicates the function populations of each state at any time instant.

We are interested here to give some analytical results for expectation and covariance over the vector semi-Markov process. We have to use the information vector $\mathscr{I}$ given before. We follow the assumption that the functions are identical and so are not distinguishable (see [26]) (that leads to $L^\lambda = L$ for $\lambda = 1, 2, \ldots, \omega$). The following theorem uses theoretical results on semi-Markov population models in a similar manner to that given by Kao [52].

THEOREM 8.5. *In a polyfunctional system, the expected number of functions operating in state q, if at time t the system were in state i, is given by*

$$E[r_q^{i,t}(X)] = \sum_{e=1}^{E}\sum_{f=1}^{F}\sum_{p=1}^{\omega'}\sum_{s=1}^{S}\sum_{k=1}^{L}\sum_{r=1}^{M_k} \cdot \left[N_{etpskrq} \sum_{\lambda=1}^{\omega} \varphi_i^\lambda(t \mid E_k^\lambda(r)) \cdot \eta_e^q(\cdot \mid \cdot) \right], \quad (8.37)$$

$$(i, q = 0, 1, \ldots, L;\ t = 0, 1, 2, \ldots, T),$$

and the variance is

$$\operatorname{Var}[r_q^{i,t}(X)] = \sum_{e=1}^{E}\sum_{f=1}^{F}\sum_{p=1}^{\omega'}\sum_{s=1}^{S}\sum_{k=1}^{L}\sum_{r=1}^{M_k} \cdot N_{efpskrq}[\sum_{\lambda} \varphi_i^\lambda(t \mid E_k(r)) \cdot \eta_e^q(\cdot \mid \cdot)] \cdot$$

$$\cdot [1 - \sum_{\lambda} \varphi_i^\lambda(t \mid E_k(r)) \eta_e^q(\cdot \mid \cdot)], \quad (8.38)$$

$$(i, q = 0, 1, \ldots, L;\ t = 0, 1, \ldots, T).$$

Following an approach similar to Howard's [8], the total number of functions that have occupied state j in the period 0 through t, and for which at the next transition each function will choose state q, is

$$^c r_q^{j,t}(X) = \sum_{j=0}^{L} \sum_{m=0}^{t} r_q^{j,m}(X). \quad (8.39)$$

THEOREM 8.6. *In a polyfunctional system with ω identical functions, the cumulative population moments (the mean and variance) of the vector semi-Markov process for which $v_{ij}(t)$ is the number of times the process enters state j through time t given that it entered state i at time zero, are given by*

$$^c r_q^{j,t}(X) = \sum_{e=1}^{E}\sum_{f=1}^{F}\sum_{p=1}^{\omega'}\sum_{s=1}^{S}\sum_{k=1}^{L}\sum_{r=1}^{M_k} \left[N_{efpskrq} \sum_{\lambda=1}^{\omega} \sum_{m=0}^{t} \cdot \bar{v}_{kj}^\lambda(t-m) \cdot p_{jq}^\lambda \cdot {}^>h_{jq}^\lambda(t-m) \right],$$

$$(8.40)$$

$$\operatorname{Var}[^c r_q^{j,t}(X)] =$$

$$\sum_{e=1}^{E}\sum_{f=1}^{F}\sum_{p=1}^{\omega'}\sum_{s=1}^{S}\sum_{k=1}^{L}\sum_{r=1}^{M_k} \cdot \left[N_{efpskrq} \cdot \sum_{\lambda=1}^{\omega} \sum_{m=0}^{t} \check{v}_{kj}^\lambda(t-m) \cdot p_{jq}^\lambda \cdot {}^>h_{jq}^\lambda(t-m) \right], \quad (8.41)$$

where [8, pp. 650], in general,

$$\bar{v}_{ij}(t) = \sum_{m=0}^{t} f_{ij}(m)[1 + \bar{v}_{jj}(t-m)]$$

$$\check{v}_{ij}(t) = \overline{v_{ij}^2}(t) - (\overline{v_{ij}(t)})^2 \tag{8.42}$$

$$\overline{v_{ij}^2}(t) = \sum_{m=0}^{t} f_{ij}(m)\overline{v_{jj}^2(t-m)} + \overline{2v_{ij}(t)} - \overline{\xi f_{ij}(t)}.$$

Proof. The proof is immediate from the definition of the cumulative population moments [8, pp. 488], where the underyling semi-Markov process is governed by the information vector $\mathscr{I}$ through its components $N_{efpskrq}$.

The above theoretical results are widely used in stochastic decision-making applied to complex systems safety. The probabilistic figures obtained from this analysis can improve the design alternative for systems operating in an uncertain environment. The hypercorrelation index is instrumental in designing complex multifunctional systems and eventually in adequate management policies. The numerical values for $E[\cdot]$ and $\text{Var}[\cdot]$ can help the maintenance engineer plan his activities over the system components from the standpoint of staff and auxiliary spare equipment and devices.

8.5. Maintenance policies in polyfunctional technical systems experiencing semi-Markov deterioration. General formulation and "policy iteration" solution

Management decisions concerning maintenance are still empirical. In view of a detailed management model, we must consider (1) the dynamics of the system given by the system states (countable and finite), (2) the decision process following some maintenance strategies, i.e. repair, replace or even "do nothing", and (3) some other technological and management constraints which concern the system. In the analysis of complex systems for optimal maintenance strategies, we must have an appropriate effectiveness index regardless of which an objective function has to be defined. An extensive discussion for choosing such an index is given in ref. [9] by White and Armitage. In our approach, we hall choose the effectiveness index to be the "system availability" which describes properly two distinct properties of complex systems: reliability and maintenance. By "system availability" we understand the quality or the state performance of the system or some of its functions to be available (i.e. functions that may be used for the accomplishment of one or a set of objectives). In the present analysis we are concerned with polyfunctional systems and maintenance decisions toward individual structure functions $\Phi_{\lambda}(X)$,

($\lambda = 1, 2, \ldots, \omega$), which can be modelled by a polydesmic semi-Markov decisions process [8]. In the way the model is built we need to define the gain for each individual function λ, e.g. $g^\lambda \cong E$ [Cash flow from $\Phi_\lambda(X)$ — Maintenance Cost — Penalties induced to the system], where $E[\cdot]$ indicates the expected value operator. System availability subject to each function λ is given by

$$\mathcal{A}_\lambda = g^\lambda / \text{Revenue from a totality available function} = \frac{g^\lambda}{\overline{q}_0^\lambda}. \tag{8.43}$$

For the polyfunctional system with ω independent functions, the system availability measure is

$$\mathcal{A} = \prod_{\lambda=1}^{\omega} \mathcal{A}_\lambda. \tag{8.44}$$

Let the stochastic behaviour of function λ generate a semi-Markov process; the deterioration dynamics is given by p_{ij}^λ and $h_{ij}^\lambda(m)$. The "trapping state" L^λ describes zero performance level of $\Phi_\lambda(X)$. As we deal with ω-function system, the process will be characterized by ω independent chains, corresponding to each function. The expected gain for function λ is q_i^λ and reflects the relative worth of the i-th state. The value structure of q_i^λ will include the revenue of the system for function λ. Let us consider that the previously defined system availability is a subjective measure of system effectiveness. The return from the process when this is visiting a state i is R_i^λ and the penalty for each visited state j is given by r_j^λ. The penalty cost in state j is independent of the last state occupied before the transition occurs. In complex systems with big social and economic impact (e.g. an electric power network or a nuclear power station), penalties are used for sake of high availability to stimulate a continuous and perfect operation of the system. So far, we have not introduced the maintenance cost for function λ of the system. We can say that a fixed cost c_i^λ per unit time is given when the system is in a particular state i, which finally corresponds to a state configuration for the vector X. In our approach, we are not concerned with staff maintenance modelling and subsequent optimization. A survey with extended results of this problem are given in ref. [111]. The cost structure could be summarized as $\overline{R}_i^\lambda(k) = R_i^\lambda - c_i^\lambda(k)$, where k is a pertinent maintenance alternative for the system in state i, that is $k \in K_i$, r_j^λ is a fixed penalty cost incurred when the system subject to function λ enters state j, $v_i^\lambda(t)$ denotes the expected cumulative return if the system starts in state i and the process operates for a period t following a fixed policy, and $v_i^\lambda(0)$ represents the terminal reward of the system being in state i at the end of time period t, when function λ is under observation (in this chapter, the system was assumed to be completely observable).

In what follows, the main results given by Howard [7, 8] and Blackwell [96] will be used as an optimization tool for the maintenance problem. The linear programming solution for semi-Markov decision processes was introduced by Mine and Osaki. Further results can be found in ref. [10]. We shall assume that the optimization procedure is for an infinite time horizon, when the complex system by its ω functions is to operate for a very long time period compared with the mean time between transitions and the holding time in any particular state. We can define a set of maintenance decision alternatives as "do nothing", "inspect with repair or replacement" of the components belonging to a particular function λ. A particular maintenance decision alternative will be called k and in each state there will be a set K_i of decisions such that the decision space $\Omega^\lambda(t)$ is given at any time by the cartesian product $\Omega^\lambda(t) = K_0^\lambda \otimes \ldots \otimes K_{L^\lambda}^\lambda$.

Let $^>w_i^\lambda(t)$ be the complementary cumulative probability distribution for the defined previously waiting time. The contribution to the total expected reward $v_i^\lambda(t)$

$$^>w_i^\lambda(t)\,[\bar{R}_i^\lambda(t) + v_i^\lambda(0)] \tag{8.45}$$

as the system does not leave state i. However, it may be possible that the system leaves state i for visiting other state j at time τ. Then it is clear from the way that we have defined the cost structure of the decision process that the system can "earn" the penalty r_j^λ and rate $\bar{R}_i^\lambda$ for some period of time τ. Then the value $v_j^\lambda(t - \tau)$ represents the value of the reward lottery for the time τ. The transition reward lottery is then given by

$$\sum_{j=0}^{L} p_{ij}^\lambda \sum_{m=0}^{t} h_{ij}^\lambda(m)[r_j^\lambda + \bar{R}_i^\lambda \tau + v_j^\lambda(t-m)], \tag{8.46}$$

and, subsequently, we can write

$$v_i^\lambda(t) = {}^>w_i^\lambda(t)[\bar{R}_i^\lambda(t) + v_i^\lambda(0)] + \sum_{j=0}^{L} p_{ij}^\lambda \sum_{m=0}^{t} h_{ij}^\lambda(m)[r_j^\lambda + \bar{R}_i^\lambda \tau + v_j^\lambda(t-m)], \tag{8.47}$$

$$(i = 0, 1, \ldots, L^\lambda).$$

As was emphasized before, the maintenance analysis at this level of approach is for a long time horizon, $t \to \infty$. For real systems, the holding time distribution mean has a finite value, and then $\lim_{t \to \infty} {}^>w_i^\lambda(t) \to 0$; $\lim_{t \to \infty} t\,{}^>w_i^\lambda(t) \to 0$, $(i = 0, 1, \ldots, L^\lambda)$.

From the above results, it is clear that

$$v_i^\lambda(t) = \sum_{j=0}^{L^\lambda} p_{ij}^\lambda \sum_{m=0}^{\infty} h_{ij}^\lambda(m)[r_j^\lambda + \bar{R}_i^\lambda \tau + v_j^\lambda(t-m)] \tag{8.48}$$

or

$$v_i^\lambda(t) = \sum_{j=0}^{L^\lambda} p_{ij}^\lambda r_j^\lambda + \bar{R}_i^\lambda \bar{\tau}_i^\lambda + \sum_{j=0}^{L^\lambda} p_{ij}^\lambda \sum_{m=0}^{\infty} h_{ij}^\lambda(m)\, v_j^\lambda(t-m), \tag{8.49}$$

where $\bar{\tau}_i^\lambda$ is the mean waiting time in state i for function λ. The immediate expected earning rate of function λ at state i is

$$q_i^\lambda = \bar{R}_i^\lambda + \frac{1}{\bar{\tau}_i^\lambda} \sum_{j=0}^{L^\lambda} p_{ij}^\lambda r_j^\lambda, \quad (i = 0, 1, \ldots, L^\lambda), \tag{8.50}$$

and, subsequently,

$$v_i^\lambda(t) = q_i^\lambda \bar{\tau}_i^\lambda + \sum_{j=0}^{L^\lambda} p_{ij}^\lambda \sum_{m=0}^{\infty} h_{ij}^\lambda(m) v_j^\lambda(t-m). \tag{8.51}$$

Using a geometric transformation [8], we can show that for large t, $v_i^\lambda(t)$ has an asymptotic value given by

$$v_i^\lambda(t) = g^\lambda t + v_i^\lambda, \quad (i = 0, 1, \ldots, L^\lambda), \tag{8.52}$$

where a monodesmic process gain g^λ is independent of the state. Since v_i^λ are relative values, the value for state L^λ is set to zero. A "policy iteration" solution to this problem was given in ref. [8] and a rigorous theoretical demonstration could be found in ref. [10]. For any stationary policy we need to solve a set of linear equations (the value iteration phase)

$$v_i^\lambda + g^\lambda \bar{\tau}_i^\lambda(k) = q_i^\lambda(k) \bar{\tau}_i^\lambda + \sum_{j=0}^{L^\lambda} p_{ij}^\lambda(k) v_j^\lambda, \tag{8.53}$$

for $g^\lambda, v_1^\lambda, \ldots, v_{L^\lambda-1}^\lambda$ setting the value of $v_{L^\lambda}^\lambda = 0$ and k is an index for the chosen maintenance alternative.

For the next state of the algorithm (policy improvement routine) using the values of v_i^λ the values of the test quantity $G(i,f)$ must be maximized for each i such that

$$q_i^\lambda(k) + \frac{1}{\bar{\tau}_i^\lambda(k)} \left[\sum_{j=0}^{L^\lambda} (p_{ij}^\lambda(k) - \delta_{ij}) v_i^\lambda \right] \tag{8.54}$$

for all k. If $G(i,f)$, i.e. the test quantity is empty for all i, then the corresponding decision strategy is optimal. Mine and Osaki [10] proved that it is only the policy improvement routine phase which yields the improved strategy whose average return is greater than the previous one. Howard [8] and Mine and Osaki [10] proved that the algorithm is working in a finite number of steps. Using the above algorithm, we obtain an optimal term maintenance strategy for any function λ, $(\lambda = 1, 2, \ldots, \omega)$.

However, as in real practical conditions it is seldom possible to implement optimal solutions, we need to apply some suboptimal control actions. In the subsequent approach we shall give a linear programming (LP) solution to the maintenance problem under the assumptions given above. From the LP we shall take the advantage of suboptimal solutions, and we can define the m-th suboptimal solution as a limit control decision for our maintenance purposes, regardless of any function λ of the system. Then we can use the same LP for all ω-functions with maximum saving of memory on the computer. The LP needs to print out at least from the m-th suboptimal solution (simplex tablou); the results can be used later in real maintenance management decisions when the system is under a feasible set of constraints. For an LP formulation, we shall adopt the results given in refs. [10] and [54].

8.6. Linear programming formulation for replacement strategies

We consider the problem of maximising the expected total reward (given by the system availability) of a semi-Markov decision process through a set of transient states. For the sake of simplicity, assume that under every policy for each function λ, state L^λ is a trapping state. Once a set of optimum policies were found, the second-to-optimum, third-to-optimum, m-th-to-optimum policies are found in succession (see fig. 8.3). For further theoretical discussion on suboptimal policies in MDP, see refs. [75, 76, 77, 78].

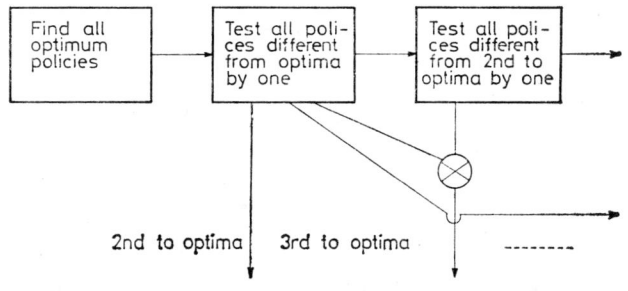

Fig. 8.3

Consider a maintenance policy ("do nothing" or "do replacement") in the policy space given by the cartesian product $\Omega^\lambda(\cdot)$. The corresponding relative value equation has been given above:

$$v_i^\lambda(t) = [q_i^\lambda(t) - g^\lambda]\bar{\tau}_i^\lambda(k) + \sum_{j=0}^{L^\lambda} p_{ij}^\lambda(k)\, v_j^\lambda(k), \qquad (i = 0, 1, \ldots, L^\lambda). \quad (8.55)$$

Since state L^λ is a trapping (absorbing) state under every policy, then

$$p_{L^\lambda j}(k) = \delta_{L^\lambda j} \quad (8.56)$$

and the L^λ-th relative value is

$$v_{L^\lambda}(k) = [q_{L^\lambda}^\lambda - g^\lambda]\bar{\tau}_{L^\lambda}(k) + v_{L^\lambda}(k). \quad (8.57)$$

Whence $g^\lambda = q_{L^\lambda}$.

Then we can write

$$v_i^\lambda(k) = [q_i^\lambda(k) - q_{L^\lambda}]\bar{\tau}_i^\lambda(k) + \sum_{j=0}^{L^\lambda} p_{ij}^\lambda(k)\, v_j^\lambda(k), \qquad (i = 0, 1, \ldots, L^\lambda). \quad (8.58)$$

In a matrix formulation, the above relation becomes

$$[I - P^\lambda(k)]\, \boldsymbol{v}(k) = \boldsymbol{s}(k), \quad (8.59)$$

where
$$P^\lambda(k) = [p_{ij}^\lambda(k)], \quad (i,j = 0, 1, \ldots, L^\lambda)$$
$$\mathbf{s}(k) = [(q_1^\lambda(k) - q_{L^\lambda})\bar{\tau}_1^\lambda(k), \ldots, (q_{L^\lambda-1}^\lambda(k) - q_{L^\lambda})\bar{\tau}_{L^\lambda-1}^\lambda(k)]^T.$$

The quantities $v_i(k)$ are the total expected reward generated before trapping, given that the initial state was i and policy k was in effect.

The maximization problem is given by $\sum_{i=0}^{L^\lambda} \pi_i v_i^\lambda(k)$, where $\pi_i = \Pr\{S(0) = i/\mathscr{E}\}$ and is the probability that the initial state was i and $\sum_{i=0}^{L^\lambda} \pi_i = 1$, $(\lambda = 1, 2, \ldots, \omega)$.

Let us define the initial state vector $\boldsymbol{\pi} = [\pi_0, \ldots, \pi_{L^\lambda}]$ for function λ. In view of the above maximization problem for maintenance strategies, we can write

$$\begin{aligned}
\boldsymbol{\pi}\,\boldsymbol{v}(k) &= \boldsymbol{\pi}[(I - P^\lambda(k))^{-1}\mathbf{s}(k)] \\
&= \boldsymbol{\pi}[D(k)\,\mathbf{s}(k)] \qquad (8.60)\\
&= (\boldsymbol{\pi} D(k))\,\mathbf{s}(k) \\
&= (\mathbf{s}(k))^T \boldsymbol{x}(k),
\end{aligned}$$

where $\boldsymbol{x}(k) \geqslant \boldsymbol{\theta}$ since $\boldsymbol{\pi} \geqslant \boldsymbol{\theta}^T$ and $D(k) \geqslant 0$, $((\boldsymbol{x}(k))^T = \boldsymbol{\pi}\, D(k))$ and $\boldsymbol{\theta}$ is a row vector with all elements equal to zero. Multiplying the above relation by $[I - P^\lambda(k)]$, we obtain

$$[I - P^\lambda(k)](\boldsymbol{x}(k))^T = \boldsymbol{\pi} \qquad (8.61)$$

and by transposing the above relation

$$[I - P^\lambda(k)]\,\boldsymbol{x}(k) = \boldsymbol{\pi}^T. \qquad (8.62)$$

Hence, the problem to be solved may be formulated as the following LP

$$\max: (\mathbf{s}(k))^T \boldsymbol{x}(k)$$

such that $[I - P(k)]^T \boldsymbol{x}(k) = \boldsymbol{\pi}^T$; $k \in \Omega^\lambda(\cdot)$; $\boldsymbol{x}(k) \geqslant \boldsymbol{\theta}$ and $\boldsymbol{\theta} = [0, 0, \ldots, 0]^T$.

From the above formulation, we can find the m-th optimal solution to the problem, using the basic properties (simplex method) of LP.

8.7. Systems with a complex maintenance; an LP formulation

In the previous section an LP formulation has been given for maintenance strategies in polyfunctional systems experiencing a semi-Markov behaviour, where for each function only two decision actions were appropriate (i.e. "do nothing" or "replace" the components belonging to X). However, a more realistic case is that of complex maintenance strategies (i.e. "do nothing", "inspect and repair" or "inspect and replace"). A policy iteration solution to this problem was given previously. In the remainder of this section, we deal with an LP formulation for the above case. The semi-Markov chain which describes such a process is included in the completely ergodic case, where all states form a single ergodic class under every policy, communicate and are recurrent. Using (8.54) and introducing the notation

$$r_i^\lambda(k) = q_i^\lambda(k)\,\bar{\tau}_i^\lambda(k), \tag{8.63}$$

we can write

$$v_i^\lambda + g^\lambda(k)\bar{\tau}_i^\lambda(k) = r_i^\lambda(k) + \sum_{j=0}^{L^\lambda} p_{ij}^\lambda(k)\,v_j^\lambda, \ (i = 0, 1, \ldots, L^\lambda) \tag{8.64}$$

or, in matrix formulation,

$$[I - P^\lambda(k)]\,\boldsymbol{v}(k) + g^\lambda(k)\,\bar{\boldsymbol{\tau}}^\lambda(k) = \boldsymbol{r}^\lambda(k). \tag{8.65}$$

Let $\pi_\lambda(k)$ be a normalized left eigenvector of $P^\lambda(k)$ with eigenvalue 1 and multiplicity 1. Then, we can write

$$g^\lambda(k) = \frac{\pi_\lambda(k)}{\pi_\lambda(k)\,\bar{\tau}^\lambda(k)}\,\boldsymbol{r}^\lambda(k) = \boldsymbol{e}^\lambda(k)\,\boldsymbol{r}^\lambda(k), \tag{8.66}$$

where $\boldsymbol{e}^\lambda(k)$ is the limiting entrance probability vector.

Following ref. [10], we can consider that $\boldsymbol{x}^\lambda = \boldsymbol{e}^\lambda(k)$. Then the maximization problem is

$$\max_k : (\boldsymbol{x}^\lambda)^T\,(\boldsymbol{r}^\lambda(k))$$

such that

$$\begin{bmatrix} (I - P^\lambda(k))^T \\ \hline (\bar{\boldsymbol{\tau}}^\lambda(k))^T \end{bmatrix} \boldsymbol{x}^\lambda = \begin{bmatrix} \boldsymbol{0} \\ \hline 1 \end{bmatrix},$$

$$\boldsymbol{x}^\lambda \geqslant \boldsymbol{0}.$$

A review of model parameters for the above problem is given in fig. 8.4.

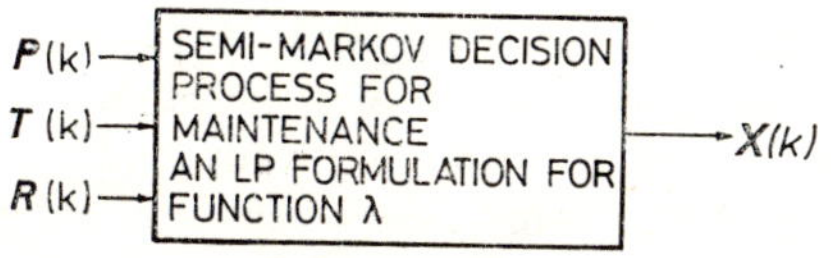

```
P(k) ──▶ ┌─────────────────────────┐
         │ SEMI-MARKOV DECISION    │
         │ PROCESS FOR             │
T(k) ──▶ │ MAINTENANCE             │──▶ X(k)
         │ AN LP FORMULATION FOR   │
R(k) ──▶ │ FUNCTION λ              │
         └─────────────────────────┘
```

Fig. 8.4

8.8. Degree of decomposition for maintenance strategies in large-scale technical systems

Owing to the special decomposition by ω-independent functions, where each function is described by an SMDP, we can build the angular structure given by the matrix

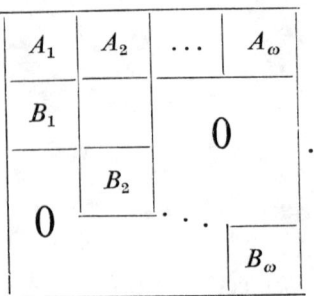

Let us suppose that we can build such a block-matrix

$$\begin{array}{|c|c|} \hline B_2 & 0 \\ \hline 0 & B_3 \\ \hline \end{array},$$

i.e. $\Phi_2(X) \cup \Phi_3(X)$, and we have a finite number of such block matrices. We can have a second level decomposition for each function λ at the level of a module (see fig. 8.5).

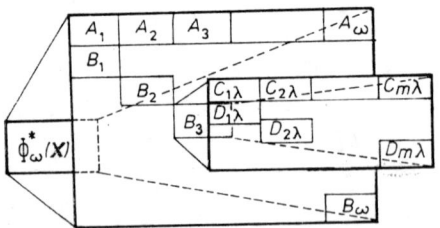

Fig. 8.5

For technical reasons, we assume that in the corresponding states of the process we have to deal with the same number of maintenance alternatives. For a given decomposition rule, a feasible question is "how many composite problems is it reasonable to solve by an LP?". This aspect does not refer implicitly to the maintenance problem of technical systems but rather to the problem-solving in Markovian decision processes with a large number of states. We shall follow Veress et al. [97]

in order to give a more formalized representation of the problem. A tied group of subproblems (i.e. structure functions) which is solved as one problem is called a composite problem p^c such that

$$p^c = \tilde{p} \circ p, \qquad (8.67)$$

where $\tilde{p}$ is the composing problem and p is a component. The sign "$\circ$" denotes a mapping composition. We deal here with a multicomponent problem, with one composing (master) problem and ω component problems. Following an approach similar to that of Masen [97], we investigate decomposition rules for polyfunctional technical systems under a semi-Markov behaviour for the composite problem which gives the minimum computing time. For a well-defined problem we can have $1, 2, \ldots, \alpha$ composite problems. For any given composite problem, we must define (1) the number of LP, $\varkappa$, to obtain the overall solution to the problem and (2) the number φ of functions within the polyfunctional system in a given $\varkappa$-LP model; however, the triplex $\{\alpha, \varkappa, \varphi\}$ has to be defined. The model characteristics, such as (1) maximum number of variables, are

$$\text{MAX } V = \begin{cases} \sum_{\lambda=1}^{\omega}\left(\sum_{i=0}^{L^\lambda} k_i^\lambda + L^\lambda + 2\right) - \omega & \text{for } \varphi = \omega, \varkappa = 1, \\ \max_{s_i: s_i \in s}\left\{\int_{\lambda: \lambda \in s_i}\left(\sum_{i=0}^{L^\lambda} k_i^\lambda + L^\lambda + 1\right)\right\} & \text{for } \omega > \varkappa, \varphi > 1 \\ \max_{s_i}\left\{\sum_{i=0}^{L^\lambda} k_i^\lambda + L^\lambda + 1\right\} & \text{for } \varphi = 1, \varkappa = \omega, \end{cases} \qquad (8.68)$$

and (2) the number of constraint equations are given by

$$\text{MAX } C = \begin{cases} \sum_{\lambda=1}^{\omega}(L^\lambda + 1) & \text{for } \varkappa = 1, \varphi = \omega \\ \max_{s_i: s_i \in s}\left\{\int_{\lambda: \lambda \in s_i}(L^\lambda + 1)\right\} & \text{for } \omega > \varkappa, \varphi > 1, \\ \max_{s_i}\left\{\sum_{\lambda: \lambda \in s_i}(L^\lambda + 1)\right\} & \text{for } \varkappa = \omega, \varphi = 1, \end{cases} \qquad (8.69)$$

where s is a set of all composite problems (number of subproblems), s_i represents a subset for a particular combination of the composite problems, and s_i^j denotes the j-th component of s_i.

The following relations are immediate.

$$s_i \subseteq s: s_i^j, \quad (j = 1, 2, \ldots, \varphi)$$
$$\varphi = \text{card } \{s_i\}, \quad \varphi \neq \emptyset$$
$$a = \text{card } \{s\}, \quad a \neq \emptyset$$
$$s = \{s_i; i = 1, 2, \ldots, a\}$$
$$s_i = \{s_i^j; j = 1, 2, \ldots, \varphi\}, \quad 1 \leq \varphi \leq \omega$$

(see also fig. 8.6). The sign "∫" denotes general summation. It is difficult to give a general picture for MAX V and MAX C (the model characteristics) variation because they are finally depending on the number of Markovian states, the number of decision alternatives for maintenance strategies, the number of functions which the system must perform, alternative of decomposition, etc. From fig. 8.7, we can see how MAX C increases for a fixed number of decomposition alternatives.

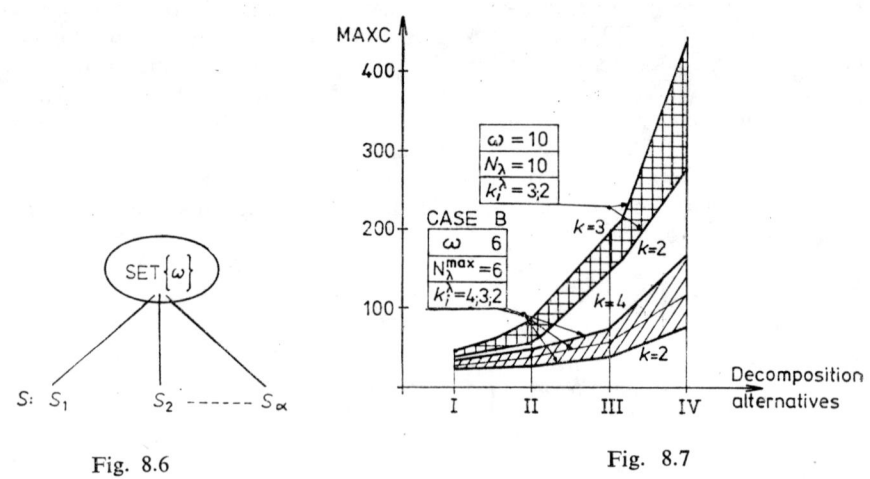

Fig. 8.6   Fig. 8.7

An example has been investigated for which (see tables 8.1—8.4) the number of functions $\omega = 6$; the number of composite problems $\varphi = 4$; the number of decision alternatives in each state of the SMDP is $k = 3$; and the number of states (maximum 6, minimum 4).

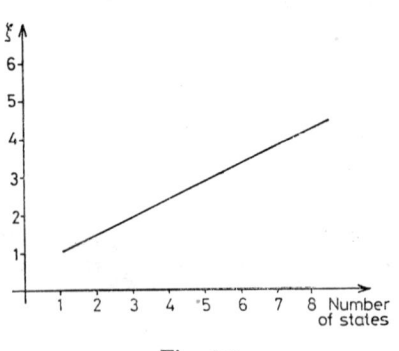

Fig. 8.8

In counting the number of MAX C, the slack variables for LP were taken into account. In tables 8.2 and 8.3 relative values for the computation time of SMDP are given on an ICL 1905 computer. In fig. 8.8 we show a pictorial representation for the variation of the number of states (performance index) for any function λ and the index ξ as the ratio between the number of states and the programming time for the computer. Clearly, ξ varies linearly with the increasing cardinality of performance index for any function λ. From the above computational experience, we may conclude that it is possible to give a guide rule on how to decompose an SMDP given by an angular LP-model, when a decomposition algorithm (functional, modular or item decomposition) is available. A direct empirical rule toward an optimal degree of decomposition for SMDP by an LP-model is to solve ω-independent LPS. The size of the problem increases with the levels of performance index for $\Phi_\lambda(X)$, $\lambda = 1$, $2, \ldots, \omega$ and also with the pertinent number of maintenance decision alternatives for each state i.

SEMI-MARKOV MAINTENANCE AND SAFETY MODELS

Table 8.1

Alternative	PARAMETERS					Compilation time (seconds)	Programming time (seconds)	Total mill time
	Number of states for SMDP	Number of decision alternatives in each state	Number of constraint equations	Columns and slack variables	Real variables in LP			
I	4	3	5	18	12	1	32	48
	6	3	7	26	18			
	5	3	6	22	16			
	4	3	5	18	12			
	6	3	7	26	18			
	5	3	6	22	16			
II	6+4	3	12	43	30	1	97	106
	5+5	3	12	43	30			
	6+4	3	12	43	30			
III	6+5+4	3	18	64	75	1	195	210
	6+5+4	3	18	64	45			
IV	6+5+4 +6+5+4	3	36	127	90	1	657	677

Table 8.2

Alternative	Compilation time (seconds)	Computation time (main programme) (seconds)	Programming time (Reference case 1 = 1)	Total mill time (Reference case 1 = 1)	Total mill time / Programme-time
I	1	1	1 (Reference)	1 (Reference)	1.5
II	1	2	3.03125	2.2084	1.093
III	1	5	6.09375	4.375	1.077
IV	1	16	20.53125	14.104	1.030
Average time	1	6	7.664	5.42185	—

Table 8.3

Alternative	Average number of states	Number of states (Reference 1 = 1)	Time (Table 6.2)		$\xi = \dfrac{\text{No. of states}}{\text{Programme time}}$
			A	B	
I	22	1	1	1	1
II	43	1.954	3.03	2.2	1.550
III	64	2.910	6.09	4.37	2.092
IV	127	5.772	20.5	14.10	3.551

Table 8.4

Number of maintenance alternatives in each state	$(a, \varphi) - \max \varphi$			
	(10, 1)	(5, 2)	(2, 5)	(1, 10)
$k = 3$	411	83	206	42
$k = 2$	311	63	156	32

8.9. An implicit enumeration program for optimization and management of maintenance strategies in polyfunctional systems

For practical reasons, we shall consider that the combination level for a ω function system is given by the minimization of the maintenance cost. We shall also consider the case when each function can be associated with a priority index. We can define a boolean decision variable $x_{\lambda j k} = 0, 1$ for each function λ, ($\lambda = 1, 2, \ldots, \omega$), number of states $j = 0, 1, \ldots, L^{\lambda}$, and optimal and suboptimal solutions $k = 1, 2, \ldots, m_{\lambda}$ (see fig. 8.9). In the remainder of this section we shall give an implicit enumeration program, or a 0—1 program, for optimization and management of maintenance strategies in polyfunctional technical systems. The objective function for the maintenance strategy is to minimize the cost for ω function maintenance, when the maintenance engineer must take a decision concerning the whole system.

Let us define the coefficients $a_{\lambda j k}$ to be the maintenance cost for the components belonging to function λ, when the performance index is given by state j, the k-th maintenance decision alternative is given by state j and the k-th maintenance decision alternative may be followed for a finite set of optimal and suboptimal solutions given by the SMDP. Then we can write the objective function as given by

$$\min_k : \sum_{\lambda=1}^{\omega} \sum_{j=0}^{L^\lambda} \sum_{k=1}^{m_\lambda} a_{\lambda jk} x_{\lambda jk}^t.$$

An appropriate set of constraints might be defined such as technological constraints, management and economical constraints, model constraints.

The set of constraints will be introduced in the sequel.

(1) The maintenance capital budget is limited subject to each function λ. If the meaning of $a_{\lambda jk}$ is defined as above and if d_λ represents the amount of money available for the maintenance of function λ, then

$$\sum_{j=0}^{L^\lambda} \sum_{k=1}^{m_\lambda} a_{\lambda jk} x_{\lambda jk} \leq d_\lambda,$$

$(\lambda = 1, 1, \ldots, \omega)$.

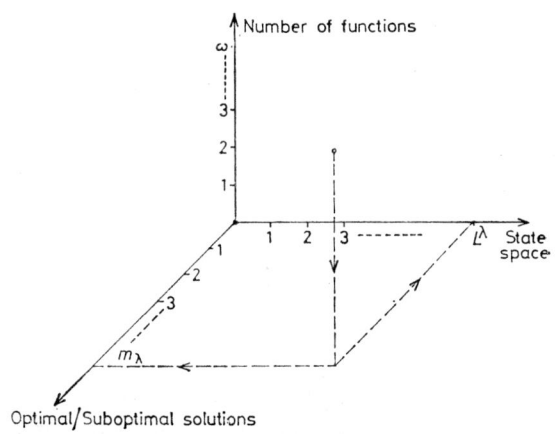

Fig. 8.9

(2) A limited number of functions a could be maintained at the same time:

$$\sum_{\lambda=1}^{\omega} \sum_{j=0}^{L^\lambda} \sum_{k=1}^{m_\lambda} x_{\lambda jk} \leq a.$$

(3) At least one decision alternative must be chosen when the ω-function system can be in a finite number of states:

$$\sum_{j=0}^{L^\lambda} \sum_{k=1}^{m_\lambda} x_{\lambda jk} \geq L^\lambda, \quad (\lambda = 1, 2, \ldots, \omega).$$

(4) For technological reasons, a function λ is not allowed to operate under a given performance level. If $e_{\lambda jk}$ represents function λ under the conditions given above, and h_λ is the minimal performance level for which function λ is allowed to operate, then

$$\sum_{j=0}^{L^\lambda} \sum_{k=1}^{m_\lambda} e_{\lambda jk} x_{\lambda jk} \geq h_\lambda, \quad (\lambda = 1, 2, \ldots, \omega).$$

(5) Some priorities among the ω-functions of the system are considered. A positive constant α_λ, $(\lambda = 1, 2, \ldots, \omega)$, indicates priorities for each function and is defined as

$$\sum_{\lambda=1}^{\omega} \alpha_\lambda = 1, \quad \alpha_\lambda \geq 0.$$

Then we can write

$$\sum_{\lambda=1}^{\omega} \sum_{j=0}^{L^\lambda} \sum_{k=1}^{m_\lambda} \alpha_\lambda x_{\lambda jk} \leq 1.$$

(6) The last constraint to the problem is related to the priority to follow an optimal or a suboptimal maintenance strategy for each function λ. We can define a priority coefficient $\xi_{\lambda k}$ to be the probability that for function λ, the k-th-to-optimal solution should be used.

$$\sum_{k=1}^{m_\lambda} \xi_{\lambda k} = 1, \quad (\lambda = 1, 2, \ldots, \omega),$$

$$\xi_{\lambda k} \geq 0, \quad (\lambda = 1, 2, \ldots, \omega, \ k = 1, 2, \ldots, m_\lambda).$$

However, the constraint for the priority to an optimal solution is

$$\sum_{\lambda=1}^{\omega} \sum_{j=0}^{L^\lambda} \sum_{k=1}^{m_\lambda} \xi_{\lambda k} x_{\lambda jk} \leq 1.$$

8.10. The implicit enumeration (0−1) program model for maintenance management strategies

An epitome of the general 0−1 program model for maintenance management strategies takes the form

$$\min_{k} : \sum_{\lambda=1}^{\omega} \sum_{j=0}^{L^\lambda} \sum_{k=1}^{m_\lambda} a_{\lambda jk} x_{\lambda jk}$$

such that

(1) $\sum_{j=0}^{L^\lambda} \sum_{k=1}^{m_\lambda} a_{\lambda jk} x_{\lambda jk} \leq d_\lambda, \quad (\lambda = 1, 2, \ldots, \omega),$

(2) $\sum_{\lambda=1}^{\omega} \sum_{j=0}^{L^\lambda} \sum_{k=1}^{m_\lambda} x_{\lambda jk} \leq a$

(3) $\sum_{j=0}^{L^\lambda} \sum_{k=1}^{m_\lambda} x_{\lambda jk} \geq L^\lambda, \quad (\lambda = 1, 2, \ldots, \omega).$ \hfill (8.70)

(4) $\sum_{i=0}^{L^\lambda} \sum_{k=1}^{m_\lambda} e_{\lambda jk} x_{\lambda jk} \geq h_\lambda, \quad (\lambda = 1, 2, \ldots, \omega),$

(5) $\sum_{\lambda=1}^{\omega} \sum_{j=0}^{L^\lambda} \sum_{k=1}^{m_\lambda} a_\lambda x_{\lambda jk} \leq 1$

$$(6) \quad \sum_{\lambda=1}^{\omega} \sum_{j=0}^{L^{\lambda}} \sum_{k=1}^{m_{\lambda}} \zeta_{\lambda k} x_{\lambda jk} \leqslant 1$$

$$(7) \quad x_{\lambda jk} = 0, 1 \text{ for all } \lambda, j, k.$$

Bounds

$$a_{\lambda} \geqslant 0$$

$$\sum_{\lambda=1}^{\omega} a_{\lambda} = 1$$

$$\zeta_{\lambda k} \geqslant 0$$

$$\sum_{k=1}^{m_{\lambda}} \zeta_{\lambda k} = 1, \quad (\lambda = 1, 2, \ldots, \omega).$$

A pictorial representation for the maintenance management model given above is shown in fig. 8.10.

Model characteristics. The number of problem variables is given by

$$M \text{ MAX } V = \sum_{\lambda=1}^{\omega} [(L^{\lambda} + 1) + m_{\lambda}], \tag{8.71}$$

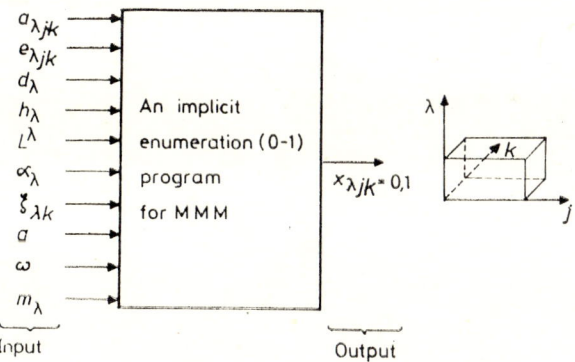

Fig. 8.10

and the number of constraint equations by

$$M \text{ MAX } C = 3(\omega + 1), \tag{8.72}$$

where ω is the total number of functions; L^{λ} represents the cardinality of functions λ and m_{λ} is the m-th optimal solution for function λ. Equations (8.71) and (8.72) are plotted in figs. 8.11 and 8.12, as a function of ω and for several values of L^{λ} and m_{λ}.

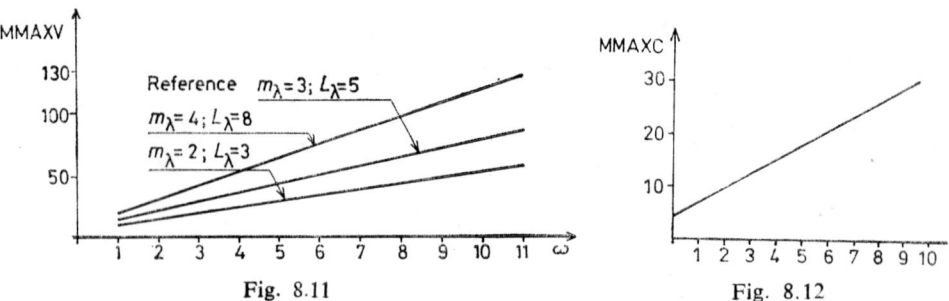

Fig. 8.11   Fig. 8.12

Further comments. If a complex technical system can be decomposed into a finite number of functions, then an SMDP may be used for each function λ in order to model optimal and suboptimal maintenance strategies. A $0-1$ program model was given for the case when the maintenance engineer must take management decisions concerning the whole system. The decomposition scheme of the system at the functional level could be extended to the module or item level. The model efficiency *versus* an appropriate decomposition scheme is given in fig. 8.13. However, for hypothetical cases, the model cost should increase exponentially when it is possible to build models (MDP) with 10^6 states. Alternative system models for maintenance strategies in complex technical systems are given in fig. 8.14. From these figures, it is obvious that the problem of optimal maintenance strategies may be classified as an intermediate scale programming.

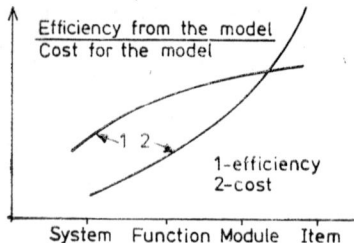

Fig. 8.13

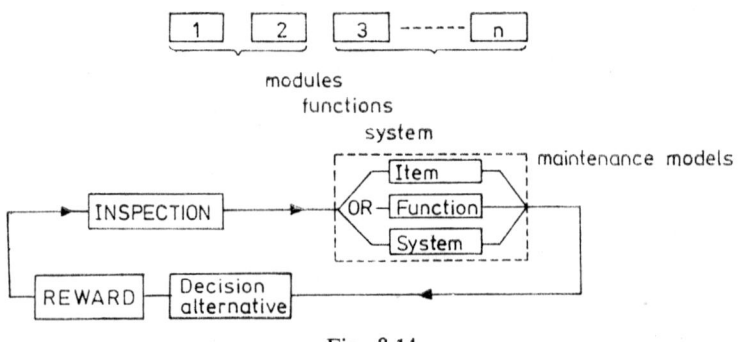

Fig. 8.14

Part III

Applied studies

Chapter 9

Applications of systems engineering in maintenance-safety models and medical diagnosis

In this chapter, we shall apply the theoretical results obtained so far to large scale engineering systems. The examples presented cover energy systems engineering, referring in particular to nuclear reactors (BWR) and power stations, as well as medical diagnosis. Following the severe social and economic side-effects of the improper functioning of a nuclear reactor, much attention has been given of late to safety in system operation. For analysis, we shall consider the class of light water-cooled reactors and the Boiling Water Reactor (BWR) in particular. The following general description of a BWR system will stress maintenance strategies for the pressure vessel and safety analysis of a nuclear power station equipped with a BWR.

9.1. Description of a BWR

The nuclear BWR acts as a recirculation boiler with steam separators and dryers situated in the top section of the reactor pressure vessel (fig. 9.1). Jet pumps situated around the reactor core and driven by small external pumps ensure the recirculation flow of water from the steam separators and feedwater returning from the turbine condenser. The BWR uses enriched uranium fuel and zirconium alloy cladding with square lattice fuel element. Rods inserted hydraulically between the fuel assemblies achieve control of the reactor. A special type of pressure suppression containment is adapted to deal with the consequences of the primary circuit rupture. For safety reasons, the BWR is equipped with a safeguard system, Emergency Core Cooling System (ECCS), consisting of four independent sub-systems, High Pressure Core Injection (HPCI), High Pressure Circulation System (HPCS), Low Pressure Core Injection (LPCI), Automatic Depressurization System (ADS). An integrated design concept toward nuclear reactor safety equips the BWR with redundant components.

Much of the current controversy is concerned with probabilistic *vs.* deterministic safety analysis for nuclear reactors (see Palmer [103]), but a lot of interest is taken in the probabilistic approach which could be finally correlated with the attitude of the population towards risk in a nuclear reactor operation (see Farmer [106]).

A very controversial piece of equipment within a BWR system is the pressure vessel. Before performing a complex safety analysis on a nuclear power station with a BWR system, we shall first investigate optimal repair and inspection policies for the pressure vessel of a BWR system under the assumption that the maintenance engineer is a risk-sensitive person; a Markov decision model as described in Chapter 6 will be used in the sequel.

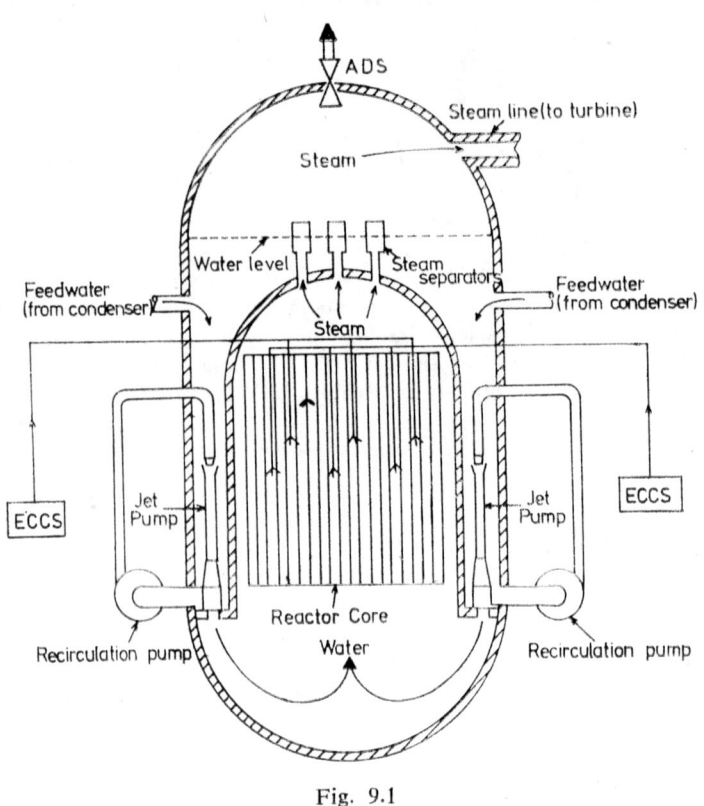

Fig. 9.1

9.2. Risk-sensitive Markovian policies for detection and maintenance of fatigue cracks

As is well known (see Bonhomme [30], Farmer [106], Nichols [110]), the fatigue failure mechanism for cylindrical vessels is an inherently stochastic process. Much attention has been given to the study of pressure vessels of the nuclear power units. The pressure vessel of a LWR (Light Water Reactor) has received special interest of late. For further modelling, we shall make the following assumptions [30]:
- the fatigue crack is assumed to develop in the central part of a wide plate from an alternating stress in the axial direction,

— the crack initiates at the surface and then grows in depth and length until the plate is penetrated,
— plate penetration is considered a failure of the plate.

For nuclear reactor vessels, Bonhomme [30] defined the states in a Markov chain as operating and failure states. The major index description for the state of the vessel is given by the depth of the crack:

State 1 — no detectable crack,
State 2 — crack size 0.0093 — 0.014 inch,
State 3 — crack size 0.014 — 0.022 inch,
State 4 — crack size 0.022 — 0.041 inch,
State 5 — crack size 0.041 — 0.10 inch,
State 6 — crack size 0.10 — 0.50 inch,
State 7 — crack size > 0.50 inch, failure.

The deterioration process ($k = 0$) is given by the rate transition matrix

$$P^0 = \begin{bmatrix} q & p & & & 0 \\ & q & p & \ldots & \\ 0 & & & q & p_1 \end{bmatrix}, \qquad (9.1)$$

where $q = \exp(-\lambda \Delta t)$ and $p = 1 - q$. In the above equation, λ is the failure rate constant associated with the deterioration process of the pressure vessel. The system states are arranged in the order of their increasing deterioration. After an inspection has been performed and a failed state has been observed, a repair action must be taken so that the system may become "as-good-as-new", with the following additional assumptions:

— the time for repair is considered negligible,
— the effect of a repair action is to reset the unit immediately into state 1.

The probability transition matrix following a repair action ($k = 1$) is given by

$$P^1 = \begin{bmatrix} q & p & & \\ \vdots & \vdots & & 0 \\ q & p & & \end{bmatrix}. \qquad (9.2)$$

The probability of crack initiation is estimated to follow a Weibull distribution [30]. Three different values for crack initiation probabilities have been considered: 2.2475×10^{-5}, 0.49×10^{-4} and 0.41×10^{-3}, respectively. The influence of the cost parameters were assessed by exploring a range of values for the ratio maintenance and inspection cost *versus* failure cost. The following additional assumptions were made:

— in state 7 an inspection procedure is always available and no cost is incurred for this inspection;

- for the detection of the crack sizes corresponding to states 3, 4, 5 and 6, a second inspection procedure by a non-zero cost is available;
- a crack with a depth less than 0.014 inch cannot be detected; a large-value real number was assessed as the inspection cost corresponding to this situation;
- the repair cost was assumed to be independent of crack size and considered to be of the same order of magnitude as the inspection cost;
- the relation between the repair and inspection cost was assumed to be constant;
- the decision-maker is a risk-sensitive person characterized by an exponential utility function.

In the above model we have considered a perfect inspection (the underlying Markov process is completely observable). After an inspection has been performed, in each state of the process, we can leave the system to a further deterioration ($k = 0$) or perform a repair activity ($k = 1$). The value for the inspection-maintenance lottery is given by the relative value $Y = (\mathrm{IN} + \mathrm{RP})/H$, where IN is the inspection cost, RP is the repair cost and H is the failure cost (penalty). Under the assumptions given above, the value of Y_i for each state ($i = 1, 2, \ldots, 7$) can be written as $Y^1 = 2.5\, K_i/H$, where K_i, H_i, corresponds to a repair cost activity ($k = 1$) for the pressure vessel in state i.

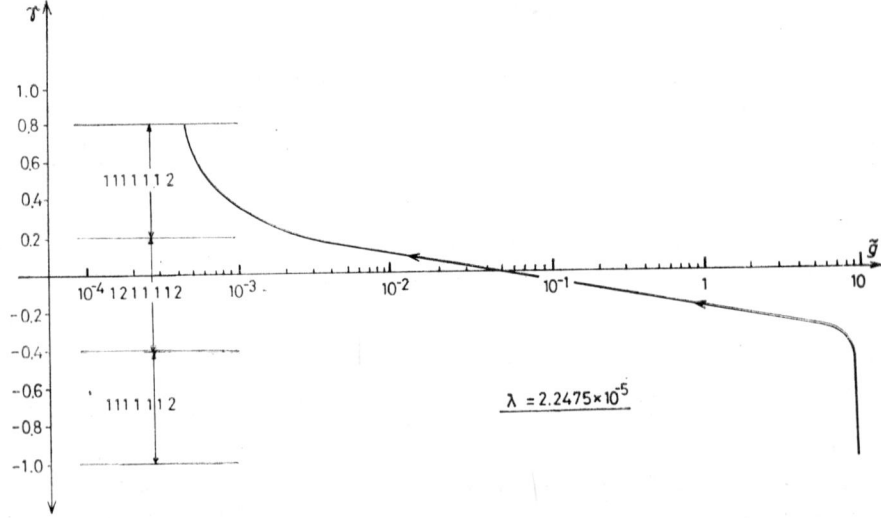

Fig. 9.2

Computer simulation results for risk-sensitive maintenance/inspection strategies, using a model similar to that given in Chapter 6 are presented in fig. 9.2 for $\lambda = 2.2475 \times 10^3$. From an examination of this figure, it is clear that the certain equivalent gain for the Markov process, which finally corresponds to some appropriate availability figures for the pressure vessel system in a repair environment, decreases dramatically from the risk-preference to the risk-averse decision maker. The biggest change in the value of $\tilde{g}$ occurs when the risk-aversion coefficient γ changes in sign, which implies in fact a change in the risk-aversion attitude of the decision-maker. For $\gamma \in [1, -0.4]$, no repair action is taken until an ultimate deterioration state has been achieved. For $\gamma \in [-0.4, 0.2]$, a limiting repair state is state 2 and for $\gamma > 0.2$ the decision-maker (risk-averse) becomes "myopic" in his choice to repair detectable states until the system reaches a final deterioration level; some other interesting results have been also obtained.

— For increased values of λ, corresponding to larger crack initiation probabilities (transition from state 1 to state 2), maintenance policies for a risk-averse decision-maker ($\gamma > 0$) do not seem to reach an optimal solution (this is explained later in this Chapter).

— The variation of the certain equivalent gain is from $\tilde{g} = 10$ for $\gamma = -0.2$ and reaches $\tilde{g} = 1$ for $\gamma = -0.2$ with slight variations in between when the value of λ changes (see fig. 9.3).

— The switching point for the value of $\tilde{g}$ was generally observed to be $\gamma = -0.6$, when the certain equivalent gain decreases by a factor of two to ten.

A second simulation procedure has been carried out, this time for optimal inspection scheduling over a long-time horizon. Three distinct inspection policies have been taken into account for a fixed value of the repair/failure cost ratio. If K_2 represents the repair cost for the system corresponding to degrading states 3, 4, 5, 6, H is the failure cost (penalty) for the system and a is the inspection cost for the k-th type of inspection procedure, then we can write the relation $a^k = b^k K_2$, where b^k is a cost coefficient for the k-th ($k = 1, 2, 3$) type of inspection. Three values

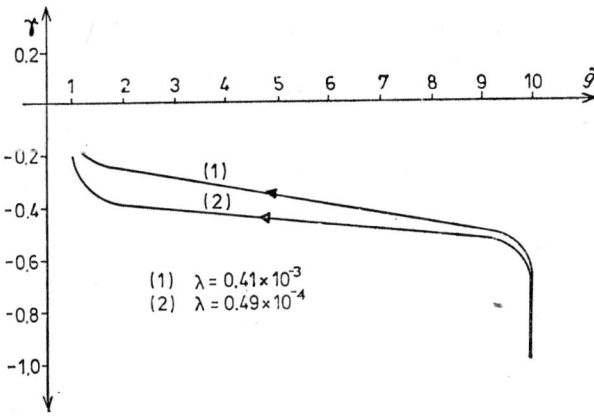

Fig. 9.3

for b^k have been chosen and $a = K_2/H$ (the ratio of maintenance and failure cost) took four different values (table 9.1). The decision coefficient corresponding to states 3, 4, 5, 6 was

$$a_{\text{total}}^{3,4,5,6} = a(1 + b^k), \quad (k = 1, 2, 3).$$

For states 1, 2 and 7, for reasons derived from the above assumptions, we can write $a_{total}^{1,2,7} = 1$; the inspection decision vector has the form (11 3333 1). Computer simulation results showed that the certain equivalent gain for the above problem remains almost constant for $\gamma > 0$, $\gamma < 0$, but a dramatic decrease in its value occurred when the decision-maker was characterized as a risk-averse person (fig. 9.4).

Table 9.1

a	k	b_k	$a_{total}^{3,4,5,6}$
10^{-4}	1	1.5	1.5×10^{-7}
	2	4.3	5.3×10^{-7}
	3	7.5	8.5×10^{-7}
4×10^{-4}	1	1.5	6×10^{-3}
	2	4.3	2.12×10^{-3}
	3	7.5	3.7×10^{-3}
10^{-3}	1	1.5	1.5×10^{-3}
	2	4.3	5.3×10^{-3}
	3	7.5	8.5×10^{-3}
2×10^{-3}	1	1.5	3×10^{-3}
	2	4.3	0.0106
	3	7.5	0.017

A risk-preference decision-maker has been shown to prefer cheap inspection procedures for a fixed maintenance cost. As soon as $\gamma > 0$ (risk-aversion), state 5 became a limiting state for more expensive inspection procedures. Other computer results showed that the inspection strategies do not converge to an optimal solution when the decision-maker is risk-averse.

The Markov model used to analyse maintenance policies for the nuclear reactor pressure vessel indicates that if the decision-maker is not prepared to take risks then he cannot formulate a reactor maintenance policy, nor can he operate the reactor at all. The fact that the algorithm does not converge to an optimal maintenance management policy, but oscillates between a set of alternative policies, reflects a state of decision in the mind of the decision-maker, due to the intimate nature of the risk-aversion phenomenon. The implication is that a totally risk-averse person will not operate a nuclear reactor (or any other system involving risk). Maintenance

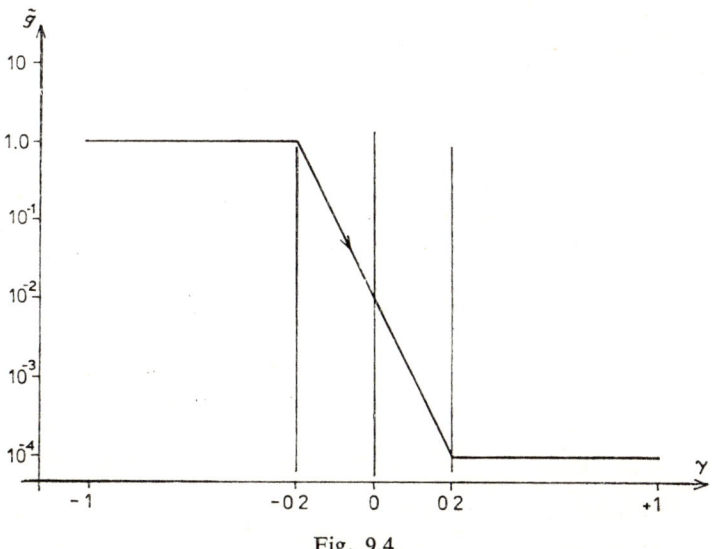

Fig. 9.4

actions of systems which by definition imply risk (i.e. a nuclear reactor) have not to be investigated by classical operational methods. However, realistic models, which take into consideration the "man-machine" interface by incorporating risk-sensitivity decision actions and a probabilistic system analysis, give more appropriate results which can be implemented in the "real-world" operation. The computational results of the present investigation show that the RSMDP [35] does not necessarily reach an optimal solution for the class of decision processes which by definition are risk processes.

9.3. A probabilistic safety analysis for a power station equipped with a BWR

A short description of a BWR system was given at the beginning of this chapter. In this section we shall report the computational results of the safety analysis for a BWR system, using SPTM as described in Chapter 4. We have first to identify the initiating events which eventually could lead the system to unsafe operational states. A complete list of the initiating events $E(i)$ considered for a BWR is given in table 9.2. A pictorial representation for the safety transfer func-

Table 9.2. BWR initiating events

$E(i)$	Initiating events	Class of initiating events
1	Pressure vessel rupture	Internal mechanical initiating events
2	Recirc. line break	
3	Steam line break (inside)	
4	Steam line break (outside)	
5	Feedwater line break (inside)	
6	Feedwater line break (outside)	
7	Small pipe break	
8	Recirc. pump seizure	
9	Damage to fuel and core structure	
10	Rod ejection	
11	Spontan. struct. failure	
12	Loss of heat sink	
13	Power distrib. failure	
14	Refuelling accid. (fuel drop)	
15	Turbine trip	
16	Rod withdrawal	
17	Coolant flow decrease	
18	Coolant flow increase	
19	React. press. valve failure	
20	All other mech. fails.	
21	Internal fire	Internal fire
22	Contr. room oper. error	Internal human error
23	Other plant pers. error	
24	Earthquake	External natural disaster
25	Storm (causing pow. loss)	
26	Flood	
27	Power loss (no storm)	Ext. connect. with syst.
28	Plane crash	External human accidents
29	Area-wide fire or expls.	
30	Intentional damage by humans	External intentional damage

tions (path description) is given in figs. 9.5—9.10. Three event trees were drawn after a very careful and detailed analysis of the reactor operational status and subsystems interaction — all these toward the safety goals.

The reactor is assumed to have been in steady state, full capacity operation for one year (including normal maintenance).

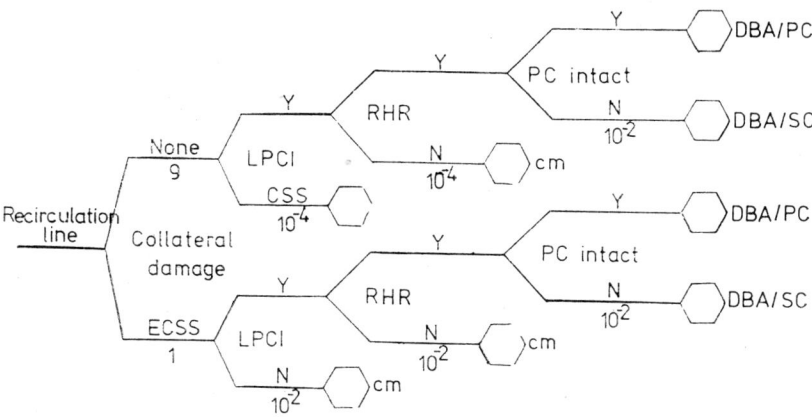

Fig. 9.5

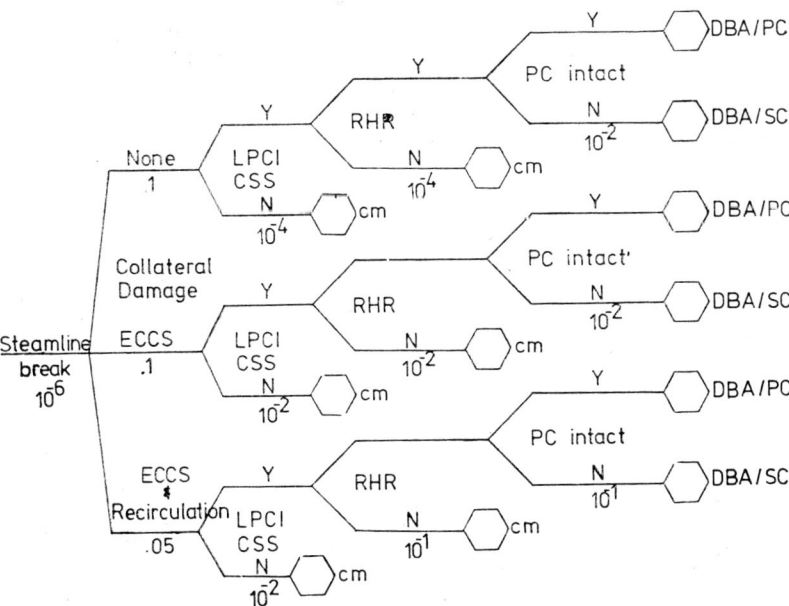

Fig. 9.6

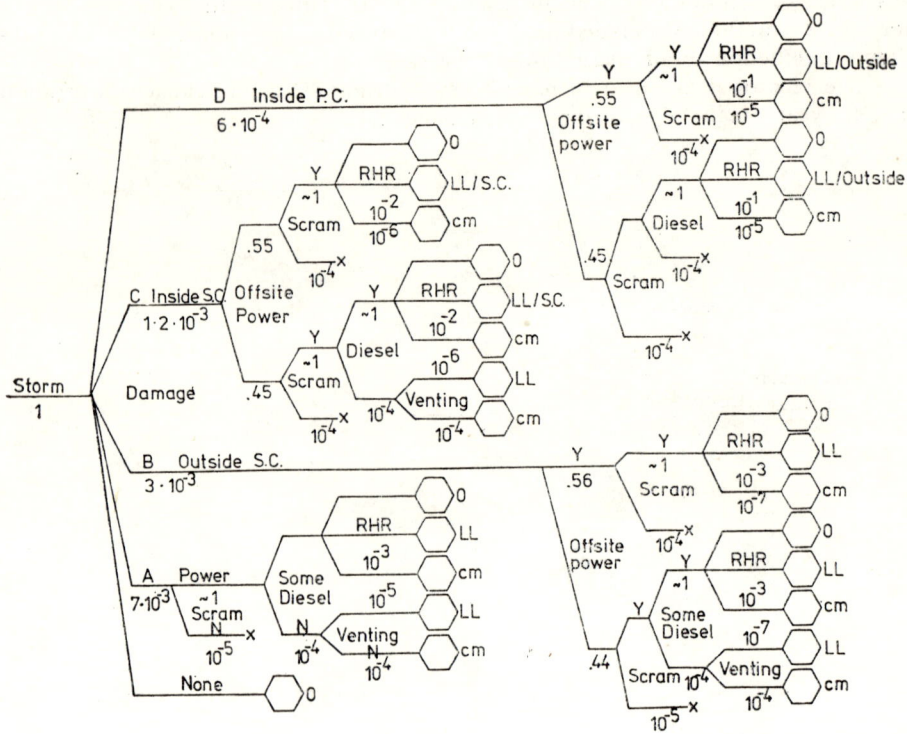

Fig. 9.7

Fig. 9.8

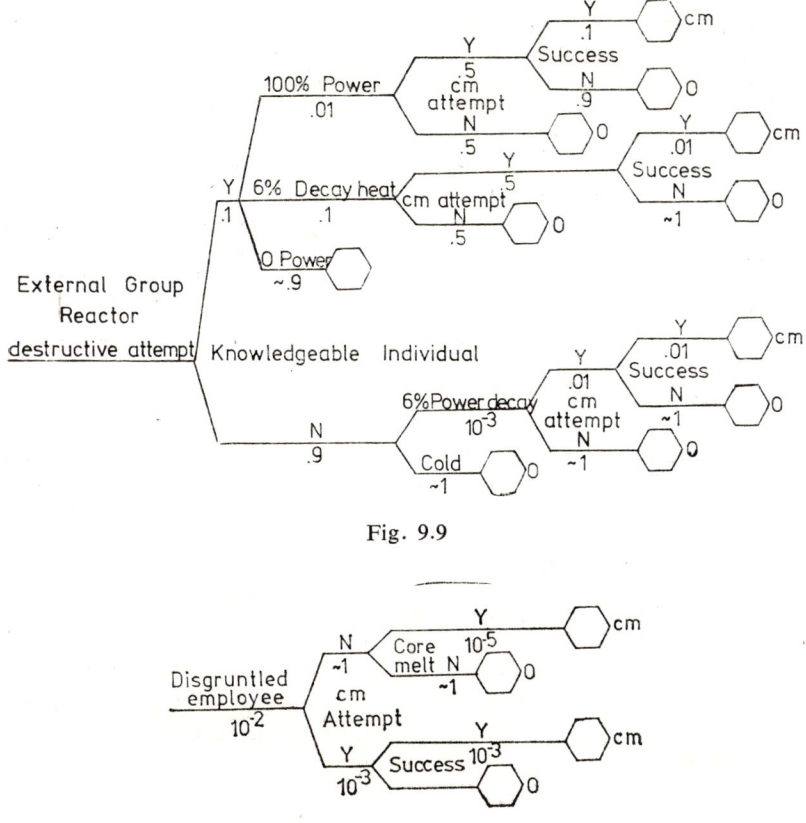

Fig. 9.9

Fig. 9.10

Description of initiating events and their causes. A brief description of initiating events and their causes will be given in the sequel.

Pressure vessel rupture, as a possible initiating event, could be caused by a non-ductile rupture or by a human-active-induced mechanical failure. The probability of such an event, after experts consultation and assessment, was 10^{-6} per reactor year, and the discriminator description is the core melting for the nuclear reactor.

Break of the recirculation line could be caused by collateral damage (secondary damage resulting from faults elsewhere in the system) elsewhere, earthquake, reactor overpressure, intentional damage with missiles and "spontaneous" causes.

Break of the steam line inside the primary containment could be caused by collateral damage.

Feedwater line is an important subsystem for nuclear reactor operation. It can break due to earthquake or structural failure, fire, line overpressure, collateral damage elsewhere.

Recirculation pumps failure could be caused by loss of power either because of the equipment failure or an operator error, mechanical failure, etc.

Turbine trip: the turbine-generator block is tripped off from the power network either manually or automatically. The trip considered in the present analysis is not viewed as the result of any other internal mechanical accident. The causes for the turbine trip without reactor scram could be: overspread due to electrical load loss, overpressure in the reactor or control error, steam line pressure rise due to reactor activity or a valve malfunction (e.g. stop valve, bypass valve), turbine operator error, mechanical (turbine and generator) failure, earthquake, vibrations, autotrip due to neutron flux, electrical load loss (network pulled down or other internal reasons), oversensitive instrumentation which can lead to autotrip or operator trip.

Internal fire is another initiating event towards an unsafe nuclear reactor in operation. The sources of these events may be power distribution system (control room, transformers, human errors, short or overheats of an electrical motor), hydrogen due to storage generators cooling systems, human error (e.g. smoking), diesel fuel (generator building, storage tanks).

Loss of outside power could be caused by network pulled down by another plant or power distribution equipment failure, localized loss of transmission line or tie due to plane crash, storm, intentional damage, controller error or line crew error.

Probabilistic event trees. The purpose of a probabilistic event tree is to enumerate and ssociate probabilities with the various accidents equences that can result from a given initiating event.

Table 9.3. Accident sequence events (BWR)

$f_{(j)}$	Primary failure events	$f_{(j)}$	Primary failure events
1	Correct functioning of the safety system	16	Power distrib. 13 kV failure
		17	Refuelling accident (critical mass)
2	Reactor scram	18	Diesel failure
3	Rod run in — rods inserted	19	Off-site power loss
4	Shut down mode	20	Fire damage
5	Relieve value failed operated	21	RPS (Reactor protection syst.) alarm failure
6	Heat sink		
7	SLC (Standby liquid control system)	22	PRS signal failure
8	FWIS (Feedwater injection system)	23	Operator action failure
9	HPCI	24	Intentional damage
10	LPCI	25	Damage A
11	RHR (Reactor heat removal)	26	Damage B
12	RCIC	27	Damage C
13	CSS (Core spray system)	28	Damage D
14	Coolant damage	29	Flood damage
15	Power distrib. 4 kV failure	30	Control system failure

Table 9.4. Initiating events and their causes

No.	Causal events	Initiating events *
1	Earthquake	all but 28, 30
2	Reactor overpressure	1−7, 9, 10, 12, 15, 18−20
3	Intentional damage	all but 24−26
4	Structural failure	1−20
5	Operator error	all but 24−26, 28, 29
6	Stop/By pass valve closure	2, 4, 15, 20
7	Internal fire	1−23
8	Pipe overpressure	3−7, 12, 15, 17−19
9	Power loss distrib. (internal)	all but 24−26, 28−30
10	Mechanical failure	1−20, 21−23
11	Vibration	1−13, 20, 21, 15
12	Spontaneous failure	1−23
13	Valve closure	−
14	Rupture	1−3, 5, 7−10, 12−23
15	Pump failure	1−21
16	Off site power loss	1−23, 27
17	Instrument failure	10, 12, 13, 15−18, 20, 22, 23
18	Vehicular	4, 6, 12, 13, 17, 18, 20, 21, 23, 27
19	Natural events (flood and storm)	1−29
20	Plane crash	1−23, 28
21	Mischief	1−23, 27−30
22	External fire	all but 24−26, 28, 30

* see Table 9.2.

Table 9.5

Accident source description		Causal events (See Table 8.4)
(A)	(B)	
Internal	Mechanical failure	2, 4, 6, 8, 10, 11, 12, 13, 14, 15
	Human errors	3, 5
	Aggregation (regarding the system)	9, 17, 21
External	Nature effect	1, 19
	Human errors	18, 20, 21, 22
	Aggregation	16

At each stage of analysis, efforts have been directed toward those portions of the master tree that appeared most important under current probability assessments. In places where current assessments indicated stochastic dominance, the appropriate portions of the tree have been trimmed. If some assessments are later changed in the light of new information or supportive analysis, it may be revealed that the trimming was unjustified. In this case, however, we can resume the development of the branch or branches in question with no effort wasted. A number of benefits result from this process of model refinement in response to sensitivity analysis; many questions regarding the level of model detail were found to be relatively unimportant.

The accident sequence events for BWR are given in table 9.3. The correlation between the causal and initiating events is presented in table 9.4. Table 9.5 lists the relations between causal events given in table 9.4 and accident source description, which results from rolling back the master event tree.

Results and conclusions for the nuclear reactor safety. The analysis has been carried out for a relative large master tree in the SPTM. A substantial number of combinations between initiating events and final consequences (curies released to environment) have been analysed. The results are presented in tables 9.6 and 9.7. From the above tables, we can see the probability distribution function of radioactive releases due to different classes of initiating events. Owing to an internal mechanical class of failure (see table 9.2 for initiating events belonging to this class), the safety discriminator with the biggest probability is given by low level releases outside the primary containment. Fuel failure releases have the probability of occurrence 0.5. Design basis accidents outside primary and secondary containment, which could produce radioactive releases, have the probability 10^{-4} and $3 \cdot 10^{-8}$ per reactor year, respectively. The most undesirable event (the core melt) has the probability of occurrence $1.6 \cdot 10^{-6}$, according to the model accepted here.

A close analysis of table 9.6 indicates that according to the data bank used, the most probable initiating event which would lead to the core melt event is intentional damage, with a probability of 10^{-6} per reactor year. External accidents and external natural initiating events (e.g. floods, storms) induce the lowest probability

Table 9.6. Release lottery: Pr {Release/$\mathscr{E}$}

Category	Safety discriminator					
	FF	LL(PC)	LL(SC)	DBA (PC)	DBA (SC)	CM
I. Internal mechanical	.5	1		10^{-4}	$3 \cdot 10^{-8}$	$1.6 \cdot 10^{-6}$
II. Internal fire				$3 \cdot 10^{-8}$		$3 \cdot 10^{-8}$
III. Internal human error		$2 \cdot 10^{-4}$				$2 \cdot 10^{-7}$
IV. External natural		10^{-5}	10^{-5}			10^{-9}
V. External systems interaction		10^{-6}		10^{-10}		10^{-7}
VI. External accidents		10^{-6}	10^{-7}			$5 \cdot 10^{-9}$
VII. Intentional damage						10^{-6}

Table 9.7. Curies released to environment (Ci)

Releases	Primary containment	Secondary containment	Outside
FF (.5)			1 Ci
LL (PC) (.1)	10 Ci	$<.1> \Big\langle \begin{array}{c} \text{—0} \\ 10^{-2} \\ \text{—10} \end{array}$	$<10^{-3}> \Big\langle \begin{array}{c} \text{—0} \\ 10^{-2} \\ \text{—}10^{-1} \end{array}$
LL (Sc) $(3,5 \cdot 10^{-5})$		10	.1
DBA (P.C) $(5 \cdot 10^{-6})$	10^6	$<10^4> \Big\langle \begin{array}{c} 10^{-1} \text{—0} \\ 10^{-2} \text{—}10^3 \\ \text{—}10^6 \end{array}$	$<101> \Big\langle \begin{array}{c} 10^{-1} \text{—0} \\ 10^{-2} \text{—}10 \\ \text{—}10^4 \end{array}$
DBA (S.C) $(8 \cdot 10^{-8})$		10^6 $\quad$ 10^5	10^3
C.M. $(4.4 \cdot 10^{-6})$	10^8	10^6	10^5
Releases	446	4.46	.94

values for the core melt discriminator, $5 \cdot 10^{-9}$ and 10^{-9}, respectively. Internal human operational errors and the interaction of the nuclear power station with external systems (e.g. the power network or other power stations in the system, either nuclear, thermal or hydropower stations) could induce the probability of the magnitude 10^{-7} to the core melt event for an operational reactor, as was mentioned at the beginning of this chapter. Internal human errors could lead the system to low level releases, outside the primary containment with a sensitive probability value. These facts proved that under real world nuclear operation, human errors may change the safe operational status of the reactor.

On a careful analysis of the results given in the above tables, the systems analyst and the design decision-maker can find the class of initiating events which is most likely to reach an unsafe nuclear reactor status, and therefore, by performing appropriate design changes and implementing protective systems, he may decrease the probability of the different events which characterize safety discriminators.

The results given in table 9.7 only reassess the real danger of nuclear reactors (BWR) to the population safety, by producing figures for released curies (Ci) to the environment. From an overall interpretation of the above figures, it is clear that the value of Ci released decreases almost exponentially by passing primary and secondary containment of the BWR system. The mean values of Ci released per reactor year in the "worst" scenario are 446 Ci outside the primary containment, 6.46 Ci outside the secondary containment and 0.94 Ci outside the reactor building.

We do not have to take these figures as fixed values. A sensitivity analysis to the input probability and other model data can be performed.

More comments will be given now about the reference lottery.

Let us assume that the probability of death per reactor year is 10^{-6} for every person living close to the reactor. Suppose that our comparison is now with the accident induced by driving a car. From statistics, the auto accidental death rate is 1/4000 (data refers to the U.S.A.). As an average, people are driving 10^4 miles/year then the probability of death per mile/year/person is 2.5×10^{-8}. If $x =$ mileage given 10^{-6} death/year from the nuclear reactor accident, then we can write (2.5×10^{-8}), $x = 10^{-6}$, or $x = 40$ miles/year. A reference event having 10^{-6} probability of death per year is 40 miles of extra driving. In other words, we can say that living next to a nuclear reactor increases the probability of death per year by the same amount as an extra 40 mile trip per year.

Some other results of this investigation are given in figs. 9.11 and 9.12. The probability figures (per reactor year) in fig. 9.11 indicate different levels of radioactive release, such as fuel failure (FF), low level (LL) of primary and secondary containment (PC, SC), design basis accidents (DBA), core melt (CM). Figure 9.12 shows the accident source events contribution (%) in the "worst" scenario and as a discriminator the core melt is given. The internal mechanical class of events has the largest percent in a core melting event (54.5%), followed by intentional damage event (34%), human errors (7%), external systems interaction (3.5%), internal fire (1%).

From the above results we can deduce the "weak" points in design, operation and maintenance of a nuclear reactor system and that the design and maintenance of the plant affect the accident state probability distribution function.

The following areas may prove fruitful for further research concerning the applied models in the field of systems engineering. At least theoretically, new models — such as partially observable semi-Markov decision processes — have a good potential for a broad set of applications in systems engineering.

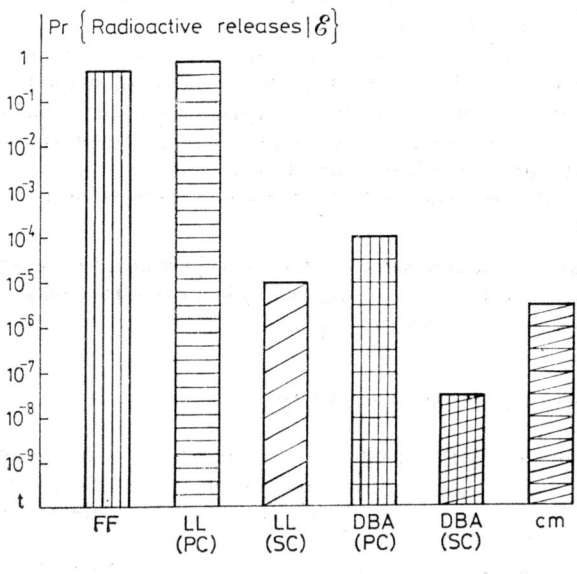

Fig. 9.11

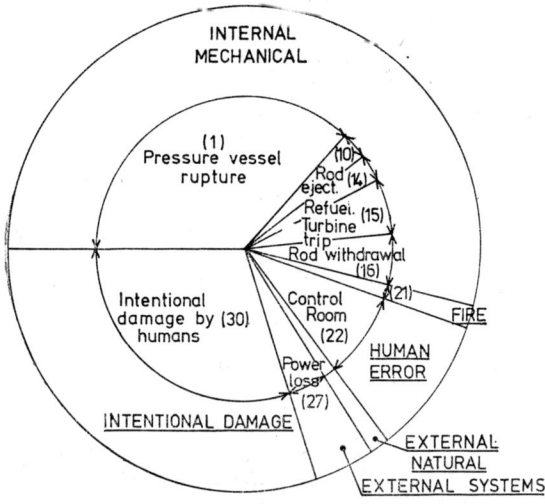

Fig. 9.12

In the present approach, we considered the inspection time equal to unity. More realistic assumptions for finite horizon models are required when the inspection time has some finite value, governed by a probability law distribution function and the observation process is probabilistically coupled with the Markovian deterioration of the "core process". Markovian-models with fuzzy goals could be developed for maintenance strategies in multilevel performance coherent systems.

The risk-sensitive MDP are far from being satisfactorily investigated. However, much was done for problem-solving and optimization techniques for partially or generally observable MDP. Practical applications of such models to real systems maintenance strategies could provide more insight for the "real world" problems. However, the most difficult problem of probability assessment has to be viewed before any computational effort is developed.

Finally, let us mention, as further research aspects, several comments given by Turban [17]:

"The problem of implementation of this 'most difficult subject matter' is intricate and complex in the area of maintenance. Hence, inquiry into this subject holds both challenge and promise for the researcher and for industry".

9.4. Computer systems safety

The present application evaluates the operational performances for an informational system using electronic computers by means of fault tree analysis. The system is made up of an informational subsystem for data collection and display and a computing subsystem which performs the computational functions (see fig. 9.13). In the proposed configuration process, microcomputers of the FELIX-MINI generation are used.

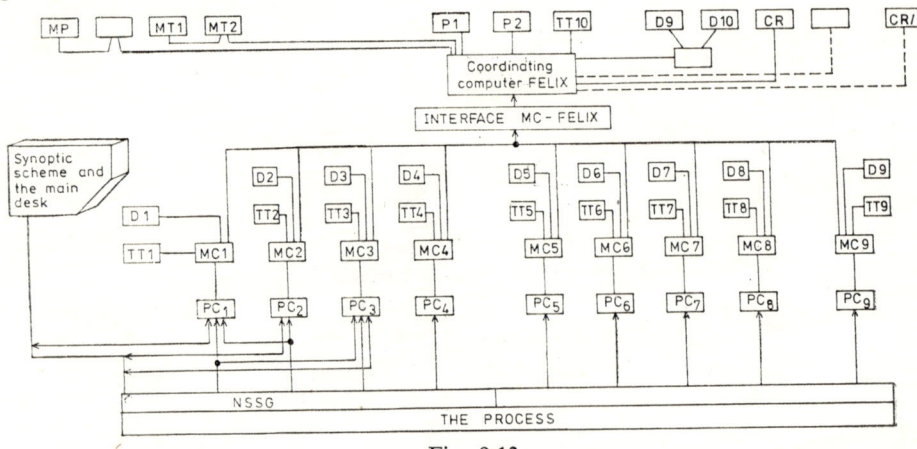

Fig. 9.13

The informational subsystem consists of eight process minicomputers. Every minicomputer explores directly from the process a number of maximum 256 analogous data and 256 discrete data. By the analogous data explored through the computer software, there are realized the validity operation, numerical filtering, linearization, compensation, conversion and a comparison with technological limits.

On the teletype associated with each minicomputer, the fixed values of technological limits are indicated in accordance with the analogous explored measures. These messages indicate the fault events within the measurement chain and messages concerning the commuting of the state of any explored binary measure (the events protocol).

On the display associated with every minicomputer the corresponding values of 16 analogous or binary parameters are visualized. This protocol is given in a predetermined protocol.

Each minicomputer is associated to a zone on the synoptic scheme from the control room. Optical indicators are powered by the binary outputs (maximum 256) of the process multiplexor.

On the request of the top coordinating computer, which is formulated *via* a special hardware interface, the minicomputer sends individual values for the parameters or the entire vector of the existing values to the data base at the moment of request.

The computing subsystem is made up of a top high level computer and an autonomous process minicomputer. The top high level computer is of a FELIX-MINI type, 64 K memory. One of the line printers is used for periodic protocol registration, and the second line printer is for post-failure and other protocols, on request. On the D9 and D10 displays, we can visualize the computed parameters, which must be known in a normal activity. Teletype TT 10 is used as a main workstation to the system, for the operating system as well as a conversational peripheral equipment for the off-line programmes. The magnetic disks DM 1 and DM 2 are used within the operating system as well as by the applicative programs within the system.

Magnetic tapes BM 1 and BM 2 are used for data storage on a long-time basis and for other computing facilities. The minicomputer MC 9 is used as an auxiliary computing system. The main and the auxiliary computing systems work at the same time. The results of both systems must coincide.

Definition of the top event. Vital input parameters for the nuclear power station operation were divided into groups of 40 elements. Each group of parameters is simultaneously processed by the two above mentioned computers.

The fault state of the system is defined by the simultaneous failure of two minicomputers (two out of eight) which affects the same group of final input parameters. Under this state, the system becomes incapable of performing its designed functions.

Failure mode analysis and the side-effects of the failure event. The lowest reliability performance for this on-line computer system is associated with the process hardware interface which can end it in a faulty state due to the deterioration of transistors or other reasons.

A minicomputer system can also malfunction due to the electronic or mechanical components of the peripheral devices (e.g. line printers).

The displays and teletype failure directly affect the system availability. Direct replacement of failed components is the best maintenance policy in this case.

The other periperal equipments of the computer configuration do not directly affect the availability of the vital parts of the analysed informational system.

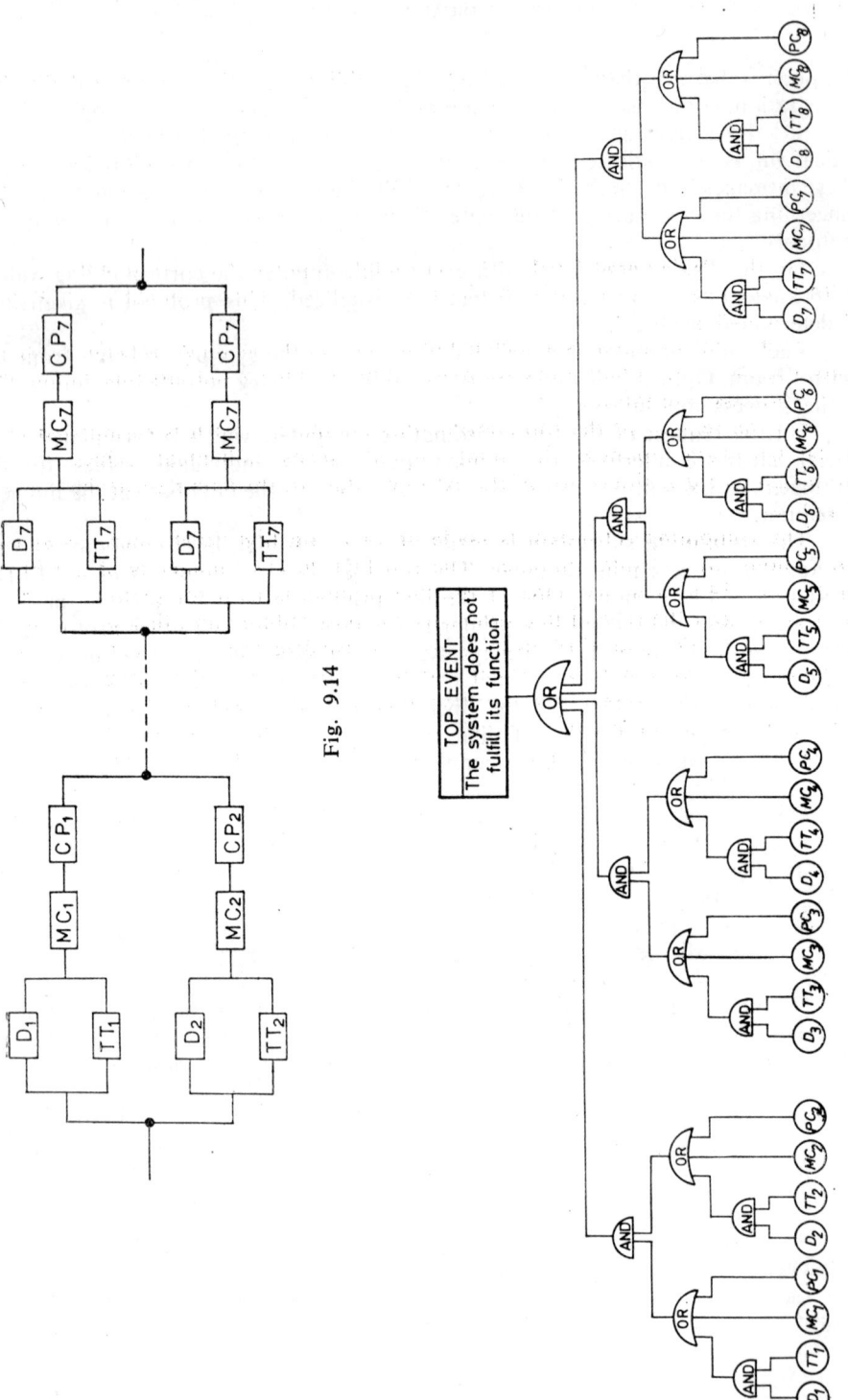

Fig. 9.14

Fig. 9.15

MAINTENANCE SAFETY MODELS AND MEDICAL DIAGNOSIS 313

Fault tree drawing. The fault tree is drawn with respect to the previously defined top event. In the further analysis have been included only those components which have a significant contribution to the undesired top event occurrence (see figs. 9.14 and 9.15).

Primary events quantification. Failure and repair notes for the system components are given in table 9.8. The statistical data were obtained either by consulting experts in computer systems or by using experimental data.

Table 9.8

Number	Component	Failure rate	Repair time
1−8	D1 −D8	30×10^{-6}	2
9−16	TT1−TT8	50×10^{-6}	3
17−24	MC1−MC8	7×10^{-6}	0.5
25−32	CP1−CP8	15×10^{-6}	6

Table 9.9

Failure mode number	Failure mode matrix Identifiers for primary events		
1	7	3	0
2	7	4	0
3	8	3	0
4	8	4	0
5	15	11	0
6	15	12	0
7	16	11	0
8	16	12	0
9	23	19	0
10	23	20	0
11	24	19	0
12	24	20	0
13	31	27	0
14	31	28	0
15	32	27	0
16	32	28	0
17	6	5	3
18	6	5	4
19	7	2	1
20	8	2	1
21	14	13	11
22	14	13	12
23	15	10	9
24	16	10	9
25	22	21	19
26	22	21	20
27	23	18	17
28	24	18	17
29	30	29	27
30	30	29	28
31	31	26	25
32	32	26	25

Table 9.10. Characteristics for components number 1, 5, 9, 13, 17, 21, 25, 29

Time $T(h)$	Failure probability at time t Q	Failure per second w	Failure intensity $\lambda(h^{-1})$	Failure in the interval $0-T$	Failure probability in the interval $0-T$
0.0	0.0	0.300000D−04	0.300000E−04	0.0	0.0
0.200000D−00	0.599998D−05	0.299998D−04	0.300000E−04	0.599998D−05	0.599998D−05
0.400000D−00	0.119999D−04	0.299996D−04	0.300000E−04	0.119999D−04	0.119999D−04
0.600000D−00	0.179998D−04	0.299994D−04	0.300000E−04	0.179998D−04	0.179998D−04
0.800000D−00	0.239997D−04	0.299993D−04	0.300000E−04	0.239997D−04	0.239997D−04
0.100000D−01	0.299995D−04	0.299991D−04	0.300000E−04	0.299995D−04	0.299995D−04
0.200000D−01	0.599982D−04	0.299982D−04	0.300000E−04	0.599982D−04	0.599982D−04
0.400000D−01	0.599964D−04	0.299982D−04	0.300000E−04	0.119995D−03	0.119993D−03
0.600000D−01	0.599964D−04	0.299982D−04	0.300000E−04	0.179991D−03	0.179984D−03
0.100000D−02	0.599964D−04	0.299982D−04	0.300000E−04	0.299984D−03	0.299955D−03
0.100000D−03	0.599964D−04	0.299982D−04	0.300000E−04	0.299982D−02	0.299550D−02
0.600000D−03	0.599964D−04	0.299982D−04	0.300000E−04	0.179989D−01	0.178390D−01
0.120000D−04	0.599964D−04	0.299982D−04	0.300000E−04	0.359978D−01	0.353597D−01
0.180000D−04	0.599964D−04	0.299982D−04	0.300000E−04	0.539967D−01	0.525679D−01
0.270000D−04	0.599964D−04	0.299982D−04	0.300000E−04	0.809951D−01	0.778063D−01
0.360000D−04	0.599964D−04	0.299982D−04	0.300000E−04	0.107993D−00	0.102372D−00
0.450000D−04	0.599964D−04	0.299982D−04	0.300000E−04	0.134992D−00	0.126284D−00
0.540000D−04	0.599964D−04	0.299982D−04	0.300000E−04	0.161990D−00	0.149559D−00
0.620000D−04	0.599964D−04	0.299982D−04	0.300000E−04	0.185989D−00	0.169726D−00
0.700000D−04	0.599964D−04	0.299982D−04	0.300000E−04	0.209987D−00	0.189416D−00

Table 9.11. Characteristics for failure mode no. 1 with 7 and 3 as identifiers for primary events

Time $T(h)$	Failure probability at time t	Failure per second	Failure intensity (h^{-1})	Failure in the interval $0\text{-}T$	Failure probability in the interval $0\text{-}T$
0.0	0.0	0.0	0.0	0.0	0.0
0.200000D−00	0.204263D−11	0.200089D−10	0.200089D−10	0.200089D−11	0.200089D−11
0.400000D−00	0.817053D−11	0.400177D−10	0.400177D−10	0.800354D−11	0.800354D−11
0.600000D−00	0.183837D−10	0.600264D−10	0.600264D−10	0.180079D−10	0.180078D−10
0.800000D−00	0.326821D−10	0.800351D−10	0.800351D−10	0.320141D−10	0.320141D−10
0.100000D−01	0.492257D−10	0.982247D−10	0.982247D−10	0.498401D−10	0.498401D−10
0.200000D−01	0.196903D−09	0.196448D−09	0.196448D−09	0.197176D−09	0.197176D−09
0.400000D−01	0.787612D−09	0.392891D−09	0.392891D−09	0.786515D−09	0.786515D−09
0.600000D−01	0.123064D−08	0.491110D−09	0.491110D−09	0.167052D−08	0.167052D−08
0.100000D−02	0.123064D−08	0.491110D−09	0.491110D−09	0.363496D−08	0.363496D−08
0.100000D−03	0.123064D−08	0.491110D−09	0.491110D−09	0.478348D−07	0.478348D−07
0.600000D−03	0.123064D−08	0.491110D−09	0.491110D−09	0.293390D−06	0.293390D−06
0.120000D−04	0.123064D−08	0.491110D−09	0.491110D−09	0.588056D−06	0.588056D−06
0.180000D−04	0.123064D−08	0.491110D−09	0.491110D−09	0.882722D−06	0.882721D−06
0.270000D−04	0.123064D−08	0.491110D−09	0.491110D−09	0.132472D−05	0.132472D−05
0.360000D−04	0.123064D−08	0.491110D−09	0.491110D−09	0.176672D−05	0.176672D−05
0.450000D−04	0.123064D−08	0.491110D−09	0.491110D−09	0.220872D−05	0.220872D−05
0.540000D−04	0.123064D−08	0.491110D−09	0.491110D−09	0.265072D−05	0.265071D−05
0.620000D−04	0.123064D−08	0.491110D−09	0.491110D−09	0.304361D−05	0.304360D−05
0.700000D−04	0.123064D−08	0.491110D−09	0.491110D−09	0.343649D−05	0.343649D−05

316

Table 9.12. Characteristics for the system — upper limits

Time $T(h)$	Failure probability at time t	Failure per second	Failure intensity (h^{-1})	Failure in the interval $0\text{-}T$	Failure probability in the interval $0\text{-}T$
0.0	0.0	0.0	0.0	0.0	0.0
0.200000D−00	0.774416D−10	0.774428D−09	0.774428D−09	0.774428D−10	0.774428D−10
0.400000D−00	0.309775D−09	0.154891D−08	0.154891D−08	0.309777D−09	0.309777D−09
0.600000D−00	0.697011D−09	0.232346D−08	0.232346D−08	0.697014D−09	0.697014D−09
0.800000D−00	0.123916D−08	0.309806D−08	0.309806D−08	0.123917D−08	0.123917D−08
0.100000D−01	0.193624D−08	0.387271D−08	0.387271D−08	0.193624D−08	0.193624D−08
0.200000D−01	0.774591D−08	0.774687D−08	0.774687D−08	0.774603D−08	0.774603D−08
0.400000D−01	0.309808D−07	0.154937D−07	0.154937D−07	0.309866D−07	0.309866D−07
0.600000D−01	0.625040D−07	0.220066D−07	0.220066D−07	0.684869D−07	0.684869D−07
0.100000D−02	0.625001D−07	0.220059D−07	0.220059D−07	0.156512D−06	0.156512D−06
0.100000D−03	0.624996D−07	0.220058D−07	0.220058D−07	0.213704D−05	0.231704D−05
0.600000D−03	0.624996D−07	0.220058D−07	0.220058D−07	0.131399D−04	0.131399D−04
0.120000D−04	0.624996D−07	0.220058D−07	0.220058D−07	0.263434D−04	0.263431D−04
0.180000D−04	0.624996D−07	0.220058D−07	0.220058D−07	0.395469D−04	0.395461D−04
0.270000D−04	0.624996D−07	0.220058D−07	0.220058D−07	0.593522D−04	0.593504D−04
0.360000D−04	0.624996D−07	0i220058D−07	0.220058D−07	0.791574D−04	0.791543D−04
0.450000D−04	0.624996D−07	0.220058D−07	0.220058D−07	0.989626D−04	0.989577D−04
0.540000D−04	0.624996D−07	0.220058D−07	0.220058D−07	0.118768D−03	0.118761D−03
0.620000D−04	0.624996D−07	0.220058D−07	0.220058D−07	0.136373D−03	0.136363D−03
0.700000D−04	0.624996D−07	0.220058D−07	0.220058D−07	0.153977D−03	0.153965D−03

Fault tree quantification. The fault tree evaluation was completed using the ARBOR-computer program. For a given fault tree topology, the computer program calculates the failure modes of the entire computing system, for a series of time-dependent probabilistics of the primary event. The failure mode matrix corresponding to the above analysed computing system is presented in table 9.9.

Presentation of computer results. The probabilistic characteristics of the system components are presented in tables 9.10—9.12. Tables 9.10 and 9.11 show the probabilistic characteristics of the most representative failure modes of the system. When the system behaviour is time-dependent, the appropriate probabilistic characteristics are those given in table 9.12.

9.5. Dynamic decision models for clinical diagnosis

The complex process of medical diagnosis has traditionally relied on the experience and judgement of the clinician. With the increased application of systemic techniques to medical problems in general, it is timely to consider application to diagnosis and treatment situations as, inherently, clinicians are using some form of pattern recognition techniques, relationships and decision rules in the diagnosis-treatment process.

Decision theory and dynamic probabilistic models are well established and several decision models have been applied to medical problems. Equally, pattern recognition techniques turned out to be useful in medical diagnosis and classification where large quantities of information have to be processed. Unlike the real clinical situation, however, there have been few attempts at combining these techniques in an overall diagnostic decision model.

In an attempt to remedy this deficiency, partially observable Markovian decision processes (Markov or semi-Markov) are here combined with cause-effect models as a probabilistic representation of the diagnostic process. Pattern recognition techniques are used in a first stage of disease classification within which the patient states can be identified. The methodology is given for combining the patient state of health, the clinician's state of knowledge of the cause-effect representation from the observation space (measurements), feature selection using pattern recognition techniques and, finally, the treatment decisions to be used in restoring the patient to a more desirable state of health. A cost functional for the decision process has then to be optimized according to some pre-assigned objective functions (social return from the patient, state of health or treatment cost for the patient), when the process has an infinite-time horizon.

Model formulation. Quantitative approaches to clinical diagnosis and decision-making offer the premise of an improved understanding of the relationships between process variables and an improved quality of patient care. However, successful model formulation must afford parallelism with the "real world" clinical diagnosis-treatment process.

By examining the real decision process of clinical diagnosis and its associated infrastructure, it can be readily seen that the clinician, with his *a priori* medical knowledge, has to make observations on the patient who behaves as a dynamic system under conditions of uncertainty. The clinician has to recall his professional knowledge, identify patterns and classify features as a result of making these observations and measurements.

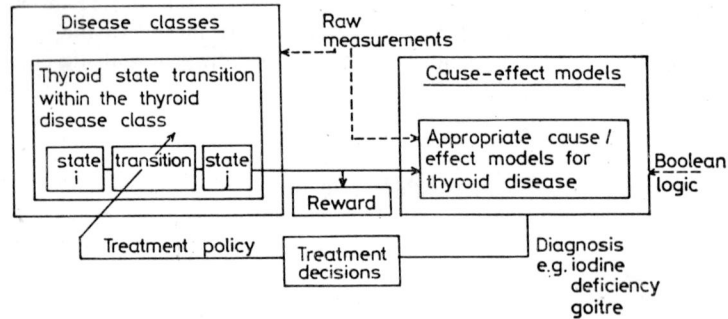

Fig. 9.16

The whole process is not a one-stage operation but rather a hierarchical system of pattern recognition, identification and decision-making. In an advanced conceptual model of computer-aided medical diagnosis (fig. 9.16), we can identify the dynamic state space of the patient (i.e. states of health for a particular class of disease), a cause-effect model and clinical decisions bringing about a change in the patient's state of health in accordance with the new information vector (state of health — measurements — *a priori* knowledge). These decisions are evaluated on the basis of combined OR techniques.

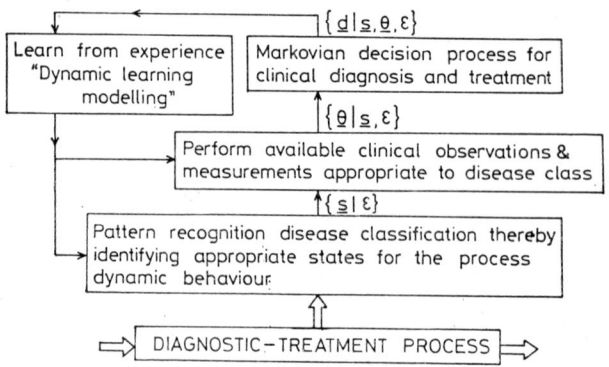

Fig. 9.17

The model presented here uses hierarchical modelling techniques and faces two distinct stages in the process of medical diagnosis — treatment (fig. 9.17). First, pattern recognition techniques can be used to identify the state space of the patient. Then Markovian (Markov or semi-Markov) models can be used when the states of the process are only partially observable and when we dispose of a set of cause-effect models which describe the probabilistic relationship between the patient's symptoms (observable effects of a possible disease) and the cause to which these observations can be ascribed [139].

Pattern recognition. The cause-effect model provides a mapping from a set of input measurements to a diagnosis/primary event. In principle, the cause-effect model alone is sufficient for medical diagnosis. In practice, the number of possible measurements may be of the order of 20,000 and the number of possible diagnoses of the order of 6000. A single cause-effect model to achieve such a mapping would clearly be unwieldy and probably impossible to design.

Similarly, a pattern recognition system provides a mapping from a set of input measurements to the appropriate pattern class (diagnosis/primary event). Again, the complexity of the medical diagnosis problem renders a one-step pattern recognition solution unrealistic. Both cause-effect models and pattern recognition have advantages worth mentioning. With the former, problem knowledge is readily incorporated into the decision structure. How to introduce such problem knowledge into pattern recognition systems is currently not well understood.

Pattern recognition techniques are ideally suited to extracting information from the large clinical data base. Such a system is therefore used to operate on the raw measurements, together with any features derived from those measurements in order to produce a broad classification into disease classes (e.g. respiratory disease, circulatory disease, thyroid disease, etc.). This classification is then extended to the cause-effect model, together with the raw measurements and derived features for a finer classification. This will comprise identification of the primary cause and the associated treatment decision recommended.

In this way, the information to be handled by the pattern recognition system is reduced to a feasible level and the demands made upon the cause-effect model are similarly eased. Effectively, the pattern recognition system guides the cause-effect model to an appropriate area of uncertainty. This could be realised, for instance, by having a multiplicity of cause-effect sub-models, one for each specific disease class assigned by the pattern recognition system.

The problem of the number of possible measurements which may be taken can be eased in several ways. First, the clinician can neglect certain measurements after a preliminary examination of the patient. Second, features and complex features can be defined, using problem knowledge, as functions of raw measurements. Whilst not reducing the number of measurements to be taken, it does reduce the number to be handled by the pattern recognition system and the cause-effect model, since these are needed to examine the features and complex features only. Third, sequential pattern recognition techniques can be employed.

With these techniques it is possible to decide whether the measurements already taken carry sufficient information to allow a classification and, if they do not, which measurement should be taken next optimally to increase that information. The optimal choice amongst the remaining measurements can incorporate the relative cost and difficulty of taking each particular one. Similarly, the cause-effect model can incorporate the facility of calling for particular additional measurements to achieve the final diagnosis.

State dynamics. Let us consider that after a careful analysis, it has been decided that there are N patient states within the appropriate disease class and that a finite set of M observations is always available to the clinician. For instance, in the case of thyroid disease the patient states would be hyperthyroid, euthyroid and hypothyroid with the observations corresponding to the levels of the thyroid hormones, thyroxine and tri-odothyroxine and that of thyroid-stimulating hormone. Unlike the situation in preventive medicine in which decisions have to be taken in order to maintain the patient within a fully satisfactory state, in this paper patient state transitions are considered to occur (e.g. the patient makes the transition from the euthyroid state to the hyperthyroid state within the class of thyroid disease).

The dynamics of patient health evolution is considered to be a semi-Markov process, where the probability rate transition

$$p_{ij}^k = \Pr\{S(t+1) = j/S(t) = i, \ S(t-1) = l, \ldots, S(0) = m, u(t) = k,$$
$$\mathscr{E}(t)\} = \Pr\{S(t+1) = j/S(t) = i, u(t) = k, \mathscr{E}(t)\},$$
$$(i = 1, 2, \ldots, N; j = 1, 2, \ldots, N; \ k = 1, 2, \ldots, K_i),$$

where K_i is the maximum number of decision alternatives available in state i, k is a pertinent treatment alternative, S is the patient state, $\mathscr{E}(t)$ is the current state of knowledge of the system and $u(t)$ is the decision set at time t.

$$p_{ij}^k \geqslant 0, \quad \sum_{j=1}^{k} p_{ij}^k = 1, \quad (i = 1, 2, \ldots, N; k = 1, 2, \ldots, K_i).$$

Let the random integer-valued variable τ_i be the holding time in state i. Then the probability cumulative distribution function for the semi-Markov process (the holding-time mass function) can be defined as $h_{ij}(m) = \Pr\{\tau_{ij} \geqslant m\}$, assuming that the process is completely observable (i.e. the clinician can make a certain diagnosis of the patient state), $f_{ij}(n)$ is defined as the first passage time for the semi-Markov discrete process. This is the probability that the first passage from state i to state j will require at least n time units and, as such, is a measure of how long it takes to reach a given state from another state. The analytical relationship between p_{ij}, h_{ij} and f_{ij} is given by

$$f_{ij}(n) = \sum_{\substack{r=1 \\ r \neq j}}^{N} \sum_{m=0}^{n} p_{ir} h_{ir}(m) f_{rj}(n-m) + p_{ij} h_{ij}(n)$$

$$f_{ij}(0) = \begin{cases} 1 & \text{if } i = j \\ 0 & \text{otherwise}. \end{cases} \tag{9.3}$$

For a time-dependent Markov process, the initial state vector can be written as

$$\pi(t) = [\pi_i(t), \ i = 1, 2, \ldots, N] = [\Pr\{S(t) = i/\mathscr{E}(t)\}; \ i = 1, 2, \ldots, N].$$

The initial state vector depends on the past history of treatment decisions, and measures the probability of the patient being in each state as a result of a particular treatment decision. For each treatment there is a corresponding initial state vector. A set of all possible initial state vectors can be defined as below

$$\Pi = \text{set}\left(\pi_i : \pi_i \geq 0; \sum_{i=1}^{N} \pi_i = 1\right).$$

The clinical observation space. The real-world process implies medical observations which will be treated as a noisy information channel. By performing a set of independent observations, $\theta = 1, 2, \ldots, M$, the clinician obtains information about the patient's state. These observations may be classified as cheap or expensive. The random process of clinical observation is described by the probability

$$r_{j\theta}^k = \Pr\{Z(t+1) = \theta \mid S(t+1) = j, u(t) = k, \mathcal{E}(t)\},$$
$$(j = 1, 2, \ldots, N; \; \theta = 1, 2, \ldots, M; \; k = 1, 2, \ldots, K_j)$$

and

$$0 \leq r_{j\theta}^k \leq 1, \; \sum_{\theta=1}^{M} r_{j\theta}^k = 1, \quad (j = 1, 2, \ldots, N; \; k = 1, 2, \ldots, K_j),$$

where $Z(t+1)$ is an event representing an observable output at time $(t+1)$. Thus, $r_{j\theta}^k$ defines the probability of achieving observation θ, given that state j results from a particular treatment k.

Following an observation and clinical judgement, a cause-effect model can be built in which all the logical combinations (such as AND, OR, TRUE, FALSE) of the causes are correlated with the observation outputs. The pattern recognition model used in the first step of the hierarchical modelling process is serviceable in building a more complete and accurate cause-effect model. If $Y(t+1)$ is an event representing a cause and ξ_n is a primary event in a cause-effect model, then the model is characterized by

$$l_{\theta\xi_n} = \Pr\{Y(t+1) = \xi_n / Z(t+1) = \theta, S(t+1) = j, \mathcal{E}(t)\} \tag{9.4}$$

$\theta = 1, 2, \ldots, M; \; \xi_n = 1, 2, \ldots, N_{A_n}$, where N_{A_n} is the number of causes which could give rise to a given observation and

$$0 \leq l_{\theta\xi_n} \leq 1, \quad l_{\theta\xi_n} + Cl_{\theta\xi_n} = 1, \tag{9.5}$$

where C is the complementary function.

In eqns. (9.4) and (9.5), $l_{\theta\xi_n}$ specifies the probability of a particular cause, given an observation θ and the present state being achieved as a result of a specific treatment history. In the case of thyroid disease, causes would include thyrotoxicosis, toxic nodule, iodine deficiency goitre and Hashimoto's disease.

Overall system dynamics. From the above it can be seen that the partially observable, semi-Markov clinical diagnosis-decision process is given by the quadruplex

$$\gamma = \{\mathcal{S}, \mathcal{O}, \mathcal{E}, \mathcal{D}\}, \tag{9.6}$$

where $\mathcal{S}$ is the state space, $\mathcal{O}$ is the observation space, $\mathcal{E}$ represents the primary events (causes) of the process and $\mathcal{D}$ is the decision space.

The input parameters of the model are

$$\gamma^* = \{P, H(m), R_\theta, L_{\xi_n}, k\}, \tag{9.7}$$

where $P = [p_{ij}]$, $H(m) = [h_{ij}(m)]$, $R_\theta = \text{diag}_{\theta=\text{const}} R = \text{diag}_{\theta=\text{const}}[r_{j\theta}]$, $L_{\xi_n} = [l_{\theta\xi_n}]$ and k is a treatment decision.

Extending the results of Gheorghe [139] to semi-Markov processes, and using the first passage time, we can write the new system dynamics

$$T_j(\pi|\theta, \xi_n, n, k) = \frac{\sum\limits_{i=1}^{N} \pi_i f_{ij}^k(n) r_{j\theta}^k l_{\theta\theta_n}^{lk}}{\sum\limits_{i=1}^{N}\sum\limits_{j=1}^{N}\sum\limits_{\theta=0}^{M} \pi_i f_{ij}^k(n) r_{j\theta}^k l_{\theta\xi_n}^k} \tag{9.8}$$

or, in vector formulation,

$$T(\pi|\theta, \xi_n, n, k) = [T_j(\pi/\theta, \xi_n, n, k); j = 1, 2, \ldots, N] =$$

$$= \frac{\pi F^k(n) (R_\theta^k \triangledown L_{\xi_n}^k)}{\{\theta, \xi_n/\pi, n, k\}} = \frac{\pi F^k(n) X_d^k}{\{d|\pi, n, k\}}, \tag{9.9}$$

where $X^k = (R_\theta^k \triangledown L_{\xi_n}^k)$ ($\triangledown$ being in fact a logical multiplication operator). X^k is therefore the probability of a particular cause having given rise to a specific state which was achieved as a result of decision alternative k.

$$\{\theta, \xi_n|\pi, n, k\} = \{d|\pi, n, k\} = \pi F^k(n) X_d^k \mathbf{1}$$

is the probability that the observed output is θ given that the process started under the initial probability state vector π was holding for n time units in state i before making the transition to state j due to treatment decision k; d is an index for the cause-effect model. For instance, if there were two causes and two observations, d would have four elements corresponding to each combination of cause and observation.

As X^k is a non-square matrix, a square matrix X_d^k is defined as $\text{diag}_{d=\text{const}} X^k$.

Equation is valid since dimensionally $\{\theta, \xi_n|\pi, n, k\} = \{d|\pi, n, k\}$ involves the matrix multiplication $(1 \times N)(N \times N)(N \times N)(N \times 1)$ leading to a scalar quantity and $\mathbf{1} = \underbrace{[1, \ldots, 1]}_{N}$.

The case presented here is more general than previous model formulations [128]. Specialization to a Markov process leads to the results given by Gheorghe [139]. If (i) $R_\theta = I$, then the process degenerates to a completely observable semi-Markov decision process; (ii) $H(1) = I$, then we have a partially observable Markov process with auxiliary cause-effect models. A pictorial representation of the patient state dynamics for the case $N = 3$ is given in fig. 9.18.

MAINTENANCE SAFETY MODELS AND MEDICAL DIAGNOSIS

If the process could be represented using conditional transition probabilities (first the system selects a new state j, before making any real transition), then the first time passage is given by relation

$$f_{ij}(n) = \sum_{\substack{r=1 \\ r \neq j}}^{N} \sum_{m=0}^{n} c_{ir}(m) f_{rj}(n-m) + c_{ij}(n), \qquad (9.10)$$

where $c_{ij}(m) = p_{ij} h_{ij}(m)$ for $i, j = 1, 2, \ldots, N$ and the general equation of the system dynamics is given by eqn. 9.8.

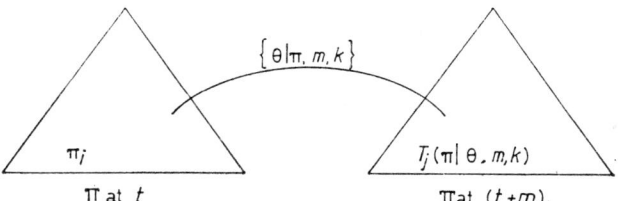

Fig. 9.18

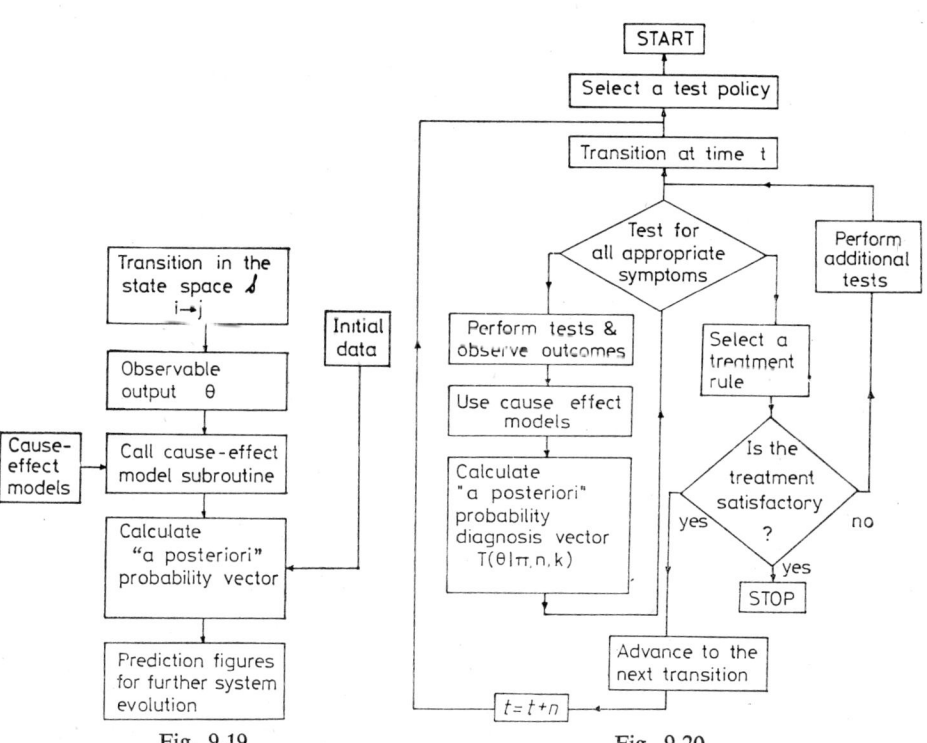

Fig. 9.19   Fig. 9.20

Figure 9.19 provides a logical diagram for the process presented above and fig. 9.20 shows an extension of this in a computer-aided diagnosis-treatment scheme.

Decision alternatives in clinical disorders. To improve the patient's state of health or even to maintain him in an acceptable state, it is necessary to carry out a set of treatment decisions. From the viewpoint of the model, we can ascribe rewards for the implementing of alternative treatments to patients in each state, given the current observations.

If in the above decision process y_{ij}^k is the variable reward (per unit of holding time) and b_{ij}^k is the fixed reward, then the expected value of the process for the case of a completely observable semi-Markov process is

$$q(\pi, k) = \sum_i \sum_j \pi_i p_{ij}^k \sum_{m=0}^{n} h_{ij}^k(m) (m y_{ij}^k + b_{ij}^k). \quad (9.11)$$

However, the reward functional for the clinical treatment-diagnosis process which has to be maximized, using a discount factor β, is given by

$$g(\pi, n) = \underset{k \in \mathscr{D}}{\text{Max}} \left[q(\pi, k, n) + \beta \sum_{\theta=1}^{M} \left(\sum_{j=1}^{N} \sum_{i=1}^{N} \pi_i p_{ij}^k r_{j\theta}^k l_{\theta \xi_n}^k \times \sum_{m=0}^{n} h_{ij}^k(m) g(T(\cdot | \cdot), n-m) \right) \right]. \quad (9.12)$$

Adopting an explicit cost $\tilde{c}_{j\theta}^k$ for the observation process under a treatment alternative k, the reward functional which has to be maximized can be expressed as

$$v_j(n) = \max_k (\gamma_{ij\theta\xi_n}^k(n) + \beta \sum_{i,j,\theta,\xi_n} \pi_j c_{ij}^k(m) r_{j\theta}^k l_{\theta\xi_n}^k \cdot v_{Tj(\pi|\theta,m,k)}(n-m)), \quad (9.13)$$

$$(j = 1, 2, \ldots, N)$$

where

$$\gamma_{ij\theta\xi_n}^k(n) = \sum_{i=1}^{N} \pi_i \sum_{j=1}^{N} \sum_{m=0}^{n} c_{ij}^k(m) \sum_{\theta=1}^{M} r_{j\theta}^k \sum_{\xi_n=1}^{N_{A_n}} l_{\theta\xi_n} \cdot (m y_{ij}^k + b_{ij}^k + \tilde{c}_{j\theta}^k).$$

The reward functional given by eqn. (9.13) is piecewise linear and convex. As the proof is similar to that given for Markov processes, it will be omitted here.

Provided that the treatment process for the partially observable system with Markovian dynamics can be represented as a decision tree, an optimization algorithm (Branch and Bound algorithm) can be built up to find the optimal treatment decision k, under conditions postulated by the model. A general formulation of Branch and Bound algorithms, as applied to partially observable Markov models, was given by Satia et al. [34] and Gheorghe [128].

Model implementation. The model presented in this section is the result of a careful analysis of the real world clinical diagnosis and treatment process. The implementation of the theoretical results in clinical systems (i.e. with small numbers of patient states and cause-effect models) can be analysed. For instance, the development of dynamic decision models for thyroid disease is part of an on-going research programme.

Large-scale diagnostic systems, incorporating the dynamic behaviour of the patient, can reveal the practical limitations of the class of models developed here. For example, large cause-effect models and a high-order decision space reveal the inadequacies of existing computational techniques, indicating the need of more OR work and improved algorithms with which to evaluate the optimal decision strategy.

The class of models given here is potentially applicable to a wide range of diagnostic systems, including those in systems engineering, and could stimulate further theoretical research. However, the analysis could be taken a stage further by applying risk-sensitive Markovian decision models to the diagnosis-treatment process and defining a utility functional over the rewards of the treatment process. The maximizing of the expected utility functional in a diagnosis-treatment process has been previously used by Ginsberg and Offensend [49] for the case of steady-state system behaviour. This use of Markovian models when the decision-maker (i.e. the clinician) is risk-sensitive and the process is only partially observable leads to a better understanding over the spectrum of optimal strategies.

List of Notation

A	— the money value to participate in the "safety game" (Chapter 7),
$A(t)$	— availability function for the system,
A, F, C	— a partition of the system state space $\mathscr{S}$,
B, C, D	— matrix representation for probability state transition using the "evolution diagram concept",
A, B, C, D	— matrix representation for the stochastic process of "failure arrival",
$\mathscr{B}$	— branching rule,
c_{ij}	— decision reward structure for a completely observable process which makes the transition from state i to state j,
c_π^k	— expected immediate reward from implementing decision k, given initial state vector π,
d	— index for cause-effect model,
DIFF	— branching criterion,
$\mathscr{E}(t)$	— the state of knowledge about the process through time t,
EP	— expected profit (Chapter 4),
$f_{ij}^\lambda(t)$	— first passage time probability for λ-th function,
F_N	— Fibonnacci number,
$\tilde{g}$	— certain equivalent gain for RSMDP,
$g(\pi)$	— expected cost on maintenance in the case of a completely observable process
$\bar{g}(\pi)$	— expected cost on maintenance for a partially observable process,
$g^*(\pi)$	— expected cost on maintenance for a probabilistic observation of the system's states and auxiliary cause-effect models,
$H(m) = [h_{ij}(m)]$	— the pmf for τ_{ij} (holding-time pmf),
$H^B(.), H^C(.), H^D(.), H^Q(.)$	— holding-time pmf matrices for the case given in Chapter 5
$H^\lambda(.)$	— as previous, but for λ-th function,
I	— the identity matrix with elements δ_{ij},
$\mathscr{I}$	— an information vector (Chapter 8),
k	— maintenance decision alternative,
$\underline{L}_{\xi_\varkappa} = [1_{\theta\xi_\varkappa}]$	— probability matrix in a cause-effect model,
$l_{\theta\xi_\varkappa}$	— the complementary value for $l_{\theta\xi_\varkappa}$
L^A	— lower bound cost vector following some policy A,
LOB(.)	— lower bound of any node,
L^λ	— number of states for the λ-th function,
L	— number of states for the Markovian process,
$\mathscr{L}^\lambda(t)$	— length-to-stay distribution for λ-th function,
M	— a diagonal matrix of mean waiting times
M	— number of independent observations in partially observable MDP,

M_λ — number of minimal paths in a structure,
$\mathcal{M}$ — a cause-effect model at time $(t+1)$,
m_λ — m-th optimal solution for λ-th function,
$O(t)$ — the observable output at time t,
$Q = [q_{ij}]$ — disutility matrix for a RSMDP (Chapter 6),
$P = [p_{ij}]$ — the probability that a Markov (semi-Markov) process that entered state i on its latest transition will visit state j on the next transition,
p — probability of death (Chapter 4),
$P^\lambda = [p_{ij}^\lambda]$ — as previous, but for λ-th function,
R — number of independent observations,
$R = [r_{j\theta}]$ — stochastic observation probability matrix,
$\mathcal{R}(t)$ — reliability function for the system at time t,
r — the largest eigenvalue for Q (Chapter 6),
$r^j(X)$ — number of functions operating at the performance level j,
$\mathcal{S}$ — index for the system state-space,
$S(t)$ — the state of the system at time t,
$\mathbf{s}_\lambda j$ — the j-th minimal path for a function λ,
$T_j(./.)$ — the j-th component in the state probability vector for a partially observable MDP,
$\mathcal{T}_j(./.)$ — the j-th component in the state probability vector for a partially observable MDP with cause-effect models,
U — the unity matrix (a square matrix with all elements = 1),
$u(v)$ — a utility function,
UPB(.) — upper bound of any node,
$\hat{\mathbf{u}} = [\hat{u}_i]$ — a normalized utility vector,
$\bar{v}$ — certain equivalent for a lottery,
v_i — expected cost for process terminating in state i as a result of maintenance alternative,
$\mathbf{v}_\lambda$ — a cut set representation,
$x_{\lambda jk}$ — a 0–1 decision variable (Chapter 8),
$X = [x_{ij}]$ — probability matrix that particular cause gives rise to a specific process state (Chapter 7),
$\mathbf{X} = (x_1, \ldots, x_n)$ — the vector of the state components,
$Y(t+1)$ — an event representing a cause-effect model,
$\mathbf{Y}_\omega$ — a set representation for a polyfunctional system,
$Z(t+1)$ — an event representing an observable output at time $(t+1)$,
$w_i(.)$ — pmf for $\tau_i \left(w_i(.) = \sum_{j=1}^{L} p_{ij} h_{ij}(.) \right)$,
$^>w_i(t)$ — complementary cdf for the waiting times

$$\left({}^>w_i(t) = \sum_{m=t+1}^{\infty} w_i(m) \right),$$

W — the losses out of a "safety game" (Chapter 4),
$\mathbf{\alpha} = [a_j]$ — the state probability vector
α — the α-vector representation for the "one-pass algorithm" (Chapters 6 and 7),
β — discount factor,

LIST OF NOTATION

$\Gamma^g(z)$	— the geometric transform for matrix $\Gamma(.)$,
γ	— risk-aversion coefficient,
$\gamma^\lambda_{ijq}(t)$	— destination probabilities for λ-th function,
δ_{ij}	— the Kronecker symbol defined as:

$$\delta_{ij} = \begin{cases} 0 \text{ if } i \neq j \\ 1 \text{ if } i = j, \end{cases}$$

$\delta(u)$	— the unit step function defined as:

$$\delta(u) = \begin{cases} 1 \text{ if } u = 0 \\ 0 \text{ if otherwise,} \end{cases}$$

ε_s	— the hypercovariance measure,
$\varepsilon_{s\lambda}$	— hypermoment of the λ-th function,
θ	— an observable output,
$\zeta_\varkappa$	— a primary event in a cause-effect model,
$\boldsymbol{\pi} = [\pi_i]$	— initial probability state vector,
Π	— the set of all possible initial state vectors,
O	— a strategy for MDP,
ρ^θ_ω	— hypercorrelation coefficient,
τ_i	— waiting time in any state $i \in \mathscr{S}$,
τ_{ij}	— the holding-time in state i before performing a transition to state j,
$\varphi^\lambda_{ij}(t)$	— the interval transition probability for the λ-th function,
$\Phi_\lambda(X)$	— the structure function representation for function λ,
$\Phi^*_\omega(X)$	— the total structure function for a polyfunctional system,
$\psi_{gi}(X)$	— a structure function for a module of the system,
ω	— total number of functions of a polyfunctional system,
ω'	— equivalent function representation for a polyfunctional system,
$\Omega^\lambda(.)$	— cartesian product for the policy space (Chapter 8),
$E[.]$	— expected value notation,
$\text{Var}[.]$	— variance; notation,
$\Pr\{./.\}$	— conditional probability notation,
$\square$	— congruent matrix multiplication,
"$\succeq$"	— sign for lexicographical ordering,
"card"	— cardinality of the polyfunctional system (a measure of the number of operational functions at any given moment),
$\emptyset$	— the empty set,
$\text{set}[.]$	— a set representation,
"$\cup$"	— a notation for the union of two or more sets,
"$\cap$"	— intersection of two or more sets,
$\int$	— notation for a general summation.

References

1. BARLOW, R. E., PROSCHAN, F., *Mathematical theory of reliability*, J. Wiley & Sons, New York, 1967.
2. DE MERCADO, I., "Reliability prediction studies of complex systems having many failed states", IEEE T − Reliability, vol. R-20, November, 1971, pp. 223−231.
3. GIRAULT, M., *Stochastic processes*, Springer Verlag, Berlin 1966.
4. OSAKI, S., "Systems analysis by Markov renewal processes", Jour. of Oper. Research, Japan, vol. 12, December 1970, pp. 127−189.
5. DOLAZZA, E., "System states analysis and flow graph diagrams in reliability", IEEE T − Reliability, vol. R-15, August 1966, pp. 58−70.
6. BRANSON, H. M., SHAH, B., "Reliability analysis of systems comprised of units with arbitrary repair - time distributions", IEEE T − Reliability, vol. R-20, November 1971, pp. 217−223.
7. HOWARD, R. A., *Dynamic programming and Markov processes*, M.I.P. Press, Cambridge, Mass., 1960.
8. HOWARD, R. A., *Dynamic probabilistic systems*, vols. I and II, John Wiley, New York, 1971.
9. JARDINE, A. K. S., *Operational research in maintenance*, Manchester University Press, New York, 1970.
10. MINE, H., OSAKI, S., *Markovian decision processes*, Elsevier, New York, 1970.
11. RAU, J. G., *Optimisation and probability in systems engineering*, Van Nostrand Reinhold Co., New York, 1970.
12. JORGENSON, D. W. et al., *Optimal replacement policy*, North-Holland Co., Amsterdam, 1967.
13. SONDIK, E. J., "The optimal control of partially observable Markov processes", Ph. D. Dissertation, Stanford University, EES Department, California, 1971.
14. HOWARD, R. A., "The foundations of decisions analysis", IEEE T − Systems, Science and Cybernetics, vol. SSC-4, no. 3, September 1968, pp. 211−220.
15. SMALWOOD, R. D., "A decision analysis of model selection", IEEE T − Systems, Science and Cybernetics, vol. SSC-4, no. 3, September 1968, pp. 333−342.
16. VESELY, W. E., "A time-dependent methodology for fault tree evaluation", Nuclear Engineering and Design, vol. 13, 1970, pp. 337−360.
17. TURBAN, E., "The use of mathematical models in plant maintenance decision making", Management Science, vol. 13, no. 6, February 1967, pp. B-342−358.
18. KOLESAR, P., "Minimum cost replacement under Markovian deterioration", Management Science, vol. 12, no. 9, May 1966, pp. 694−707.

19. KOLESAR, P., "Randomised replacement rules which maximise the expected cycle length of equipment subject to Markovian deterioration", Management Science, vol. 13, no. 11, 1967, pp. 867–876.
20. MUNFORD, A. G., SHAHANI, A. K., "A nearly optimal inspection policy", Oper. Res. Quart., vol. 23, no. 3, 1972, pp. 373–380.
21. STAPLETON, R. C. et al., "Technical change and the optimal life of assets", Oper. Res. Quart., vol. 23, no. 1, 1972, pp. 45–60.
22. QUAYLE, N. J., "Damaged vehicles — replace or repair", Oper. Res. Quart., vol. 23, no. 1, 1972, pp. 83–87.
23. KLEIN, M., "Inspection — maintenance — replacement schedules under Markovian deterioration", Management Science, vol. 9, 1966, pp. 25–32.
24. ROLFE, A. J., "Markov chain analysis of a situation where cannibalization is the only repair activity", N.R.L.Q., vol. 17, 1970, pp. 151–159.
25. SIMON, R. M., "The Reliability of Multicomponent systems subject to cannibalization", N.R.L.Q., vol. 19, no. 1, March 1972, pp. 1–14.
26. O'BRIEN, T., "Reliability of multifunctional systems", Ph. D. Dissertation, Depart. of Industrial Engineering, New York University, 1970.
27. KAO, E. P. C., "Optimal replacement rules when changes of state are semi-Markovian", Oper. Research, vol. 21, no. 6, 1973, pp. 1231–1250.
28. TAKAHASHI, Y., "On the effects of small deviations in the transition matrix of a finite Markov chain", J. Oper. Research Soc. of Japan, vol. 16, no. 2, 1973, pp. 104–130.
29. TAHARA, A., NISHIDA, T., "Optimal replacement policies for a repairable system with Markovian transition of states", J. Oper. Research Soc. of Japan, vol. 16, no. 2, 1973, pp. 78–103.
30. BONHOMME, N. M., "Evaluation and optimisation of inspection and repair schedules", Ph. D. Dissertation, Carnegie-Mellon Univ., May, 1972.
31. SMALWOOD, R. D., SONDIK, E. J., "The optimal control of partially observable Markov processes over a finite horizon", Oper. Research, vol. 21, no. 5, 1973, pp. 1071–1088.
32. GHEORGHE, A. V., "Reliability and availability prediction of polyfunctional technical systems with semi-Markov structure", Technical Report, The City University, London, 1974.
33. GHEORGHE, A. V., "A Markovian decision model for clinical diagnosis and treatment as applied to the respiratory system", IEEE T — Systems Man and Cybernetics, April 1974.
34. SATIA, J. K., LAVE, R. E., "Markovian decision processes with probabilistic observation of states", Management Science, vol. 20, no. 1, 1973, pp. 1–14.
35. HOWARD, R. A., MATHESON, J. E., "Risk-sensitive Markov decision processes", Management Science, vol. 18, no. 7, 1972, pp. 356–370.
36. HOWARD, R. A., MATHESON, J. E., "Risk-sensitive semi-Markov decision processes", to be published in Management Science (private communication).
37. HOBBS, R. J., "Optimal maintenance politics for multi-state Systems Experiencing Semi-Markov Deterioration", Ph. D. Dissertation, Stanford University, May 1972.
38. PATTON, A., "Short-term reliability calculation", IEEE T — PAS, Vol. 89, April 1970, pp. 509–513.
39. PYKE, R., "Markov renewal processes — definitions and preliminary properties", *Annal. Math. Stat.*, vol. 32, Part 6, 1961, pp. 1231–1242.

REFERENCES

40. HTUN, TIN, L., "Reliability prediction techniques for complex systems", IEEE T-Reliability vol. R-15, August 1966, pp. 58–70.
41. BIRNBAUM, Z. W., ESARY, J. D., SAUNDERS, S. C., "Multi-component systems and structures and their reliability", Technometrics, vol. 13, no. 1, February 1961.
42. BIRNBAUM, Z. W., ESARY, J. D., "Modules of coherent binary systems", J. Soc.. Indust. Appl. Math., vol. 13, no. 2, June 1965.
43. HIRSCH, W., MEISNER, M., BOLL, C., "Cannibalization in multi-component systems and the theory of reliability", N.R.L.Q., vol. 15, 1968, pp. 331–359.
44. BODIN, L. D., "Optimisation procedure for the analysis of coherent structures", IEEE T-Reliability, vol. R-18, no. 2, August 1969.
45. M'PHERSON, P. K., "Systems Science and Systems Philosophy", Futures, vol. 6, no. 3, June 1974, pp. 219–240.
46. FUSSELL, J. B., "A formal methodology for fault tree construction", Nuclear Science and Engineering, 52, 1973, pp. 421–432.
47. TALBOT, S. A., GESSNER, V., *System physiology*, John Wiley, New York, 1973.
48. BALI, H. N., et al., "Mathematical model of the respiratory system", Research Memorandum, 13, Dept. of Systems and Automation, The City University, 1972.
49. GINSBERG, A. S., OFFENSEND, F. L., "An application of decision theory to a medical diagnosis-treatment problem", IEEE-T-SMC – 4, 1968, pp. 355–363.
50. ROUSSEAU, W. F., "A method for computing probabilities in complex situations", Ph. D. Dissertation, Stanford University, Ca., May 1968.
51. LEDLEY, R. S., "Practical problems in the use of computers in medical diagnosis", *Proceedings of IEEE*, vol. 57, November 1969, pp. 1900–1918.
52. KAO, E. P. C., "A semi-Markovian population model with application to hospital planning", IEEE T-SMC-3, July 1973, pp. 327–336.
53. KRISHNAN KUTTY, K. K. et al., "Hyper-correlation: A concept for system characterisations", IEEE T-SSC-5, no. 2, 1969, pp. 161–166.
54. AESBITT, D., Private Communication, June 1973.
55. GHEORGHE, A. V., "Reliability and maintenance as functions of complex systems", Research Memorandum, DSA/AVG/23, Dept. of Systems and Automation, The City University, July 1972.
56. KAMIEN, M. E., SCHWARTZ, N. L., "A direct approach to choice under uncertainty", Management Science, vol. 18, no. 8, April 1972, pp. B. 470–B. 477.
57. MCCALL, J. J., "Probabilistic microeconomics", The Bell Journal of Economics and Management Science, vol. 2, no. 2, 1971, pp. 403–434.
58. BOYD, D., MATHESON, J., "Memorandum, decision analysis group", Stanford Research Institute, 1968.
59. DERMAN, C., "Finite state Markovian decision processes", Academic Press, New York, 1967.
60. FISHBURN, P. C., "Utility Theory", Management Science, vol. 14, no. 5, 1968, pp. 335–379.
61. HOWARD, R. A., "Risk-preference", unpublished paper, Stanford University, 1973.
62. CHARLWOOD, F. J., "The development of a systems engineering method and its application to a nuclear reactor", Ph. D. Dissertation, Department of Systems and Automation, The City University, London, 1971.
63. PRATT, J. W., "Risk-aversion in the small and in the large", Econometrica, vol. 32, 1964, pp. 122–136.
64. GANTMACHER, F. R., *The theory of matrices*, vol. 2, Chapter XIII, Chelsea, 1960.
65. LUCE, R. D., RAIFFA, H., "Games and decisions", John Wiley, New York, 1957.

66. JAQUETTE, S. C., "Utility criteria for Markov decision processes", Technical Report no. 200, Sept., 1973, Depart. of Operations Research, Cornell University.
67. PORTEUS, E. L., "On the optimality of structured policies in countable stage decision processes", Research Paper 141, Graduate School of Business, Stanford Univ., 1973.
68. POLLARD, E., "Time-risk preference decision analysis", Ph. D. Dissertation, Stanford University, 1969.
69. * * * "Electronic maintenance management symposium", University of Nottingham, July 9—12, 1973.
70. * * * "Maintenance cost ratios dropping, survey shows", Factory, September 1971.
71. ECKLES, J. E., "Optimum maintenance with incomplete information", Oper. Research, September—October, 1968, pp. 1058—1068.
72. VON ALVEN, W. H., (ed.), Reliability engineering, Prentice-Hall, Inc., New Jersey, 1964.
73. BELLMAN, R., Dynamic programming, Princeton University Press, Princeton, New Jersey, 1957.
74. MANNE, A., "Linear programming and sequential decisions", Management Science, 6, 1960, pp. 259—267.
75. GRINOLD, R. C., "Elimination of suboptimal actions in Markov decision problems", Oper. Research, vol. 2, no. 3, 1973, pp. 848—852.
76. MACQUEEN, J. B., "A test for suboptimal actions in Markovian decision problems", Oper. Research, 15, 1967, pp. 559—561.
77. PORTEUS, E. L., "Some bounds for discounted sequential decision processes", Management Science, 18, 1971, pp. 7—11.
78. HASTINGS, N. A. J., MELLO, J. M. C., "Tests for suboptimal actions in discounted Markov programming", Management Science, 19, 1973, pp. 1019—1023.
79. HAASL, D., "Advanced concepts in fault tree analyses", System Safety Symposium, University of Washington, June 8—9, 1965.
80. MILLER, A. C., "The value of sequential information", Research Report no. EES-DA-73-2, January 1974, Stanford University.
81. SCHLEICHER, R. W., "Stochastic decision making applied to nuclear reactor safety", Ph. D. Dissertation, Cornell University, 1972.
82. LUSS, H., KANDER, Z., "A preparedness model dealing with N systems operating simultaneously", Oper. Research., vol. 22, no. 1, 1974, pp. 117—129.
83. O'NEIL, F. J., "A decision structure for making on-line support tradeoffs for large complex systems", Ph. D. Dissertation, Case Western University, 1973.
84. GHEORGHE, A. V., "Markovian models with logical conditions for fault isolation with insufficient observation", Rev. Roum. Math. Pures et Appl., Tome XXIV, no. 7, 1973, pp. 1047—1064.
85. RANSOM, M. N. et al., "Fault isolation with insufficient measurements", IEEE T-Circuit Theory, July 1973, pp. 716—717.
86. MITTEN, L. G., "Branch-and-Bound methods: general formulation and properties", Oper. Research, vol. 18, no. 1, pp. 24—35, 1970.
87. BALAȘ, E., "A note on Branch-and-Bound principle", Oper. Research, vol. 16, no. 2, 1968, pp. 442—446.
88. * * * "Listening for cracks" (Industrial Notes), Nuclear Engineering International, November 1973, pp. 898.
89. MATHESON, J. E., "The economic value of analysis and computation", IEEE T-SSC-4, September 1968, no. 3, pp. 325—333.

90. LAWLER, E., WOOD, D. E., "Branch-and-Bounds methods — a survey", Oper. Research., vol. 14, no. 4, pp. 699.
91. CHEN, K., PATOON, G. T., "Branch-and-Bound approach for decision tree analysis", Stanford Research Institute, SRI—188531—172, December, 1967.
92. SHORT, R. A., GOLDBERG, J., "Soviet progress in the design of fault-tolerant digital machines", IEEE T-Computers, November 1971.
93. VON NEUMANN, J., MORGENSTERN, O., "Theory of games and economic behaviour", Princeton University Press, Princeton (3rd edition, 1953).
94. BELLMAN, R., ZADAH, L. A., "Decision making in a fuzzy environment", Management Science, vol. 17, no. 4, pp. 141—164, 1970.
95. SMITH, J. L., "Markovian decisions on a partitioned state space", IEEE T-SMC, vol. SMC-1, no. 1, Ian. 1971, pp. 55—60.
96. BLACKWELL, D., "Discounted dynamic programming", Ann. Math. Stat., 36, 1965, pp. 226—235.
97. HIMMELBLAU, D. M., (editor), *Decomposition of large-scale problems*, North-Holland, Elsevier, 1973.
98. BONDER, S., "Operations research education: some requirements and deficiencies", Oper. Research, vol. 21, no. 3, pp. 797—810, 1973.
99. RODGERS, W. P., *Introduction to system safety engineering*, John Wiley, New York, 1971.
100. SELVIDGE, J., "Rare-events probabilities encoding", Ph. D. Dissertation, Harvard University, 1973.
101. SPETZLER, C. S., STAËL VON HOLSTEIN, C. A. S., "Probability encoding in decision analysis", Paper presented at the ORSA — TIMS — AIEE, 1972, Joint National Meeting, Atlantic City, N.J., 8—10 November, 1972.
102. BEVERIDGE, G. S., SCHECHTER, R. S., *Optimisation ; theory and practice*, McGraw-Hill, 1970.
103. PALMER, J. F., "International views on principles and standards of reactor safety", Nuclear Safety, vol. 14, no. 5, 1973, pp. 428—439.
104. * * * "Safety of nuclear power reactors (LWR) and related facilities", WASH 1250, AEC, July 1973, 510 p.
105. GARRICK, B. J. et al., *Reliability analysis of nuclear power plant protective systems*, HN-190, Holmes and Narver, Inc., Los Angeles, May 1967, 600 p.
106. FARMER, F. R., "Quantitative safety analysis", Nuclear engineering and design, vol. 13, 1970, pp. 183—244.
107. NORTH, W. D., "A tutorial introduction to decision analysis", IEEE T-SSC, September 1968, pp. 200—211.
108. SAVAGE, L. J., *The foundations of statistics*, John Wiley, New York, 1964.
109. RAIFFA, H., *Decision analysis. Introductory lectures on choices under uncertainty*, Addison-Wesley, Boston, 1968.
110. NICHOLS, R. W., "Some applications of fracture mechanics in power engineering", Intern. Journ. on Press, Ves. & Piping, (1), pp. 281—297.
111. BODEN, J. M., "Optimum staffing for machine maintenance", Ph. D. Dissertation, New York University, 1972.
112. RIVAS, J. R., "Man-machine synthesis of disaster-resistant operations", Operations Research, 23, 1, January 1975, pp. 2—21.
113. NEUMANN, P. C., BONHOMME, M. M., "Optimal inspected and repair schedules for multicomponent systems", IEEE T-Systems, Man and Cybernetics, January 1974, pp. 58—66.

114. NEUMAN, P. C., BONHOMME, M. N., "Evaluation of maintenance policies using Markov chains and fault tree analysis", IEEE T-Reliability, 1975.
115. HOWARD, A. R., JUDD B. R., MATHESON, E. J., *A pilot model for nuclear reactor accident safety*, Stanford Research Institute, Menlo Park, California, December, 1972.
116. SHARMA, J., VENKATESWARAN, V. K., "A direct method for maximizing the system reliability", IEEE T-Reliability, R-20, 4, November 1971, pp. 256—259.
117. ROLFE, J. A., "Markov chain analysis of a situation where cannibalization is the only repair activity", Naval Research Logistic Quarterly, 17, 1970, pp. 151—158.
118. NITU, V. et al., *Metode statistico-probabilistice utilizate în energetică*, Editura tehnică, București, 1968.
119. M'PHERSON, P. K., *System theoretic concepts of systems engineering, Proceedings of the European Meeting*, Vienna, 1973, *Advances in cybernetics and systems research*.
120. WARFIELD, HALL., *A unified systems engineering concept*, Battelle Memorial Institute, 1973.
121. KLIR, G. J., *An approach to general systems theory*, Van Nostrand-Reinhold, 1970.
122. LEE, A. M., *Systems analysis frameworks*, Macmillan, New York, 1970.
123. MESAROVIC, M. D. et al., *Theory of hierarchical, multilevel systems*, Academic Press, New York, 1970.
124. WYMORE, A. W., *A mathematical theory of systems engineering*, J. Wiley, New York, 1967.
125. FAN L. T. et al., "Optimization of systems reliability", IEEE-T, Reliability R-16, 2, September 1967, pp. 81—86.
126. NEGOIȚĂ, V., RALESCU, *Mulțimi vagi și aplicațiile lor*, Editura tehnică, București, 1975.
127. FU, K. S., *Sequential methods in pattern recognition and machine learning*, Academic Press, New York, 1968.
128. GHEORGHE, A. V., "The derivation of optimal maintenance and safety strategies for complex technical systems using probabilistic models", Ph. D. Dissertation, The City University, London, 1974.
129. HARISON MICHAEL, J., *Critique of reactor safety modeling methodology, Project Memo*, Stanford Research Institute, June 1973.
130. BARLOW, R. E., PROSCHAN, F., *Statistical theory of reliability and life testing*, Holt, Rinehart and Winston, New York, 1975.
131. CHATTERJEE P., "Fault tree analysis. Reliability theory and systems safety analysis", Operation Research Centre University of California, Berkeley, Rept., ORC 74—34 (1974).
132. HOANG, TUI, "Concave programming with linear constraints", Soviet Math. Dobl., 5, 1964, pp. 1437—1440.
133. GHEORGHE, A. V., BALI, H. N., HILL, J., CARSON, E., "Dynamic decision", *Modern trends in general systems and cybernetics*, Springer Verlag, 1977.
134. ESARY, J. D., PROSCHAN, F., "Coherent structures of non-identical components", Technometrics, 5, 1963, pp. 191—209.
135. BIRNBAUM, Z. W., "On the importance of different component system", *Multivariate analysis*— II, P.R. Rishnaiah, Academic Press, New York, 1969.
136. EAGLE, J. N., "A utility criterion for the Markov decision process", Ph. D. Thesis, Stanford University, Ca., 1975.
137. GHEORGHE, A. V., "Probabilistic modelling for safety analysis of technical systems", Rev. Roum., Sci. Techn. — Electrotechn. et Energ., 21, 1, 1976, pp. 85—93.
138. GHEORGHE, A. V., "On risk-sensitive Markovian decision models for complex systems maintenance", *Economic computation and economic cybernetics studies and research*, 1, 1976.

REFERENCES

139. GHEORGHE, A. V., BALI, H. N., HILL, J. W., CARSON, E. R., "Dynamic decision models for clinical diagnosis", Int. J. Bio-Medical Computing (7), 1976, 2, p. 81—93.
140. GHEORGHE, A. V., "Risc și securitate în proiectarea tehnologiei moderne", *Revoluția științifico-tehnică și modernizarea forțelor de producție*, Editura politică, București, 1976.
141. GHEORGHE, A. V., "Analiza aplicată a sistemelor în studiul sistemelor de energie", *Revoluția socialistă și revoluția științifică-tehnică*, Editura politică, București, 1975.
142. GHEORGHE, A. V., "Structural modelling for energy strategies management", *"Energy and physics"*, Proceedings of the third General Conference of the European Physical Society, 9—12 September, 1975, Bucharest, Romania.
143. GHEORGHE, A. V., *Ingineria sistemelor. Modele și tehnici de calcul* (Systems engineering. Models and computational techniques), Editura Academiei R. S. România, Bucharest, 1979.
144. GOLDMAN, A. G., SLATTERY, T. B., *Maintainability: a major element of system effectiveness*, John Wiley & Sons, Inc., New York, 1964.
145. SAUNDLER, G. H., *System reliability engineering*, Prentice Hall, Englewood Cliffs, N. J., 1963.
146. GOLOMB, S. W., "Mathematical models: uses and limitations", IEEE T-Reliability, vol. R-20, no. 3, August 1971, pp. 130—132.
147. BLANCHARD, B. S., LOWERY, E. E., *Maintainability — principles and practices*, McGraw-Hill Book Co., New York, 1969.
148. EVANS, R. A., "Redundancy configurations and their effect on system reliability", *Microelectronics and reliability*, Pergamon Press, 1971, vol. 10, pp. 355—357.
149. BELLMAN, R., DREYFUS, S., "Dynamic programming and the reliability of multicomponent devices", Operations Research, vol. 6, March—April 1958, pp. 200—206.
150. KETTELLE, J. D., "Least-cost allocations of reliability investment", Oper. Research, vol. 10, 1962, pp. 249—265.
151. KABAK, J. W., "System availability and some design implications", Oper. Research, vol. 7, September 1969, pp. 838—847.
152. SHERSHIN, A. C., "Mathematical optimisation techniques for the simultaneous apportionments of reliability and maintainability", Oper. Research, January—February 1970, vol. 18, pp. 95—107.
153. JENSEN, P. A., "Optimization of series parallel-series networks", Oper. Research, vol. 18, 1970, pp. 471—482.
154. LÜTTSCHWAGER, J. M., "Dynamic programming in the solution of a multistage reliability problem", The Journal of Industrial Engineering, vol. XV, no. 4, July—August 1964, pp. 168—176.
155. WOODHOUSE, C. F., "Optimal redundancy allocation by dynamic programming", IEEE T-Reliability, February, 1972, pp. 60—62.
156. PROSCHAN, F., BRAY, T. A., "Optimum redundancy under multiple constraints", in Oper. Research, vol. 13, 1965, pp. 800—814.
157. BURTON, R. M., HOWARD, G. T., "Optimal design for system reliability and maintainability", IEEE T-Reliability, vol. R-20, no. 2, May 1971, pp. 56—60.
158. LAMBERT, B. K., WALVEKAR, A. G., HIRMAS, J. P., "Optimal redundancy and availability allocation in multistage systems", IEEE-T-Reliability, vol. R-20, no. 3, August 1971, pp. 182—185.
159. TILLMAN, F. A., "Optimization by integer programming of constrained reliability problems with several modes of failure", Management Science, vol. 13, No. 11, July 1967, pp. 887—899.

160. MISRA, K. B., "A method of solving redundancy optimization problems", IEEE T-Reliability, vol. R-20, No. 3, August 1971, pp. 117—121.
161. MISRA, K. B., "A simple approach for constrained redundancy optimization problem", IEEE T-Reliability, vol. R-21, No. 1, February 1972, pp. 30—35.
162. MIZUMAKI, K., "Optimum redundancy for maximum system reliability by the method of convex and integer programming", Oper. Research, vol. 16, March—April, 1968, pp. 392—406.
163. TILLMAN, F. A., HWANG, C. L., FAN, L. T., LAI, K. C., "Optimal reliability of a complex system", IEEE T-Reliability, vol. R-19, No. 3, August 1970, pp. 95—100.
164. MCNICHOLAS, R. J., MESSER, G. H., "A cost based availability allocation algorithm", IEEE T-Reliability, vol. R-20, no. 3, August 1971, pp. 178—182.
165. BECKER, P. W., JARKLER, B., "A systematic procedure for the generation of cost-minimized designs", IEEE T-Reliability, vol. R-21, no. 1, February 1972, pp. 71—75.
166. WIDAWSKY, W. H., "Reliability and maintainability", IEEE T-Reliability, vol. R-20, no. 3, August 1971, pp. 158—164.
167. BROWN, D. B., MARTZ, JR., H. F., "Simulation model for the maintenance of a deteriorating component system", IEEE T-Reliability, vol. R-20, no. 1, February 1971, pp. 28—32.
168. HWANG, C. L., LAI, Z. C., TILLMAN, F. A., FAN, L. T., "Optimization of system reliability by the sequential unconstrained minimization technique", IEEE T-Reliability, vol. R-24, no. 2, 1975, pp. 133—136.
169. BLANKS, H. S., "A review of new methods and attitudes in reliability engineering", *Microelectronics and reliability*, Pergamon Press, vol. 12, 1973, pp. 301—319.
170. MISRA, K. B., "A fast method for redundancy allocation", *Microelectronics and Reliability*, Pergamon Press, vol. 12, 1973, pp. 385—387.
171. MISRA, K. B., "A method of redundancy allocation", *Microelectronics and reliability*, Pergamon Press, vol. 12, 1973, pp. 389—393.
172. BARLOW, R. E., PROSCHAN, F., "Availability theory for multicomponent systems", ORC 72-8, April 1972, University of California, Berkeley.
173. FUSSELL, J. B., "Fault tree analysis — concepts and techniques", July 1973, Aerojet Nuclear Company, Idaho Falls, Idaho.
174. WILSON, J. R. et al., "Evaluation of common cause failure analysis techniques", EG & G Idaho Inc., Idaho National Engineering Lab., September 1977.
175. WEISMAN, J., HOLZMAN, A. G., "Engineering design optimization under risk", Management Science, vol. 19, no. 3, November 1972, pp. 235—249.
176. DERMAN, C., LIEBERMAN, G. J., ROSS, S. M., "On optimal assembly of systems", N.R.L.Q. December 1972, pp. 569—574.
177. MERRILL, H. M., "Failure diagnosis using quadratic programming", IEEE T-Reliability, vol. R-22, no. 4, October 1977, pp. 207—213.
178. SETHI, S. P., MORTON, T. E., "A mixed optimization technique for the generalized machine replacement problem", N.R.L.Q., September 1972, vol. 19, no. 3, pp. 471—481.
179. PATTON, A. D., "Short-term reliability calculation", IEEE T-PAS, vol. PAS-89, no. 4, April 1970, pp. 509—513.
180. PATTON, A. D., "A Probability method for bulk power system security assessment, I —Basic Concepts", IEEE T-PAS, vol. PAS-91, no. 1, January/February 1972, pp. 54—61.

181. PATTON, A. D., "A Probability method for bulk power system security assessment, II — Development of probability models, for normally operating components", submitted for 1972 IEEE Winter Power Meeting.
182. PATTON, A. D., "A Probability method for bulk power system security assessment, III — Models for stand-by generators and field data collection and analysis", submitted for 1972 IEEE Winter Power Meeting.
183. BOURNE, A. J., "Measuring reliability (II)", Electronics & Power, May 1972, pp. 181—184.
184. HENSLEY, G., "Plant and process reliability", Instrument Practice, November 1971, pp. 624—629.
185. DOBRYDEN, V. A., KURT-UMEROV, V. O., "Reliability of a parallel — series stand-by structure and its optimization according to an economic criterion", Telecommun. Radio Eng. (USA), vol. 26, February, no. 2, 1971, pp. 131—133.
186. TILLMAN, F. A., LÜTTSCHWAGER, J. M., "Integer programming formulation of constrained reliability problems", Management Science, vol. 13, no. 11, July 1967, pp. 887—899.
187. BANERJEE, S. K., RAJAMANI, K., "Parametric representation of probability in two dimensions — A new approach" IEEE T-Reliability, February 1972, pp. 56—60.
188. WARFIELD, J. N., "Structuring complex systems", Battelle Monograph, no. 4/April 1974.
189. SAHAL, D., "System Complexity: Its conception and measurement in the design of engineering systems", IEEE T on Systems, Man and Cybernetics, June 1976, pp. 440—445.
190. FERDINAND, A. E., "A theory of system complexity", Int. J. General Systems, 1974, vol. 1, pp. 19—33.
191. GOTTINGER, H. W., "Complexity and information technology in dynamic systems", Kybernetes, 1975, vol. 4, 192, pp. 129—141.
192. MANN, N. R., SCHAFER, R. E., SINGPURWALLA, N. D., *Methods for statistical analysis of reliability & life data*, John Wiley & Sons., New York, 1974.
193. AVENHAUS, R., et al., *New societal equations*, International Institute for Applied Systems Analysis (IIASA), WP-75-67, June 1975, Austria.
194. OTWAY, H. J. et al., "A risk estimate for an urban-sited reactor", Nuclear Technology, vol. 12, October 1971, pp. 173—184.
195. HALPERN, J., "Evaluating Boolean function with random variables", Technion — Israel Institute of Technology, June 1973.
196. KIM, Y. H., CASE, K. E., GHARE, P. M., "A method for computing complex system reliability", IEEE T-Reliability, vol. R-21, no. 4, November 1972, pp. 215—219.
197. DODERLEIN, J. M., "Technology, risk and society", University of Washington, Seattle, November 1977 (invited lecture).
198. THOMAS, K. et al., "A comparative study of public beliefs about five energy systems", IIASA, WP-79-5, January 1979, Austria.
199. THOMAS, K. et al., "Nuclear energy: the accuracy of policy makers perceptions of public beliefs", IIASA, WP-79-4, January 1979.
200. NIEHAUS, F., SWATON, E., "Cross-cultural comparison of public attitudes towards various energy supply technologies", presented at Seminar on Energy and Safety, Sandefjord, Norway, 15—16 November, 1979.
201. *** *The acceptability of risk*, Barry Rose (Publishers) Ltd., Council for Science and Society, 1977.
202. LAMBERT, H. E., "Fault trees for decision making in systems analysis", Ph. D. Thesis, Lawrence Livermore Lab., University of California, Livermore, Oct. 1975.

203. REGULINSKI, T. L., "Elements of information theoretic methods: tutorial introduction", IEET-Reliability, vol. R-18, no. 3, August 1969, pp. 82—87.
204. EVANS, R. A., "The principle of minimum information", IEEE T-Reliability, vol. R-18, no. 3, August 1969, pp. 87—90.
205. LEVY, L. L., MOORE, A. H., "A Monte Carlo technique for obtaining system reliability confidence limits from component test data", IEEE T-Reliability, vol. R-16, no. 2 September 1967, pp. 69—72.
206. JENSEN, P. A., BELLMORE, M., "An algorithm to determine the reliability of a complex system", IEEE T-Reliability, vol. R-18, no. 4, November 1969, pp. 169—174.
207. FLEMING, J. L., "Relcomp: A computer program for calculating system reliability and MTBF", IEEE T-Reliability, vol. R-20, no. 3, August 1971, pp. 102—107.
208. OSAKI, S., "On a two-unit standby-redundant system with imperfect switchover", IEEE T-Reliability, vol. R-21, no. 1, February 1972, pp. 20—24.
209. OSAKI, S., "Reliability analysis of a two-unit standby-redundant system with preventive maintenance", IEEE T-Reliability, vol. R-21, no. 1, February 1972, pp. 24—29.
210. SCHAEZER, R. E., "What are we talking about when we talk about "Risk"? A critical survey of risk and risk preference theories", IIASA, RM-78-69, December 1978.
211. BOYER, C., LEGER, B., "Relations Elf — Aquitaine/C.E.A. — Trace des arbres de defant relatifs a GR-NE", C.E.A. — D.A.M. — C.E.S.T.A. — DO 338, 28 December 1976.
212. HOCHBERG, M., "Generalized multicomponent systems under cannibalization", N.R.L.Q., December 1973, vol. 20, no. 4, pp. 585—605.
213. LINNEROOTH, J., "A critique of recent modelling efforts to determine the value of human life", IIASA, RM-75-67, December 1975.
214. MADERTHANER, R. et al., "Perception of technological risks: the effect of confrontation", IIASA, RM-76-53, June 1976.
215. OTWAY, H. J., COHEN, J. J., "Revealed preferences: comments on the Starr benefit — risk relationships", IIASA, RM-75-5, March 1975.
216. OTWAY, H. J. et al., "Avoidance response to the risk environment: a cross cultural comparison", IIASA RR-75-14, June 1975.
217. OTWAY, H. J., et al., "Social values in risk acceptance", IIASA, RM-75-54, November 1975.
218. THOMASON, M. G., PAGE, E. W., "Boolean difference techniques in fault tree analysis", Int. J. of Computer and Information Sciences, vol. 5, no. 1, March 1976, pp. 81—88.
219. FUSSELL, J. B., VESELY, W. E., "A new methodology for obtaining cut sets for fault trees", in Trans. Amer. Nucl. Soc., vol. 15, June 1972, pp. 262—263.
220. BELLMAN, R., "Large systems", USC-113 P-75, University of Southern California, August 1974.
221. THOMA, J., "Energy, entropy, and information", IIASA, RM-77-32, June 1977.
222. ARVANITIDIS, N. V., "A generalized reliability model with applications in electric power systems", Systan Inc. SRR-67-12, December 1967.
223. BLACK, S., NIEHAUS, F., SIMPON, D., "How safe is 'too' safe?", IIASA, W.P.-79-68, June 1979.
224. LEWIS, H. W. et al., "Risk assessment review group report to the U.S. Nuclear Regulatory Commission", NUREG/CR-0400, September 1978.
225. BARLOW, R. E., PROSCHAN, F., "Theory of maintained systems: distribution of time to first system failure", QRC-74-36, December 1974, University of California, Berkeley.

226. BARLOW, R. E., "Coherent systems with multistate components", ORC-77-5, January 1977, University of California, Berkeley.
227. BARLOW, R. E., DAVIS, B., "Analysis of time between failures for repairable components", ORC-77-20, July 1977, University of California, Berkeley.
228. BARLOW, R. E., LIANG, T. Y., "Availability analysis of the Superhilac Accelerator", ORC-77-21, July 1977, University of California, Berkeley.
229. MURCHLAND, L. D., *Fundamental concepts and relations for multistate reliability systems*, Conference on Reliability and Fault Tree Analysis, Berkeley, 1974.
230. PAU, L. F., "Applications of pattern recognition to the diagnosis of equipment failures", *Pattern Recognition*, Pergamon Press, 1974, vol. 6, pp. 3–11.
231. GRAHAM, J., *Fast reactor safety*, Academic Press, New York, 1971.
232. SMITH, C. F., KASTENBERG, W. E., "On risk assessment of high level radioactive waste disposal", Nuclear Eng. and Design, 39, 1976, pp. 293–333.
233. CUMMINGS, G. E., "Application of the fault tree technique to a nuclear reactor containment system", Reliability and Fault Tree Analysis, SIAM, Philadelphia, 1975, pp. 806–825.
234. CHATTERJEE, P., "Modularization of fault trees; a method to reduce the cost of analysis", Reliability and Fault Tree Analysis, SIAM, Philadelphia, 1975, pp. 101–126.
235. KOEN, B. V., CARNINO, A., "Reliability calculations with a list processing technique", IEEE T-Reliability, vol. R-23, no. 1, April 1974, pp. 43–50.
236. OLMAN, M. D., WORRELL, R. B., "A fault tree representation designed for computer analysis", Sandia Laboratories, SC-RR-71 0615 A, July 1972.
237. WORRELL, R. B., "Set equation transformation system (SETS)", Sandia Laboratories, SLA-73-0028 A, January, 1975.
238. BILLINTON, R., KRASNODEBSKI, J., "Practical application of reliability and maintainability concepts to generating station design", IEEE PAS-Winter Meeting, New York, January, 1973.
239. POWERS, G. J., LAPP, S. A., "Computer-aided fault tree synthesis", Chemical Engineering Progress, April 1976, pp. 89–93.
240. LADANY, S. P., "Optimal design of an integrated maintenance policy – maintenance force system", Int. J. Systems Sci., 1974, vol. 1, pp. 1–10.
241. HULME, B. L., WORRELL, R. B., "A prime implicant algorithm with factoring", IEEE T – Computers, November 1975, pp. 1129–1131.
242. DERMAN, C., LIEBERMAN, G. J., ROSS, S. M., "On optimal assembly of systems", N.R.L.Q., December 1972, vol. 19, no. 4, pp. 569–574.
243. WORRELL, R. B., BURDICK, G. R., "Quantitative analysis in reliability & safety analysis", IEEE T–Reliability, vol. R-25, no. 3, August 1976, pp. 164–170.
244. SCHULLER, G. I., "On the structural reliability of reactor safety containments", Nuclear Engineering and Design, 27 (1974), pp. 426–433.
245. SHIMERS, S. M., "Modeling of human operator performance utilizing time series analysis", IEEE T-Systems, Man, and Cybernetics, vol. SMC-4, no. 5, September 1974, pp. 446–458.
246. M'PHERSON, P. K., "Systems science and interdisciplinary education", Journal of Systems Engineering, vol. 4, no. 1, August 1974, pp. 20–39.
247. TAYLOR, J. R., "Sequential effects in failure mode analysis", Riso–M–1740, Danish, A.E.C., Roskilde, Denmark.

248. EVANS, R. A., "Fault-trees and cause-consequence charts", IEEE T-Reliability, vol. R-23, p. 1, April 1974.
249. BENNETS, R. G., "On the analysis of fault trees", IEEE T-Reliability, vol. R-24, August 1975, pp. 175—185.
250. RUSSELL, J. B., HENRY, E. B., MARSHALL, N. H., "MOCUS — A computer program to obtain minimal sets from fault trees", ANCR—1156, Aerojet Nuclear Company, Idaho Falls, Idaho, USA, March 1974.
251. SEMANDERES, S. N., "ELRAFT, a computer program for the efficient logic reduction analysis of fault trees", IEEE Trans. Nuclear Science, vol. NS-18, 1971, pp. 481—487.
252. VESELY, W. E., NARUM, R. E., "PREP and KITT: Computer codes for the automatic evaluation of a fault tree", IN-1349, Idaho Nuclear Corporation, Idaho Falls, Idaho USA, August 1970.
253. CROSETTI, P. A., "Computer program for fault tree anaylsis", DUN-5508, Douglas United Nuclear, Inc., Richland, Washington USA, April 1969.
254. GHEORGHE, A. V., "On maintenance policies for multifunctional systems using semi-Markov decision models", Rev. Roum. Math. Pures et Appl., Tome XXIV, no. 9, 1979, pp. 1337—1353.
255. GHEORGHE, A. V., "Operational behaviour prediction for technical systems with semi-Markov structure", Rev. Rom. Sci. Techn. — Electrotechn. et Energ., 23, 2, 1978, pp. 231—243.
256. SCHRODER, R. J., "Fault tree simulation with importance sampling", Documentation for Computer Program AS 2798, The Boeing Co., Seattle, Washington, USA, 1967.
257. CROSETTI, P. A., "Fault Tree Simulation Computer Program", DUN-7697, Douglas United Nuclear, Inc., Richland, Washington, USA, June 1971.
258. GONGLOFF, W. C., "Common mode failure analysis", IEEE T-Power, Apparatus and Systems, vol. PAS-94, pp. 27—30, January—February 1975.
259. BURDICK, G. R. et al., "COMCON — A computer program for common-cause analysis", ANCR-1314, Aerojet Nuclear Company, Idaho Falls, Idaho USA, May 1976.
260. CÂRLAN, M., Private communication, June 1980.
261. GHEORGHE, A. V., PURICA, I. I., "The decision behaviour in energy development programs", University of Bucharest, Division of Systems Studies, GPID—UNU Project Report, 1979.
262. FUSSELL, J. B., POWERS, G. J., BENNETTS, R. G., "Fault trees — a state of the art, discussion", IEEE T-Reliability, vol. R-23, pp. 51—55, April 1976.
263. DUNCAN, G. T., "Heterogeneous questionnaire theory", SIAM J. Appl. Math., vol. 27, no. 1, July 1974, pp. 59—71.
264. DUNCAN, G T., "Optimal diagnostic questionnaires", Operations Research vol. 23, no. 1, January—February 1975, pp. 22—31.
265. ONICESCU, O. et al., *Elements of informational statistics with applications* (in Romanian), Editura Tehnică, Bucharest, 1979.

236971